THE CHEMIST'S COMPANION GUIDE TO PATENT LAW

THE CHEMIST'S COMPANION GUIDE TO PATENT LAW

Chris P. Miller
Mark J. Evans

WILEY

A JOHN WILEY & SONS, INC., PUBLICATION

Published by John Wiley & Sons, Inc., Hoboken, New Jersey.
Published simultaneously in Canada.

For general information on our other products and services or for technical support, please contact our Customer Care Department within the United States at (800) 762-2974, outside the United States at (317) 572-3993 or fax (317) 572-4002.

Wiley also publishes its books in a variety of electronic formats. Some content that appears in print may not be available in electronic formats. For more information about Wiley products, visit our web site at www.wiley.com

Library of Congress Cataloging-in-Publication Data

Miller, Chris P.
  The chemist's companion guide to patent law / Chris P. Miller, Mark J. Evans.
      p. cm.
  Includes bibliographical references and index.
  ISBN 978-0-471-78243-8 (cloth)
    1. Patent laws and legislation–United States.   2. Chemistry–United States–Patents.
  3. Chemical industry–Law and legislation–United States.   I. Evans, Mark J.   II. Title.
  KF3133.C4.M55   2010
  346.7304′86–dc22

                                                                            2010000195

Printed in Singapore

10   9   8   7   6   5   4   3   2   1

CONTENTS

A man who is his own lawyer has a fool for a client (although he won't have to pay any fees).

—Source unknown

The legal information contained in this book is just that, legal information. None of the information contained in this book is intended to constitute, nor should it be interpreted to constitute, legal advice. While we have attempted to ensure the accuracy of the material in this text, please appreciate that mistakes can happen and interpretations of law as expressed herein are merely interpretations and not necessarily correct. Readers with specific legal questions are advised to contact and consult with competent legal counsel. The views expressed herein are the authors' views alone and do not reflect the views of any organization, business, or association that the authors might have been, currently are, or will be affiliated with in any way. Furthermore, one needs to appreciate that any discussion as it relates to the law often cannot be extrapolated to situations not yet considered. A decision in a particular case depends on the particular facts of that case, and since no two cases will have exactly the same facts, the law as applied may be different despite other apparent similarities. This text reflects no more than the authors' desire to provide a narrative for the interested reader in the topic at hand. The discussion herein should not be construed as taking a particular position on any legal matter, nor is it to be construed as advocating any single position. Any appearance to the contrary is not intentional.

The hypothetical examples are included for illustrative purposes only—they should be accorded no weight as prior art against inventors' future activities. Furthermore, under no circumstances is it the authors' intent to imply that any of the compounds or compositions presented herein can be safely produced by any of the processes described herein nor used according to any of the methods described in this text.

In several instances, the authors have included fictional characters to enrich the learning experience. Any resemblance to any person, past, present, or future is entirely coincidental and unintended.

Law never is but is always about to be.

—Benjamin Cardozo

Oh boy, here we go again, yet another book on patent law, just what I've been waiting for. Frankly, that is the usual reaction that many have when seeing an article or text on patent law—it's hard to imagine too many subjects that are drier. For many of you, reading or studying patents is far down the list of your preferred tasks for a given day, perhaps somewhere between washing glassware and emptying the chemical waste cans. A natural aversion to chemical patent law is not surprising because most of your exposure to patent law has probably come from reading (or helping to write) patents and/or patent applications. In the course of reading the patent literature, you surely have been assaulted with a sometimes bewildering array of gobbledygook from which you are supposed to sift out something meaningful. Somewhat counterintuitively, the difference between the level of description, detail, and clarity in a typical scientific paper often exceeds what you might find in the related patent literature—given a choice, very few chemists will choose the patent literature as their primary source for background scientific information in a topic subject area.

Some of this difference can be attributed to careless application drafting or application drafting habits unmoored from legal reality—sometimes an overly burdened patent attorney might rely on boilerplate recitation to the exclusion of real description, and the end result is not always helpful. In that regard, patents resemble many legal documents, which are why they are usually about as much fun to read as the warranty for your new refrigerator. However, there also are more valid reasons for the sometimes peculiar language and style encountered in the patent literature. We hope these reasons will become clearer as we move along.

Before proceeding farther into this book, let's all take a collective deep breath and relax because this book is not boilerplate and, we hope, will be more lucid if not more interesting than the typical patent or patent application. Rather than going through an array of patents and explaining their parts in tedious detail so that reading the patents would have been fun in comparison, we are approaching the subject matter in a different way. The approach relied on herein is more similar to the approach used for learning in law school and includes a combination of explanation surrounding a given topic coupled with discussions centered on actual federal court opinions.

Briefly, the hierarchy of patent law in the United States begins first with the U.S. Constitution, which provides the discretionary basis for federal regulation of patent laws.[1] Acting on its discretionary authority, Congress has enacted federal patent laws that are outlined in the U.S. Code Title 35 (the Code). The Code establishes the substantive framework that federal judges must interpret and apply.[2] In contrast, patent rules are administrative in nature and are outlined in the U.S. Title 37—Code of Federal Regulations (the Regs). The Regs outline the administrative rules for conducting business at the U.S. Patent Office (USPTO).[3] Beyond the USPTO rules, there is the *Manual of Patent Examining Procedure* (MPEP). The MPEP provides detailed instruction to the USPTO examiners with the USPTO's interpretation of myriad issues related to the procurement of patents (patent prosecution) in the United States. Not only is the MPEP an invaluable legal reference for USPTO patent examiners but it is also an invaluable resource for patent practitioners and chemists who have an interest. The MPEP contains the Code, the Regs, and the very extensive explanation of the USPTO's interpretation of what the law is in regard to whether your patent should issue or not—we will refer to it often.[4]

The Code and Regs provide only the minimum level of detail, and by themselves, do not provide for their own interpretation. That is for the job of the U.S. federal court system that oversees cases and controversies related to patents, copyrights, and trademarks.[5] Through their many opinions and findings, the U.S. federal courts make and explain the law, filling in the myriad gaps left by the sparse language of the statutes by applying that law to the facts of the particular cases and controversies they adjudicate. Through a system of precedent, the federal courts both rely and build on the rules and reasoning from previous decisions. The written decisions (cases) issued

---

[1] At least in theory, those powers not explicitly granted to the federal government by the U.S. Constitution are reserved for the states. However, in contrast to much federal law, which often has a tortured or tenuous link to any explicit grant in the U.S. Constitution, Article 1 §8 of the Constitution grants explicit authority to the federal legislature to make laws governing the protection of inventions: "Congress shall have power... to promote the progress of science and useful arts, by securing for limited time to authors and inventors the exclusive right to their respective writings and discoveries."

[2] So when you see something that has 35 U.S.C. followed by a number (e.g., 35 U.S.C. 102), you will know this is a federal statute.

[3] In the case of any apparent conflict in interpretation, federal laws take priority over federal rules in terms of judicial authority. Patent rules can be typically recognized by their numbers. They begin at §1.1 and currently proceed up to §1.997. The patent rules outline the specific details and requirements that the USPTO has for the many different functions that it carries out with applicants/patentees. For example, rule §1.121 covers the USPTO's policy regarding the publication of patent applications.

[4] If you find yourself in the position of needing to know any specific detail about patent law prosecution then the MPEP is indispensable. The MPEP online can be found at the following site in the USPTO Internet domain: www.uspto.gov/web/offices/pac/mpep/mpep.htm. The MPEP is also available in paperback copy from the legal publishers West and Lexus-Nexus.

[5] To appreciate why federal court opinions are so important to understanding patent law, it is first necessary to appreciate that issues related to patents are in the subject matter jurisdiction of the federal courts. The U.S. Constitution specifically grants Congress the discretionary power to regulate patents, and Congress asserted that jurisdiction pursuant to 28 U.S.C. §1338 (a), which states "The (federal) district courts shall have original jurisdiction of any civil action arising under any Act of Congress relating to patents, plant variety protection, copyrights and trademarks. Such jurisdiction shall be exclusive of the Courts of the states in patent, plant variety protection and copyright cases." Thus patents, together with military law, copyrights, immigration, and bankruptcy, make up federal question jurisdiction as explicitly described in the Constitution and exercised by Congress.

from the courts reveal their interpretations of the law. From these cases we will learn the nuts and bolts of patent law and by thoroughly familiarizing ourselves with their facts and legal reasoning, we can become better at divining how situations relevant to our own might be decided, should they ever end up in litigation.

From a procedural standpoint, unfavorable patent decisions by USPTO examiners (e.g., rejected patent claims/applications) can be appealed to the U.S. Board of Patent Appeals and Interferences (the Board). If the outcome at the Board is unfavorable to the applicant then the applicant may appeal the decision to the Court of Appeals of the Federal Circuit (CAFC). Unfavorable decisions from the CAFC may be further appealed by petition for a writ of certiorari to the U.S. Supreme Court.[6] While patent appeals from the Board to the CACF may be made as a matter of right, appeals from the CAFC to the Supreme Court are not. Due to the Supreme Court's broad jurisdiction and very limited capacity to hear cases, only a very small percent of cases that are appealed (approximately 1%) are heard (so good luck!).[7] Patent litigation on the other hand involves the enforcement of patents that have already

---

[6]Most case opinions cited in this text will be from the CAFC and in some case the U.S. Supreme Court. All Supreme Court, CAFC, and most CCPA (CAFC's predecessor) case opinions (back to 1950) cited in this text can be found at www.altlaw.org. District court cases, unfortunately, are harder to come by without a pay service; however, cases back to 2004 can be found for free at http://cases.justia.com/federal/district-courts/. Opinions of the Board from 1997 can be found at http://des.uspto.gov/Foia/BPAIReadingRoom.jsp. Case citations as presented herein will give the case heading (e.g., *Bayer Corp. v. Schein Pharm. Co.*), the court the opinion came from, and the year of the opinion. There usually will also be a cite listing to the federal reporter (a type of journal for federal cases) from which the opinion can also be accessed, though this requires a fee. Unless you have access to the federal court reporter referenced, you should not use the court reporter citation when you search the case. So if you find the following cite: *Bayer Corp. v. Schein Pharm. Co.* 301 F.3d 1306 (CAFC 2002) and wish to find the free copy of the decision, you will not want to include the citation to the court reporter and page (301 F.3d 1306). Searching the court opinions at www.altlaw.org is generally best accomplished by searching the case heading only (*Bayer Corp v. Schein Pharm*) and by not including the cite to the federal reporter or any of the other information—if you include more information than just the case title, then you are likely to pull up the many cases that are citing that case but not the case itself. If a spot cite is made to a specific page of the reporter, then usually there will be two page numbers separated by commas after the reporter edition. The first number is the page on which the case starts in the reporter and the second number (or page range) is where the quote or specifically referenced material can be found. Quoted material in the text or notes may or may not refer to the page in the reporter, and the case opinion may or may not refer to a federal reporter citation, depending on how the authors accessed the information. If a page number is cited, that page number refers only to where the referenced material can be found in the court reporter and not to the actual case opinion that you will find at the free web locations listed here.

While little if any of what you work on in the patent arena is likely to be involved in litigation before a court, the assumption should still be made that it will because even short of litigation, the perceived validity of a patent is often the key determinant of whether a patent will actually be challenged in court. As a corollary, the value of a patent is often a function of its perceived validity, and as a result, many business deals involving patents are structured accordingly. Competent patent validity opinions prepared by patent counsel are based on legal reasoning primarily derived from federal patent case law precedence.

[7]A petition for certiorari means that the petitioner is asking the court to use its discretionary power to review the case. The Supreme Court hears most cases on petition known as a writ of certiorari. Of the thousands of writs received in a given year, fewer than 100 are granted. To accept a case on a writ of certiorari, four out of the nine justices must agree (the so-called rule of four) that the petition should be granted.

been issued and is not handled at all by the USPTO but rather is initiated at the federal district court level where appeals can be taken to the CAFC (and on the very rare occasion from the CAFC to the Supreme Court).

Like the hierarchy of legal precedence, we will generally follow the statute to the court decisions to their practical implications. While following this path, we will be guided by quotes and comments liberally extracted from the text of the actual opinions rendered in those cases. What you will come to appreciate is that the law is seldom fixed in one court decision but rather evolves both in time and to fit the particular facts of a given controversy. It is simply impossible to write the law as it exists but it is sometimes possible to make reasonable guesses as to how things might turn out for a given circumstance. From reading the cases, we might come to recognize fact patterns that in some way resemble our own situations and from that take either solace or caution.

This text also includes hypothetical examples to either supplement the level of detail on a particular topic or fill in gaps where it appears no court decisions are on point. You will notice that the text contains copious footnotes. Not only does this keep with the traditions of written court opinions and legal casebook format but it also serves the best interest of you, the reader. There is simply no other best way to honor the cohesiveness of the text while at the same time fully elaborating on a particular nuance or tangent, of which the study of law is full.[8]

Our studied goal in this effort was to make the topic both approachable and useful to the practicing chemist through the use of cases (where possible) and examples that relate to the chemical arts. You may find an emphasis in the cases toward small molecule organic chemistry with a distinct pharmaceutical bent, and this is to a large part due to the fact that federal legislation related to generic drugs encourages patent litigation. Where possible, the subject matter has been broadened through the use of additional hypothetical examples that derive from a wider diversity of chemical subject matter.

We know that getting answers to patent questions can be difficult. Many times the same question can yield different responses, the answers often mysteriously cloaked in vague and shifting terminology. This book cannot provide you with all of your answers, but we hope it will at least provide you with a better place to start. If you enjoy reading this book at least half as much as we've enjoyed writing it, then we'll consider this project a great success.

---

[8]"For every complex problem, there is a solution that is simple, neat and wrong," H. L. Mencken.

CHAPTER 1

Patent Basics

1.1 INTRODUCTION

All too often one of the biggest challenges in any major endeavor is figuring out exactly where to begin. In the endeavor of learning patent law that difficulty is magnified by the fact that patent topics are often highly interwoven so that no matter where one starts, they will inevitably be drawn into other related topics. Because of this, our conceptual framework will not always be assembled in a strictly linear fashion but may come to better resemble a mosaic of interlocking pieces. For some, the sometime frequent diversions from one topic to another may be annoying, but for others (e.g., those with attention deficit disorder) such diversions can be welcome. No matter which camp your flag flies over, the benefit to seeing the same topic introduced multiple times for different purposes adds significant value to the learning experience. Although repetition can magnify boredom, it can also reinforce learning, especially where the repetition occurs in different settings, perhaps resonating somewhat differently each time. Topics will sometimes be presented once or twice as a peripheral component when necessary to help explain the main topic, only to appear yet once again, perhaps this time as the main topic. To the extent such digressions are truly peripheral, they will be explained in footnotes, often with a reference to where in the text that topic is part of the central discussion. I strongly encourage you to read the footnotes because they do much to fill in the mosaic. You'll notice the tone in the footnotes is typically more relaxed and conversational, allowing an expansiveness in nuance and detail that is not always so readily accommodated in the text.

So where to begin? In this first substantive chapter, we'll start with some opening thoughts on property, move to patents as property, and end with, a discussion on the fundamental differences between owning a patent and having freedom-to-operate.

1.2 PATENTS AS PROPERTY

> Private property began the instant somebody had a mind of his own.
>
> —E. E. Cummings

The Chemist's Companion Guide to Patent Law, by Chris P. Miller and Mark J. Evans
Copyright © 2010 John Wiley & Sons, Inc.

Most of us have grown up with tangible personal property and are familiar with the concept of ownership of this type of property. In fact, even if we hadn't grown up with the concept of property ownership, its prerequisites and attractions seem to be almost instinctively shared. If this seems dubious to you, just picture for a moment two children fighting over a toy—each one grasping firmly to an end and shouting: "Mine! Mine!" Fortunately for lawyers, this primal behavior is not limited to children. Society's attempts to organize and control the basic desire to own property are reflected in the fact that entire areas and many subareas of the law are dedicated to the rights and duties of property owners. Since society places such great emphasis on property, it is perhaps worthwhile to consider just what it actually means to *own* property in the first place.

Perhaps a good place to start our quest for understanding the nature of property ownership is to understand its limitations, beginning with the maxim, You never can truly own anything. Despite the simplistic nature of this assertion, it provides a valuable first clue into the nature of property ownership because thinking of property ownership in the legal sense means thinking about the legal *rights* of ownership. In this vein, one would be hard pressed to find an example of an absolute set of legal rights to anything, and the legal rights associated with property ownership are no exception. For example, even though we say a person owns their automobile, he probably does not have the legal right to drive it through his neighbor's yard; his ownership does not grant unlimited rights with respect to what he owns. Similarly, forgo the property taxes on your house, and you probably will not be living there for too long—ownership often poses affirmative duties to maintain that ownership. The points here are that property ownership is limited and ownership is almost never an all or nothing proposition. Rather than viewing property ownership in a fixed, static, or absolute state, it is more appropriate to think of property as a bundle of sticks in which each stick represents a right or responsibility. Rather than owning a piece of property, one actually holds one or more rights related to that property, often accompanied by one or more obligations as well. The more absolute or complete the ownership of property by one individual or entity, the more rights from the bundle that individual or entity has. For example, real property (e.g., an apartment) may be rented or owned, singly or jointly, by two or more persons at the same time. In the latter case of joint ownership, the bundle of ownership rights might be split right down the middle with each person taking half of the entire bundle. Real property may be commercial, residential, or zoned nonhabitable, but in every case it still has at least one owner—even if that owner is the government. Although the variations in property ownership are almost limitless, the bundle of rights associated with property ownership is probably never enough to be absolute—nobody ever has all of the sticks in the bundle.

1.3 PATENT RIGHTS ARE RIGHTS TO EXCLUDE

In spite of the many shades of ownership interest and the corresponding rights to which that ownership interest represents, there is often one very important right that is common to property owners and that is the right to exclude others from the property, at least up to the extent the owner owns the property. So even though the owner may be limited in her own ability to affirmatively use or enjoy the property, she usually retains the power to exclude others from using or enjoying the property. Thus, although a person with a suspended or revoked driver's license might not be able to legally drive the car that she owns, she still can decide whether another individual may use the car. Likewise, consider a landowner who owns a piece of land but cannot legally hunt on it because she does not have a hunting license. Despite the fact that she cannot legally hunt on her own land, she still might have the power to decide whether another person who has a hunting license can hunt on her land. In both of these examples, the person owning the property does not have the legal right to enjoy that property in certain ways but she still, as the legal owner, has the power to exclude (or allow) others to use the property for those same purposes. From a commercial standpoint, this power to exclude or allow is significant because one can potentially make tremendous economic benefit of this legal right. A person without a driver's license might still own hundreds of automobiles that she rents out to others or the landowner who doesn't have a hunting license might still make significant revenue renting out her land to hunters. This aspect of property ownership is a cornerstone of the economics of commerce and property—if an owner did not have the power to exclude, the power to allow would have little value because nobody would want to pay for something she could otherwise get for free.

So what does all this have to do with patents anyway? The answer is that patent rights are in certain very important ways similar to property rights, and one might think of the claims[1] in a patent taking the place of the deed to a piece of real property or the title to an automobile; the claims of the patent define the "metes and bounds" of the property in the same way that a deed to real property or a title to an automobile describes those pieces of property. Instead of a patent owner, we typically refer to that person as the patent assignee, but the effect is the same. In the same way that a person holding legal title to a piece of real property such as land might be able to lease her land to another party, so might a holder of a patent license lease some or all of her stake in the claimed invention. However, we have already seen that holding the deed to the property does not necessarily give one unlimited rights to the property; think back to the car owner who couldn't drive her own car or the landowner who could not hunt on her own land. The same is true for patents, and this is one of the most conceptually difficult *but important* concepts in patent law to understand. Owning a patent does not grant one the right to "do" the thing that the claim describes. Owning a patent grants its owner *the right to exclude* others from practicing the claimed invention.[2]

---

[1] We will discuss claims in Chapter 5, but for now claims are a set of numbered single-sentence paragraphs located at the end of a patent that define the invention.

[2] The right to exclude others from practicing the invention means the right to exclude others from making, using, selling, offering for sale, or importing the claimed invention. 35 U.S.C. §154.

1.4   PATENTS DO NOT CONVEY FREEDOM TO OPERATE THE INVENTION

As we just learned, the right to exclude can be a very valuable right; implicit in the right to exclude is the right to not exclude. Where one party has the right to exclude others from making, using, or selling a claimed invention that others covet, then the party holding the right has something of value. Very often, this value can be equated to some cash value by which the party holding the patent rents or sells that intellectual property space to others. Alternatively, the party holding the intellectual property space may want to occupy it himself. However, whether the owner of the patent decides to rent that space to another party or attempts to occupy it himself, neither the other party nor he is guaranteed by the patent right he holds to be allowed to make, use, or sell the invention that he claims since the right of the patent holder is to exclude others from the claimed patent space—the patent holder cannot grant anything or license anything to another party that he does not hold himself. You may find it odd that a person could invent something, be awarded a patent but yet not be able to practice what is described in the patent, but yet still grant the rights under that patent to somebody else who himself may or may not have the right to practice what is claimed in the patent he has taken the rights to from the first person. The primary reason for this dichotomy is that the criteria for obtaining a patent are not the same criteria for determining whether the patent obtained can be used. In other words, the USPTO is not concerned with whether you can legally practice your claimed invention; they are solely concerned with whether your claimed invention meets the requirements of patentability.[3] This does not mean that you, the patentee do not have to be worried about whether you can practice the invention claimed in your patent; it all depends what your goals are. Do you wish to practice your invention, license it to another, or simply brag to your friends about it (or even all three)? If you wish to practice the invention described by your granted claims, you will want to be sure that you are not infringing somebody else's patent(s) in the process. The primary inquiry then is whether you have "freedom to operate" the claimed invention. If practicing your invention infringes another's patent(s), then you need to consider whether you can find a way to practice your invention that avoids his patent, or consider approaching him to see if you can license his patent(s), or perhaps you are confident that his patent is not valid and that if he tries to assert it, you can get his patent invalidated in a federal court.

In this sense, having freedom to operate is like having a shield that defends one from charges of infringement, whereas a patent acts like a sword, allowing its holder to go on the offense by excluding others from the claimed space of his own patent. In some cases, a chemist may want freedom to operate only as he may simply wish to make, use, or sell a certain composition or process. He does not care to try to exclude

---

[3]The USPTO doesn't help the assignee to enforce his patent rights either. As mentioned previously, patent litigation is taken up in the federal courts if agreeable terms between the patent assignee(s) and putative infringer(s) cannot be reached. But it is up to the patent holder to enforce his own rights, the USPTO will not do it for him.

others but simply wishes to do something without being bothered by the possibility of an infringement suit by another. In other cases, the chemist may obtain a patent to a composition or process that he is not himself interested in making, using, or selling because he knows the invention would be valuable to others who do not want to be excluded from the claimed invention. Finally, the chemist may wish to make, use, or sell the product of his invention, and he may also wish to own a patent to that invention so that he may prevent others from making, selling, or using his invention so he has the sole or exclusive right to do what he wishes to do. In some cases one needs a shield, in some cases one needs a sword, and in some cases one needs both. Not all three choices are available in all circumstances, but understanding one's needs is a necessary first step for staging the inquiry that follows.

Knowing whether and to what extent one has freedom to operate in the realm of patents is usually more complicated then understanding one's rights regarding other types of property. For example, in real property, the rights and restrictions that run with the property usually can be figured out in a *relatively* straightforward manner by a title search. When a title search is performed, the chain of title is searched back in time to make sure that clear title can in fact be passed by ensuring that each party in the chain of title had conveyed the title in a legally competent manner. In the same manner, that title search should turn up any restrictions that run with the property so that the new owner can be sure he is getting the rights to the land he thinks he is getting. If somebody besides yourself has mineral rights to your land, you might wish to know that before you buy the property lest you be unpleasantly surprised one bright shiny morning when the mining crew shows up in your front yard to begin the excavation. Likewise, before a party wishes to prosecute, license, or even purchase a patent, that party likely will want to know whether the patent he covets is sufficient for him to practice the claimed invention.

In contrast to a piece of real property and its attendant deed, a freedom-to-operate search for intellectual property generally involves a much more multivariate inquiry. The practice of any given technology often requires a large number of steps, any of which may involve processes or materials that may be the subject of different patent claims held by other parties. Holding a patent, unlike the property deed, will not give one notice of whether practicing the claimed invention will infringe another's patent—such an inquiry must be independently conducted by searching databases that catalog patent claims by various search descriptors (e.g., chemical structure, compound name, keyword).

1.5 CONTRASTING FREEDOM TO OPERATE WITH PATENTABILITY

The freedom-to-operate inquiry is much broader and more complex than the basic consideration of whether clear title to the patent is being offered—even a clean assignment or license to a particular patented invention does not mean that one has freedom to operate by making, using, or selling the claimed invention if that patent results in the infringement of one or more patents held by another party. Furthermore, the freedom-to-operate inquiry is relevant whenever one wishes to practice in an area, so it is entirely separate from the question of whether one has a patent, wishes to get a patent, can get a patent, etc. because the freedom to operate inquiry primarily concerns itself with whether one can practice in a particular area without infringing a patent or patents held by one or more outside interests. It is entirely possible that multiple overlapping patents are held by different parties such that no single party has freedom to operate. Because of this, the freedom-to-operate inquiry is in no way limited to whether the party initiating the inquiry has a patent on the technology or if she doesn't—it simply asks whether the thing she wishes to do can be done without infringing somebody else's patent. To solidify these abstract concepts, let's consider a scenario that involves issues of patentability and freedom to operate to demonstrate one commonly encountered way in which patentability and freedom to operate might interrelate in an industrial chemical setting.

In our first example, a chemist working in the area of adhesives has been stuck trying to find a biodegradable adhesive useful for joining two surfaces together in a way that will provide for a strong bond but yet be degradable when exposed to protein hydrolyzing enzymes. Several approaches were tried and failed but in an entirely *novel* and *nonobvious* (requirements of patentability that will be discussed in great detail in future chapters) way, she polymerized a non-natural amino acid to form a material with strong but biodegradable properties. Her company decides to file for a patent, and after a period of several years, the patent issues with the single claim shown in Figure 1.1.

The company is interested in producing the polymer in bulk for sale to several customers that have expressed an interest in the product. The polymer will be produced by heating the amino acid methyl ester precursor according to specially staged process conditions followed by base hydrolysis of the C-terminal ester and stoichiometric protonation with dilute acid as shown in Figure 1.2.

1. A polymer having the following formula:

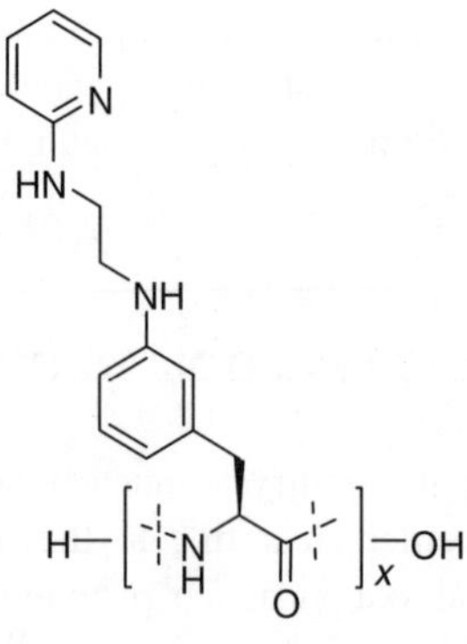

where x = an integer from 40 to 6000.

FIGURE 1.1 Claim 1 for biodegradable polymer of non-natural amino acid.

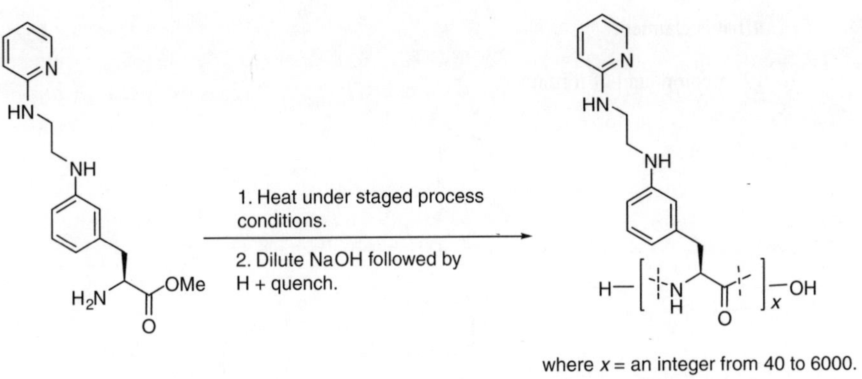

where x = an integer from 40 to 6000.

FIGURE 1.2   Process for making biodegradable adhesive.

They intend to manufacture and sell the adhesive in the United States. Before scaling up their manufacturing process, they do a freedom-to-operate search and uncover a U.S. patent filed a year before they began their work that describes and claims certain amino acids and their certain corresponding esters. Claim 17 of this earlier issued patent is of particular relevance (Figure 1.3). Notice that the genus of the compounds described in claim 17 generically covers the compound that the manufacturer's of the biodegradable polymer need to use for their own synthesis.[4] Thus if the adhesive chemists wish to prepare their biodegradable polymer according to the method they have currently planned (shown in Figure 1.2), they will infringe claim 17 of the prior issued patent. So even though the adhesive chemists own an issued patent that claims the polymer that they wish to make and sell, they cannot make it according to their current process without infringing the other party's patent. In such a case as outlined in this hypothetical, the choices available to the erstwhile manufacturer of the biodegradable adhesive include coming up with a process that does not infringe the recently issued patent, seeking a license from the patent holder (or another party that has a license to the compound and the right to sublicense), or buying the desired compound from the patent holder (or buying the compound from some other party that has a license to make and sell the amino acid methyl ester). Alternatively, the adhesive manufacturer may believe that the patent they are infringing is not legally valid or enforceable. In this scenario, they may seek out a

---

[4]The compound claim shown in Figure 1.3 contains a Markush group and so is often referred to as a *Markush claim*. Markush claiming is a very important subject in chemistry art patents and will be covered more extensively in Chapter 5. The Markush claim is referred to as a *generic* claim because it describes not a single compound but rather a whole *genus* of compounds whose scope is defined by the core structure and variables. Any particular compound within the genus is called a *species*. For now, appreciate that the proposed process would infringe claim 17 because it starts with a compound falling under the defined scope of claim 17; for their proposed starting material, X is N, Y and Y′ are each NH, and the chain is linked to the meta position of the phenyl group to which it is appended, a species covered by the genus of claim 17. In regard to the positions allowed for Y′ in the claimed Markush of the patent shown in Figure 1.3, the bond intersecting the phenyl ring between carbon atoms indicates that Y′ may be located at any position of the phenyl ring to which it is affixed.

What is claimed:

17. A compound of formula I:

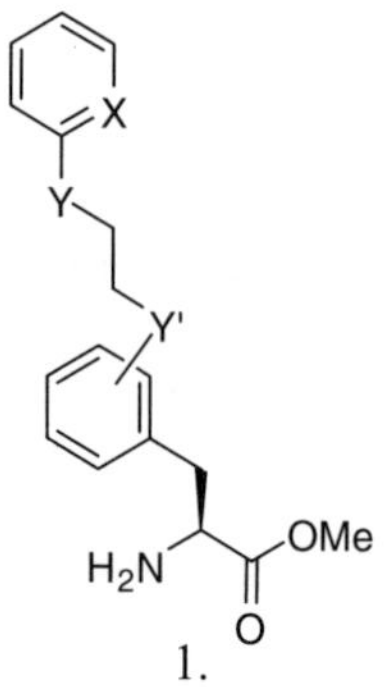

where Y and Y' are each independently selected from CH_2, O, S, and NH; and X is selected from CH or N.

FIGURE 1.3   Claim 17 from recently issued patent.

legal opinion to confirm their belief and then proceed to manufacture at risk of an infringement suit. If they are sued for infringement, they might argue many things, including that they are not actually infringing the patent or that the patent being enforced against them is invalid or unenforceable.[5] Suffice it for now to appreciate that this latter strategy is not without risk because the assignee of the prior patent might be able to get an injunction forcing our erstwhile adhesive manufacturers to stop making their present precursor and/or be forced to pay up to three times the actual economic damages from their infringing activity.[6] Add to this the high costs and the great deal of uncertainty about any outcome due to the technical and legal complexities of the subject matter itself, and one can begin to appreciate the kind of risks involved with this latter strategy.

Before leaving this section, let's briefly appreciate that the claimed polymer by itself did not lead to infringement of the patent issued to the other party but rather it was the particular intermediate used in the manufacturing process of the polymer that infringed that patent. As a result, we saw that designing around the prior art patent was a viable option since it might be possible to make the polymer of the claimed invention if one could avoid the amino acid methyl esters of claim 17. For example, perhaps it would have been possible to polymerize the amino acid itself rather than the amino acid methyl ester of claim 17. Alternatively, perhaps a different ester

---

[5]Approximately 30% of patents litigated at the federal district court level are found to be both valid and infringed.

[6]Damages for willful infringement can be awarded when the patent holder demonstrates by clear and convincing evidence that "the infringer acted despite an objectively high likelihood that its actions constituted infringement of a valid patent." For a recent holding on the requirements for willful infringement, see *In re Seagate Technology, LLC* 487 F.3d 1360 (CAFC 2007).

Claim 1. A method of providing contraception in a mammal comprising the administration of a compound of formula I,

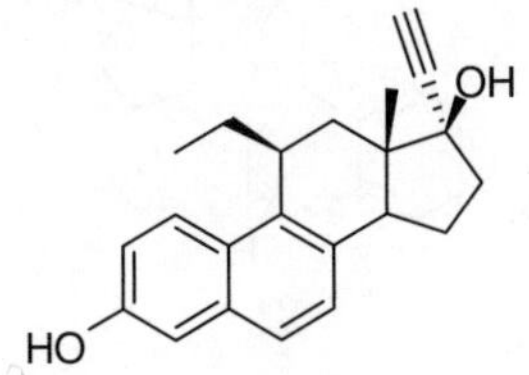

I

in combination with a progestin to a female mammal.

Claim 2. The method of claim 1 wherein said mammal is a human.

FIGURE 1.4   Claim to method using steroidal estrogen in combination with progestin.

might have been useful for the same task.[7] However, in some cases a design around is not possible because the invention that is sought to be practiced is completely encompassed by one or more patented claim(s) held by one or more other parties.

Let's consider a second hypothetical example with somewhat different subject matter, a case in which a design-around solution is not available. A pharmaceutical chemist has synthesized a steroidal compound and found that it is a particularly potent estrogenic agent in preclinical testing. The chemist's company believes the compound could be useful in combined estrogen/progestin contraceptive regimens and so decides to place the compound into clinical studies and files a patent application for a method of using the compound in combination with a progestin for use as a contraceptive. The chemist's company filed a patent application to the invention that included the two claims shown in Figure 1.4.

---

[7] These work-around solutions would avoid literal infringement of the patent in question, but infringement under the doctrine of equivalents would still be possible. According to the doctrine of equivalents, a party can be found to be infringing in situations even when their activities do not literally meet all of the elements described in the claims. Under one test for the doctrine of equivalents, an accused device or chemical can still be found to infringe if it performs substantially the same function, in substantially the same way to achieve substantially the same result. See *Gravers Tank and Manufacturing Co. v. Linde Air Products Co.*, 339 U.S. 605 (1950). The doctrine of equivalents may also be applied when there is only an insubstantial change between each element of the accused device or chemical and the claimed product. See *Warner-Jenkinson Company, Inc. v. Hilton Davis Chemical Co.*, 520 U.S. 17 (1997). The doctrine of equivalents is fairly controversial—it makes it very difficult for somebody to know simply by reading the issued patent whether her product or activity will infringe the patent. Commerce requires predictability, and ultimately, if the patentee had intended a broader reach for her claimed invention then she should have pursued those broader claims in the first place. It seems manifestly unfair and downright confusing to allow a patent to capture a broader scope than what the four corners of the claim set out. The doctrine of equivalents is subject to certain limitations and can be fairly difficult to understand in practice. Just knowing that it exists is enough for this stage of the game!

Claim 13. A compound of formula XI

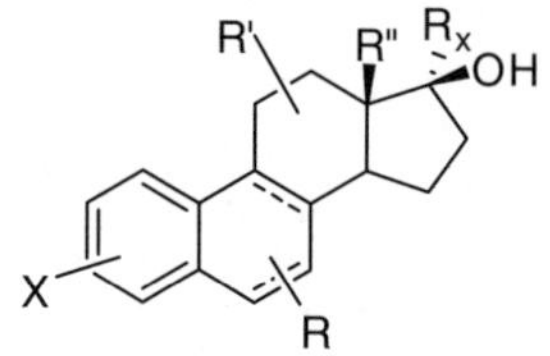

XI

where

X = H, C_{1-4} alkyl, CN, halogen, OH, SC_{1-4} alkyl, or OC_{1-4} alkyl;
R = H, C_{1-4} alkyl, CN, or OC_{1-4} alkyl;
R′ = H, C_{1-4} alkyl, CN, halogen, or OC_{1-4} alkyl;
R″ = H or C_{1-4} alkyl; and
R_x = H, C_{1-4} alkyl, C_{2-4} alkenyl, or C_{2-4} alkynyl;

and pharmaceutically acceptable salts thereof.

FIGURE 1.5   Compound claim 13 from previously issued U.S. patent.

Between the time of their filing the patent application and its issue, the company was planning several clinical studies when it came to their attention that another issued U.S. patent existed, which was filed before their chemist synthesized the compound of formula I. The patent of concern was directed to steroidal compounds, although it did not disclose their particular compound of formula I, nor did it indicate any specific utility in combination with a progestin for contraception. The patent specification (the entirety of the document) disclosed a number of Markush structures of varying scope as well as a description of the compounds as having utility for the treatment of prostate cancer. Claim 13 from the issued patent is illustrated in Figure 1.5 and it particularly concerned them.

A quick comparison of the Markush structure XI of claim 13 of the prior patent to the structure of the contraceptive steroid I of claim 1 of the chemist's patent indicates that the compound of formula I falls within its scope. In particular, consider when the compound of structure XI has X = OH, R = H, R′ = CH_2CH_3 (this is a C_2 alkyl group, thus falling within the C_{1-4} alkyl definition), R″ = hydrogen and R_x = $-C{\equiv}C-H$ (this is a C_2 alkynyl, thus falling within the C_{1-4} alkynyl definition). The two dashed lines paralleling the backbone in the steroid's B-ring indicate that those two bonds can independently be single or double bonds. The compound of formula I requires two double bonds in the B-ring of the steroid, and the presence of the two dashed lines in the B-ring of the Markush of formula XI indicate that this combination is covered in XI.

The chemist dutifully brought this prior patent to the attention of her company's patent counsel, who promptly submitted the reference to the USPTO as prior art for their consideration.[8] The USPTO found that the reference did not affect the patentability of claim 1, stating in the relevant part:

> The broad genus of the prior art patent disclosed in claim 13 does not negatively affect the patentability of applicant's pending claim 1 or claim 2. In particular, the very breadth of the prior art genus does not render the particular compound of formula I not novel, nor does it render it obvious. There is no suggestion, teaching or motivation provided by this reference to arrive at the compound of formula I nor does this reference, combined with any other prior art references teach or suggest a compound of formula I. Even if the compound of formula I were structurally obvious in view of the prior art disclosure, there still would be nothing to suggest its combination with a progestin as is require by claim1 of the pending application. Claims 1 and 2 are allowed.

The chemist and others at her company were quite pleased with the USPTO's decision, as this indicated they would soon be issued a patent to their exciting discovery, and sure enough, they were correct. The chemist mistakenly assumed that this meant they were now free to pursue the development of their compound to which they hoped would eventually lead to its successful commercial launch. What she had failed to realize, however, was that even though her company had obtained the issued patent, her company was not so free to make, use, or sell the compound described in claim 1. As the company's patent counsel explained to her and as we have already learned, a patent represents a right to exclude another from making, using, or selling the claimed invention but does not, by itself, allow one to make, use, or sell her own invention—it is a sword and not a shield. Likewise, the prior art patent grants its owner the same rights against others: It allows its owner to exclude one from making, using, or selling the subject matter of its claims. As we have already determined, the structural subject matter of claim 13 of the prior patent encompasses a compound of formula I of our chemist's claimed method. Our chemist was quick to point out to the company's attorney, however, that they were claiming and using the compound much more specifically—they required the compound to be used in combination with a progestin as a method of providing contraception, and the prior patent simply makes no mention of that use. The company's patent counsel explained to her that unfortunately for them, the prior patent was broad enough to encompass the method and the combination that they were proposing to test and eventually (they hoped) bring to market. The following conversation ensued:

> The company patent counsel explained: "In the prior patent, they claimed the compound only, with no further limitation. The fact that their claim does not specifically list any use of that compound does not mean that any use of that

[8]Patent prosecution requires a duty of candor that includes a duty to provide material information to the USPTO during the prosecution of a patent application. This duty applies to patent counsel, inventors, and anybody else substantively involved in the preparation of the application and who is associated with an inventor, assignee, or one who is obligated to assign the patent application. Failure to comply with this duty can result in a finding of inequitable conduct, which can render the application unenforceable. The subject of inequitable conduct will be discussed in detail in Chapter 2.

compound is not covered, in fact, it's quite the opposite. This means that the making, using, or selling of a compound falling within their claimed genus for essentially any purpose is an infringement of claim 13 of their patent."[9]

The chemist was understandably upset because she thought "Didn't we just get a patent covering what we wished to do?" So she asked: "How did we get a patent if what we did was already covered by that other patent?"

The company's lawyer responded, "The prerequisites for patentability are different from whether your claim is 'covered' by somebody else's claim. Patentability with respect to the prior art requires novelty and nonobviousness. Their issued claim is part of the prior art, and so it will be considered only for whether it renders our claim not novel or obvious. In the last substantive office action we read why the patent examiner thought that the prior art patent claim did not negate the patentability of our claim and for what it's worth, I completely agreed with his reasoning!"

"Well okay, I think I've got that but then haven't we already infringed their patent? I mean, isn't the mere publication of our patent with its claim that uses their compound going to tip them off that we used their compound without their permission?"

The chemist said with a worried look on her face.

"*Shhhhh*, let's ease up on that infringement talk, somebody might hear us," the company lawyer joked. "Seriously, that's a very good question and the answer is yes, no, and maybe. It's a good thing I work for the company and am not billing by the hour because you'd be running up some pretty good fees by now. Well here goes. On its face, if we did what we describe in our claim then, yes, we are literally infringing their claim. However, as to whether our published claim itself would constitute proof of the infringing act then the answer is no. For example, patents can rely on prophetic or constructive examples for support of the claimed subject matter, meaning the actual thing being claimed may not have been actually been performed as claimed. Alternatively, the thing being claimed might have been supported by work that might have taken place outside of the United States. In other words, our patent did not say where the actual work was done and any importation of *information* obtained from that work should not infringe their U.S. patent claim."[10] Pausing to catch his breath, the company patent counsel continued, "Even if one did accept the fact that we infringed

---

[9]Reading and understanding claim language will be discussed more in Chapter 5.

[10]This does not mean that you can avoid infringement by importing anything that you want from a country, where the invention is not patent protected since importing of a claimed invention is one of the ways that a patent can be infringed. However, importation of information that was obtained through an infringing use does not necessarily infringe. In the hypothetical discussion, the infringing use would have been making, using, selling, or importing the claimed compound. Simply writing the structure down and explaining how one made it would not be. Therefore, if the claimed steroid had been made and tested in a country outside the United States and the only thing transferred was information related to the compound (such as what would be included in the patent application), then that importation of information should not be an infringement of the U.S. Patent law. See *Bayer v. Housey* 340 F.3d 1367 (CAFC 2003), where the importation of information gained from a claimed process was not deemed to be an infringement of the process itself.

the patent, there are exceptions that are sometimes available to infringers to allow them to escape damages normally available to the patent holder.[11] In our particular circumstance, *maybe* we could qualify for an exception and thus escape the prior patent holder's wrath. Finally, as a practical consideration, patent litigation is very time consuming and costly. Not only is the outcome of litigation uncertain, their patent could end up being declared invalid as a final result of all of their cost and effort. But for the sake of argument let's say that hypothetically we did infringe their patent, what would the damages be?"

Answering his own question, as by now it appeared the chemist's attention had wandered elsewhere, the lawyer continued, "Chances are they would be nominal at best since our use was *very* limited and never sold or extensively practiced. If damages were to approximate a reasonable royalty negotiated between us and them in an arm's length transaction, I'd say the amount would be very small. In fact, the prior claim holders would probably be more interested in letting us develop our invention and run our clinical trials. Eventually, we would need to approach them before commercializing the product, and the more developed it is when we do finally approach them, the more they will be able to charge us on a license because the more we will have invested in the project. Does this answer your question? Is there anything else you'd like to know? Hey, hey."

The chemist, groggily coming back to her senses and not knowing how long she had drifted away realized that the lawyer was asking her something and panicking a little bit, managing to squeeze out "Yeah, sure thanks. Look, I think I just remembered that I have a reaction to turn off so I better get back to the lab real quick," before rapidly turning and walking briskly back down the hallway, careful not to turn and look back lest the lawyer might be tempted to start talking to her again.

While our somewhat worn-out chemist may have run out of questions, perhaps there still is a question or two yet worth considering. For example, if the chemist's

---

[11] There is a common law (created by courts) experimental use exception available for infringing uses that are strictly intellectual or philosophical explorations, but as a practical matter, it is applied under very limited and narrow circumstances and would not be applicable in the current circumstance in which the infringing use has an ultimate commercial goal (a new contraceptive method). However, there is a statutory experimental use privilege 35 U.S.C. §271(e)(1) originating from the Hatch-Waxman Act, which governs large sections of patent law as it relates to generic drugs in the United States. Notably, this statute has been read broadly to cover circumstances beyond generic drug manufacturers and likely could be read to provide an exception to the infringing use of the chemist's company in the instant example. The statute exempts infringing activity "solely for uses *reasonably related* to the development and submission of information under a federal law which regulates the manufacture, use or sale of drugs or veterinary biological products." In the instant case, the chemist's company was using the compound (the infringing activity) for developing a method of providing contraception with the eventual goal of submitting an application (presumably an investigational new drug [IND] application for initiating clinical trials) to the FDA for use of the contraceptive method described in their claim. The words *reasonably related* have been interpreted broadly by the Supreme Court to include inter alia activities beyond those engaged in solely by generic companies. See *Merck KGaA v. Integra Life Sciences I, Ltd.* 545 US 193 (2005). This does not mean they could then go onto sell the compound because the sale would be outside the scope of 35 U.S.C. §271(e)(1).

company cannot practice claim 1 or claim 2 of her own issued patent without infringing the prior patent, can the owner of the prior patent practice the subject matter of chemist's company issued claim without infringing the chemist's company's claim? The answer is (drum roll please)...no! Just as the prior patent owner can exclude the chemist's company from using a compound falling within the formula XI (which encompasses their compound of formula I) for any purpose including contraception, the chemist's company may exclude the owner of the prior patent from using the compound of formula I in combination with a progestin for contraception. If this seems a little bit odd and perhaps even unfair to the owner of the prior patent (after all, they were first) then you need to think again about what it is that the prior patent actually does and doesn't grant its owner. The prior patent owner was awarded their patent first and it covers any compound falling within the scope of their genus defined by formula XI for any purpose. But their patent never gave them freedom to operate, only the right to exclude others. Just as the prior patent holders were granted a right to exclude others, others can still have the right to exclude them, even when those others obtained their patents after the holder of the earlier patent.

In the example we just discussed, the first (and typically broader patent) is often referred to as the *dominant patent*, and the second, narrower patent is often referred to as a *selection patent*. Since the initial patent covers the use of a broad genus of compounds for any use, the later claim covering a single compound for a particular use might be thought of as a double selection—a selection of a compound from the prior genus as well as a selection of a single use from a full range of possible uses.[12] The practical outcome in our example is that both parties are blocked from commercializing a small portion (in the case of the prior patent holder) or all (in the case of chemist's company) of their claimed invention. In order for our chemist's company to commercialize their invention they would need to license the right to use of the prior patent from its owner. Likewise, if the prior patent owner wished to use the compound of formula I in combination with a progestin for contraception, they would need to license or purchase the patent from the chemist's company.

From a public policy perspective, the patenting of selection inventions is a good thing because most of the time such inventions reflect a refinement or often an improvement on the earlier described invention, and the patent system exists largely to encourage and reward such improvements, even when the changes are evolutionary rather than revolutionary. To each inventor should go the rewards of her invention, and it should not matter whether her invention came before or after another's; it should matter only whether she has actually made an invention. Overlapping inventions are a natural outcome of a system that encourages competition, and any attempt to limit this aspect runs contrary to the purpose of patent law. Furthermore, to the extent that

---

[12]Selection patents in the chemical arts can take many forms. Selections may be made from a prior art broader range of compounds or compound uses (as in the current example) but are not so limited. For example, selection inventions might also be made from prior art chemical processes where, for example, a broad temperature range or reaction time is disclosed and a later, narrower embodiment is discovered that provides a patentably distinct process. Selection inventions are also sometimes referred to as *improvement inventions* because the later selection may provide some unexpected result or benefit that helps overcome challenges to patentability based on assertions of obviousness of the later discovery. Obviousness challenges and rebuttals to obviousness challenges are discussed in more detail in Chapter 8.

patent law is criticized for stifling competition at the same time that it encourages it, allowing dominant patents to exist monolithically would lend further weight to such criticism.

The two hypothetical examples we just discussed served to illustrate that patentability and freedom to operate are truly different inquiries. The challenge facing a party seeking to get a patent is whether their invention meets the criteria for patentability. The challenge facing the party wishing to practice their invention or technology is whether they can do so without infringing another party's patent. As we can see from the examples, having one does not secure the other.

1.6 ASSIGNMENT AND RECORDING OF PATENTS

Despite the differences between real property deeds and a patent, the chain of title associated with real property does have its corollary in patents in which a patent holder has assigned his rights to the claimed invention to one or more other parties. When a title to a piece of real property is transferred from one party to another, that transfer is typically recorded at the municipal registrar's office. When the ownership of a patent is assigned from one party to another, that assignment should be recorded at the USPTO in the name of the new assignees.[13] This recordation has the legal effect of providing notice as to who is the holder of record, much in the same way that the transfer of a deed to real property is recorded.[14] The public notification effect of the recordation is important because it is possible that one party could attempt to assign a patent that he does not own to another party. In such a case, the rightful holder, by having his assignment recorded, serves to provide public notice of the fact that he is the rightful assignee.

[13]From the Code: 35 U.S.C. §261 Ownership; assignment:

Subject to the provisions of this title, patents shall have the attributes of personal property.

Applications for patent, patents, or any interest therein, shall be assignable in law by an instrument in writing. The applicant, patentee, or his assigns or legal representatives may in like manner grant and convey an exclusive right under his application for patent, or patents, to the whole or any specified part of the United States.

A certificate of acknowledgment under the hand and official seal of a person authorized to administer oaths within the United States, or, in a foreign country, of a diplomatic or consular officer of the United States or an officer authorized to administer oaths whose authority is proved by a certificate of a diplomatic or consular officer of the United States, or apostille of an official designated by a foreign country which, by treaty or convention, accords like effect to apostilles of designated officials in the United States, shall be *prima facie* evidence of the execution of an assignment, grant, or conveyance of a patent or application for patent.

An assignment, grant, or conveyance shall be void as against any subsequent purchaser or mortgagee for a valuable consideration, without notice, unless it is recorded in the Patent and Trademark Office within three months from its date or prior to the date of such subsequent purchase or mortgage.

[14]A patent is initially assigned to the inventor. If the invention lists more than one inventor, then each inventor is assigned an equal share in the entire invention. More will be said about this in Chapter 4, which deals with inventorship and inventorship issues.

If a legal entity is assigned an ownership interest in a patent but does not record it and that interest is subsequently assigned to a different legal entity, it is possible that the first entity could lose their interest by failing to record it. While this might not seem fair, it is entirely in the first assignee's power to make sure their right is recorded. Likewise, before purchasing or licensing a U.S. patent or patent right (including patent applications), it is incumbent on the purchaser to search the USPTO's record of assignments database to be sure the party transferring the right is the holder of record. Assignments can be viewed online by going to www.uspto.gov and searching assignments in the Public PAIR portion of the website.

An assignment of a patent right is an assignment of the complete ownership interest. In contrast, a license does not transfer the entirety of the ownership interest but rather a limited portion of that interest. For example, a license may apply for a limited time, geographical area, or field of use (where that field of use is less than actually claimed in the patent). While a license *may* be recorded in the USPTO, it does not have to be and usually is not. The reason is that the assignment recordation statute does not provide any sort of right of priority to one who has his license interest publicly recorded at the USPTO and so a purchaser of a patent takes the patent subject to any licensing rights that have been granted previously, even when those previous license grants have not been publicly recorded at the USPTO. Whether a party is acquiring the patent right by assignment or license, he needs to verify that the party transferring those rights has the legal ownership rights to do so. This verification includes not only checking the assignment register at the USPTO, as we just discussed, but also querying the patent assignee as to whether other licenses to the patents have been granted to any other party and, if so, reviewing those licenses to be sure that they do not impinge the scope of what he is attempting to acquire. This verification process is typically part of the intellectual property due diligence undertaken by the acquiring party prior to any actual transaction. Purchasing a patent in which either the transferor does not have legal title or the patent is subject to certain types of previously issued licenses can impinge on one's freedom to operate with the patented technology simply because somebody else might actually be the lawful owner of that property. This is much like buying a bogus deed to a piece of land; the erstwhile purchaser could be in for a big surprise when he attempts to take possession of the land. Nobody likes that kind of surprise.

1.7   WHY HAVE PATENTS?

Before delving further into the who, what, where, and when of U.S. patent law, as we will begin shortly, a more basic question worth considering is: Why have patents in the first place? Aren't they monopolies? Aren't monopolies bad? These are reasonable questions to ask and reflect the natural tension that has co-existed with patents ever since the Middle Ages when it appears that something akin to patents first hit the scene.[15]

---

[15] See the Wikipedia entry "Patent" for a short history of patents.

The strongest arguments for robust patent policy make the case that it encourages individuals to innovate by allowing them to maximally profit from their inventions. At the same time, patents provide the public disclosure of inventions, which allows others to further innovate based on those disclosures. Since we wish to encourage innovation and invention, it makes sense that we would want to reward the inventor. For the innovation that results in an invention, exclusivity makes a logical reward because what better way to value the invention than the market itself? If an invention has a high value then many will wish to pay for it, thus rewarding the inventor in a manner directly proportional to its value to the public.

If however, an inventor were given nothing in return for publicly disclosing her invention, there would be little to encourage innovation and even less to encourage disclosure of the innovation. Rather it might behoove people to wait for others to bring an invention to the public so they could quickly duplicate and market it. With everybody waiting and nobody inventing, not too much ends up getting accomplished in the way of innovation. Even when inventions were made in the no-patents-allowed scenario, deliberate efforts to hide the invention might take place so that others would be hindered from copying the innovation. In contrast, the patent system encourages the opposite result. By requiring a thorough disclosure of the invention in exchange for a patent, the public is effectively given a tour of the palace (only they aren't allowed to occupy it for some time, at least without paying rent).[16] Although the public is limited in terms of what they can do regarding the invention during the patent's period of exclusivity, they are free to innovate further and improve on the patented invention. Under this paradigm, the patent represents less a monopoly and more an invitation for further innovation.

But there are competing interests. For example, some have argued that patents can actually stifle innovation by preventing individuals from working in certain areas due to the presence of broad "blocking" patents in that area. Where such broad patents have been issued or when there are so many patents in a given area that a figurative minefield has been created, it can be hard for the eager inventor to find a way forward. Even licensing can become extraordinarily difficult if not impossible when there are multiple stacked patents resulting in complex and/or overly costly royalty structures. It has also been argued that certain patents, particularly in the pharmaceutical and agricultural areas, have allowed companies to price their wares in such a way that they have hurt society. For example, it has been argued that patent rights have resulted in monopoly pricing, which has left poorer nations unable to medically treat or even feed their citizens.

While the substance of these arguments can/have been dealt with in a number of ways, one should not underestimate their political and/or moral appeal, especially in the hands of erstwhile populists bent on highlighting alleged problems of patents while ignoring or obfuscating their longer term benefits. Weakening patent laws for short-term benefits can be appealing but needs to be actively guarded against. Ultimately, strong patent rights are granted by society, and they can be taken away by society; it is incumbent upon both governments and industry/inventors to continue

---

[16]U.S. disclosure requirements for chemical patents will be discussed in Chapter 9.

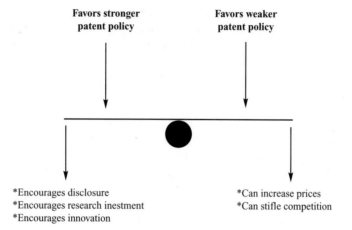

Favors stronger
patent policy

Favors weaker
patent policy

*Encourages disclosure
*Encourages research inestment
*Encourages innovation

*Can increase prices
*Can stifle competition

FIGURE 1.6   Patent policy as a balance of competing interests.

to work together in such a way to ensure that the current patent system reflects a proper balancing of the different policy interests. One might think of these competing interests as weights on opposite ends of the seesaw shown in Figure 1.6, where the fulcrum represents the balance of interests at any given time.[17]

[17]Given the important implications of patent policy in view of the balance of interests just discussed, it could be argued that we *should not* have just one set of patent laws to cover the many areas of technological endeavor, especially since each of these areas has their own particular costs and risks. In technical areas where a tremendous amount of research and development and risk are inherent, stronger patent policy should be the norm. Without the guarantee of strong patent policy in those areas, less investment is likely to be made because the return on investment might not justify the risk. If certain areas of risky research would not take place but for the incentive of patent exclusivity, then patents are serving an enabling function and should be more strongly encouraged. Such enhanced patent laws might include, for example, longer periods of exclusivity and enhanced measures of enforcement. Pharmaceuticals are one such area where the costs of development are extremely high, the development cycle extraordinarily long, and the chances of success vanishingly slim. The average cost to develop a single drug has been estimated to be as much as a billion dollars and the risk of failure is very high. Without a strong guarantee of market exclusivity for these chemical inventions, there would be much less effort put into pharmaceutical research and development. Having a relatively short time to recuperate research-and-development (R&D) costs requires higher pricing of the few successful drugs that do make it, leading to greater economic unpredictability for the companies involved and increased public backlash against the high prices. To some extent, this has been recognized and a partial solution in the way of nonpatent-based market exclusivity is offered in most countries. In the United States, this type of nonpatent exclusivity for new molecular entities is limited to 5 years, which is seldom seen as adequate inducement by itself, so the pharmaceutical industry still relies on the strength of its patents to provide a return on its investment. Conversely, in an area where the R&D cycle is normally very quick and highly competitive even in the absence of patent rights, strong patent laws could actually act to stifle innovation by blocking research that would otherwise take place on its own accord. Unfortunately, patent laws passed to rein in the latter could be too broad based and adversely affect the industries that need stronger patent laws to survive and flourish.

CHAPTER 2

The Patent Process

2.1 AN OVERVIEW OF THE PATENT PROCESS IN THE UNITED STATES

U.S. patent applications are filed with the USPTO, located in Alexandria, Virginia, which is very close to Washington, D.C. It is interesting that the USPTO is one of the few government operations that, at least up until recent years, actually produced more revenue for the government than it used. It is not surprising that such governmental economy could not last for long, and the situation has apparently been corrected such that the USPTO now appears to be approximately revenue neutral in recent years (Figure 2.1).[1] The USPTO is an extraordinarily busy place, with the number of patent applications increasing each and every year without respite (Figure 2.2).[2] Due to the volume of work that it handles, the USPTO is a relatively large operation, employing thousands of patents examiners across a number of specialties. The great majority (if not all) of patent examiners in the chemical, materials, and pharmaceutical sciences have bachelor's degrees and usually advanced degrees in the relevant disciplines. The individual examiner's work is reviewed and supervised by senior examiners. Depending on the type of decision being made by the examiner, it can be either appealed or petitioned. Appeals go to the Board of Patent Appeals and Interferences (the Board) whose membership is made up of administrative patent judges appointed by the USPTO director. Appeals heard before the Board are typically decided by a three-judge panel.[3]

---

[1] Figure 2.1 taken from United States Patent and Trademark Office "Performance and Accountability Report – Fiscal Year 2007".

[2] Figure 2.2 taken from United States Patent and Trademark Office "Performance and Accountability Report – Fiscal Year 2007".

[3] In special circumstances, the Board may decide an issue with an expanded panel. Expanded panels might be used when there are conflicting decisions from different panels of the Board, a Board panel contains substantial differences of opinion, it's a case of first impression, or an issue is of high precedential importance. If you are interested, for more details, see the website www.uspto.gov/web/offices/dcom/bpai/sop1.pdf.

The Chemist's Companion Guide to Patent Law, by Chris P. Miller and Mark J. Evans
Copyright © 2010 John Wiley & Sons, Inc.

Net (Cost)/Income (dollars in millions)	FY 2004	FY 2005	FY 2006	FY 2007
Earned revenue	$1,239.0	$1,372.8	$1,594.4	$1,735.7
Program cost	(1,289.2)	(1,424.0)	(1,514.2)	(1,769.6)
Net (cost)/income	$ (50.2)	$ (51.2)	$ 80.2	$ (33.9)

FIGURE 2.1   Revenue and cost figures for the USPTO (2004–2007).

Patent Examining Activity	2003	2004	2005	2006	2007
Applications filed, total	**355,418**	**378,984**	**409,532**	**445,613**	**467,243**
Utility	331,729	353,319	381,797	417,453	438,576
Reissue	938	996	1,143	1,204	994
Plant	785	1,212	1,288	1,103	1,047
Design	21,966	23,457	25,304	25,853	26,626
Provisional applications filed	**92,517**	**102,268**	**111,753**	**121,471**	**132,352**

FIGURE 2.2   Patent applications filed with USPTO by year and type.

Patent applicants may represent themselves (pro se applicants) before the USPTO, though it is more common to have a patent agent or attorney assisting in the process.[4] To represent clients before the USPTO, the agent or attorney needs to be registered

---

[4] In the case of *Nilssen et al. v. Osram Sylvania, et al.* 504 F.3d 1223 (CAFC 2007), the CAFC upheld the District Court's finding that plaintiff's asserted patents were invalid due to inequitable conduct committed by the patentee during the prosecution of the patents. It is interesting that the patentee had replaced his own legal representatives during the prosecution of the patents to represent himself before the USPTO. In regard to the adequacy of his representation the CAFC panel opined:

However, this case presents a collection of such problems, which the district court evaluated thoroughly and considered, including making credibility findings, and it concluded that the record and testimony indicated repeated attempts to avoid playing fair and square with the patent system. Mistakes do happen, but inadvertence can carry an applicant only so far. Thus, we cannot find that the Court's holding of unenforceability was an abuse of discretion. Perhaps some of the errors were attributable to Mr. Nilssen's representing himself during the prosecution of his patents. It surely was true that he knew more about the subject matter of his inventions than most, or even any, attorney. That is almost always the case with an invention, particularly one dealing with complex subject matter. However, the patent process is a complicated one, one that requires both technical and legal credentials in order to effectively prosecute patents for inventors. The same credentials are generally required to prosecute patents on one's own inventions. Mr. Nilssen, while apparently gaining considerable knowledge of the patenting process, thought he didn't need professional patent help. The result of this case, regrettably, proves that he was wrong.

Ironically, although the plaintiff avoided paying his own attorney's fees by representing himself, the CAFC upheld the District Court's decision requiring him to pay the defendant's litigation costs (approximately $6 million). For another round of this case, see *Nilssen et al. v. Osram Sylvania, et al.* No. 2007-1198, 1348 Slip. Op. (CAFC, June 17, 2008).

before the USPTO, which usually requires passing an entrance examination, having an accredited degree in an appropriate technical area or enough scientific college or university credits to meet the established criteria, and being able to satisfy certain personal moral fitness criteria.[5] The difference between a patent agent and a patent attorney is that the patent agent is qualified before only the USPTO, whereas a patent attorney is qualified before the USPTO as well as before one or more state bars.[6] When matters of representation are limited strictly to patent preparation and prosecution issues only, a patent agent may very well be competent to handle the situation. However, when broader matters are contemplated (e.g., licensing of intellectual property), it is likely one will need a patent attorney.

Normally, the patent applicant or assignee will sign a power of attorney identifying his representative(s), who will then conduct the patent filing and prosecution on the applicant's behalf. The applicant's representative will sign off on each item of correspondence with the USPTO and include in each signing his unique USPTO registration number.[7] Ultimately, it is the legal assignee that has the authority in determining how the process should proceed. The assignee may or may not be the inventor(s), depending on whether the inventors have already assigned their rights to another party. Should he desire, the assignee can switch representatives during the process if he is not satisfied with his representation before the USPTO.

The patent application process begins with the filing of a patent application in the USPTO, but it does not end there. From the time of the filing to the grant of a patent and sometimes beyond, a stream of correspondence between the applicant's representatives and the USPTO will begin. The primary goal of the patent applicant will be to get his invention "allowed" such that one or more patents with claims to the invention will be issued. The process of getting the claims allowed is commonly referred to as *patent prosecution*. This process is subject to a very large number of rules and formalities, which are outlined in great detail in the *Manual of Patent Examining Procedure* (MPEP).[8]

Although the patent process begins with the filing of a patent application with the USPTO, the types of patent applications can vary according to whether the application is for a design, plant, or utility patent. As chemists, we probably will be concerned

---

[5]If you are interested in the requirements for practicing before the USPTO, further information can be found at www.uspto.gov/web/offices/dcom/olia/oed/grb.pdf.

[6]To practice in a state court, the attorney typically will need to be a law school graduate of an appropriately accredited law school as well as meet that state's bar requirements, including both a satisfactory score on the state bar exam and an acceptable moral fitness evaluation. Often times, an attorney already licensed to practice law in one state may waive into another state by meeting some alternative set of requirements.

[7]Once registered to practice before the USPTO, an agent or attorney is assigned an unique registration number. These numbers are assigned sequentially. A list of each and every current member of the patent bar can be found at the USPTO's website (www.uspto.gov/web/offices/dcom/olia/oed/roster/index.html). In 2009 there were approximately 9,000 registered U.S. patent agents and 29,000 registered U.S. patent attorneys.

[8]If you find yourself in the position of needing to know any specific detail about patent law prosecution then the MPEP is indispensable. The MPEP online can be found at the following page at the USPTO, website www.uspto.gov/web/offices/pac/mpep/mpep.htm. The MPEP is also available in paperback copy from the legal publishers West and Lexus-Nexus ($70–$80).

only with the latter of these three choices, and so all references to patents from here on, unless designated otherwise, will refer to utility patents. Patent applications can be filed directly with the USPTO or may enter the USPTO indirectly through a patent filed internationally under the Patent Cooperation Treaty (PCT). Due to the important differences between these two processes for filing a U.S. patent application, they will each be detailed separately.

The initial filing of a patent application in the United States can be either as a regular (nonprovisional) patent application or a provisional patent application.[9] The differences between these two types of patent applications are numerous (and significant), with the largest difference being that a provisional patent application will never be directly examined and it becomes abandoned automatically 12 months after filing. Provisional patent applications are not examined for patentability, so they do not need to include any claims (although they often do). Given these facts, you might be surprised to learn that a fair percentage of patent applications filed directly in the United States are filed initially as provisional patent applications (Figure 2.2), but once you learn the advantages of filing a provisional patent application, you will come to see why they are so popular. In addition to decreased filing costs and formalities,[10] a provisional patent application establishes a filing date for the subject matter described in that provisional patent application but does not start the nonprovisional patent-life clock ticking.[11] Nonprovisional utility patents filed in

---

[9]Provisional applications are described in 35 U.S.C. §111(b).

[10]Almost every action taken on behalf of a patent applicant with the USPTO has a fee associated with it. The fees are staged according to a two-tier system in which small-entity applicants have lower fees than other than small-entity applicants. A party may apply for small-entity status if their organization size (employees, affiliates, associates, etc) does not exceed 500. However, small-entity status is lost when the patent is licensed or assigned to an organization that does not qualify for small entity status (including the government). Mistakenly claiming small-entity status when it does not properly apply can result in a finding that the patent is unenforceable. (See Title 13 of the Code §121.801–§121.805 for further details.) USPTO fees in 2009 for a small entity provisional patent filing was $110 and for a non-small entity was $220 (with additional charges for applications over 100 pages in length). In contrast, a regular patent application for a small entity was $165 and for a non-small entity the cost was $330. In addition, the regular application is subject to separate charges for searching, examination, and independent claims in excess of 3 or total claims in excess of 20. As a general rule, the small-entity fees are half that required for non-small entities.

[11]To be afforded a filing date, the provisional patent application needs only a written description of the invention, complying with all requirements of 35 U.S.C. §112 1st paragraph and any drawings necessary to understand the invention; the requirements of drawings are described in 35 U.S.C. §113. To be complete, a provisional application *must also* include the filing fee and a cover sheet identifying

- The application as a provisional application for patent.
- The name(s) of all inventors.
- Inventor residence(s).
- Title of the invention.
- Name and registration number of attorney or agent and docket number (if applicable).
- Correspondence address.
- Any U.S. government agency that has a property interest in the application.

the United States expire 20 years from their filing date, absent special circumstances; the day the nonprovisional patent application is filed is the day the 20-year clock starts counting down. For this reason alone, anything that can delay the filing of the nonprovisional patent application without prejudicing the rights or opportunities of the party filing the patent application is very valuable. When a provisional patent application is filed, the effective filing date for the subject matter contained in the provisional patent application is the date it is filed, much in the same way that it is in a regular patent application. Within 12 months of filing the provisional patent application, the applicants have the opportunity to do one of a couple of things.

The first and easiest thing that might be done is nothing at all. The applicant may simply decide not to pursue the invention, and after 12 months, the provisional application is automatically abandoned without any affirmative action by the applicant required; it's as if the patent application never existed at all.[12] In this regard, it must be reemphasized that a provisional patent application is not examined for patentability by the USPTO and, accordingly, will never give birth to an issued patent.[13]

If the option just discussed does not appear particularly productive to you, then you may find the next option to be more to your liking. Under this option, the provisional patent application is filed and then relied on as a priority document by a later-filed, nonprovisional patent application. Under this scenario, the non-provisional patent application claims the benefit of (borrows) the filing date of the earlier-filed provisional patent application with respect to the information disclosed in the provisional patent application(s) and to the extent that information provides support for what is claimed in the regular application.[14] In a contest for priority, the filing date is particularly critical in the rest of the world (outside the United States) because the filing date establishes who gets awarded priority of invention. In the United States, as we will learn in the next chapter, the priority of invention (within certain limits) is awarded to the first to invent and not the first to file, so the advantages of an earlier filing are not quite as acute but still can be very relevant.[15]

---

[12]35 U.S.C. §111(b)(5).

[13]This does not mean that a provisional patent application will never be publicly available. In the case in which a later-filed nonprovisional patent application is published by the USPTO and that later-filed patent application claims priority to the provisional application, then that provisional patent application is available to the public as part of that patent application's file history.

[14]The term *support* in this context refers to the requirements of 35 U.S.C. §112, which requires adequate written description of the claimed invention, enablement, and best mode. Much more will be said about these requirements in Chapter 9, but for now it is necessary only to appreciate that the provisional application is good on the date it is filed to the extent that it provides enough detail so that a person of ordinary skill in the art can make and use the later-claimed invention without having to resort to undue experimentation and that a credible utility for the claimed invention is alleged. The *best mode* means that the application includes the inventor's best mode of practicing the claimed invention.

[15]Only one party can be awarded a patent to an invention (the same invention cannot be patented twice), and the prize usually goes to the party that can establish that it was the first party to invent the contested subject matter. This is a consequence of the United States being a first-to-invent country (more will be explained regarding this facet of U.S. law in Chapter 3). The first-to-invent system makes a provisional patent application less critical than if the United States were a first-to-file country like the rest of the world. However, filing a provisional patent application in the United States is still advantageous because any

U.S. nonprovisional patent applications terms are normally 20 years from the date of their filing. A provisional patent application can establish the filing date of the subject matter disclosed in the application but does not start the 20-year clock ticking. One way of looking at the provisional patent application is that it effectively extends a 20-year patent term up to a 21-year patent term.[16]

The temporal relationship between a provisional and nonprovisional patent application claiming priority to that provisional patent application is illustrated in Figure 2.3. To better appreciate how the provisional patent process works at the USPTO, let's consider a first hypothetical example (hypothetical 2.1). A synthetic medicinal chemist working at the Wish I Were Here pharmaceutical company has been synthesizing compounds for a neuroscience program targeting schizophrenia. In the course of his work, he makes the following discoveries, and his company takes the following actions. On June 3, 2008, he synthesizes the carbazole-containing structure A (Figure 2.4 shows the structures of the compounds and the respective patent filings in this example).

After establishing that the compound is active in well-established biological models, a provisional patent application (number 61/136,899) is filed on July 15, 2008, describing the synthesis and biological activity of the compound A.

Not content to rest on his laurels, the chemist continues to work on the project by optimizing the properties of his initial lead compound. Every time he makes a new analog with good activity, he writes up his experimental section and heads down to the company's patent attorney who promptly drafts a new provisional patent application containing information relating *only* to the new analog in it. On September 29, 2008, he synthesizes the molecule with the structure B. After establishing that the compound is active in well-established biological models, a provisional patent application is filed on December 29, 2008, describing the synthesis and biological activity of the compound B.

---

prior art cited by an examiner will need to have an effective date (antedate) the provisional patent filing date (assuming the provisional patent application adequately describes the invention) . If a reference cited by the patent examiner antedates the earliest filing of the nonprovisional patent application, the patent applicant would need to demonstrate invention before the effective reference date rather than potentially being able to rely on the earlier-filed provisional application to establish the earlier invention date. Perhaps more important, filing a provisional patent application reserves a priority date for the rest of the countries of the world should the applicant decide to file outside the United States. Moreover, the first-to-invent rule in the United States may not continue in its present form; it has been the subject of much criticism, and it is possible that it could be legislatively modified or eliminated in the near future. Current versions of pending federal legislation (e.g., Patent Reform Act of 2009) propose to convert the United States from a first-to-invent system to a first-to-file system, which, of course, makes the filing of a provisional patent application all the more important.

[16]The use of *patent term* in this context does not imply that there is 20 or 21 years of enforceable patent life; with some limited exceptions, a patent is usually issued before it is enforced against infringers. If one considers that patent prosecution can easily take a few years before any claim issues from the filed patent application, than the enforceable patent life can be markedly shorter than the 20-year patent term. In the context used in this chapter, the patent term refers to the period from the filing date of the earliest patent application relied on for its priority date in the chain (the first-filed provisional patent application in this instant).

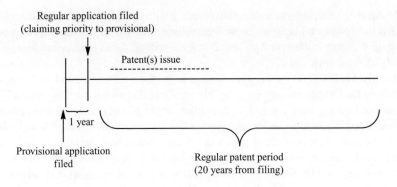

Total patent priority period equals 20 plus 1 years.

FIGURE 2.3 Timeline showing the effect of provisional filing followed by a regular filing in the United States.

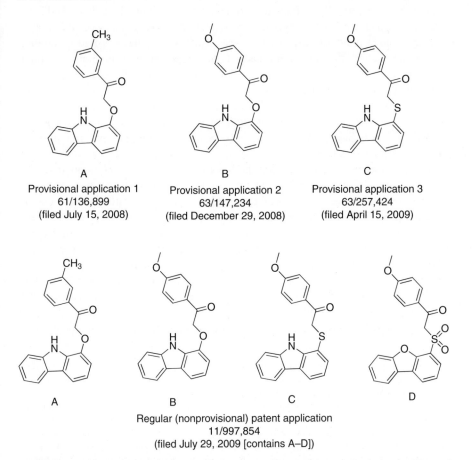

FIGURE 2.4 Compound structures, filing dates, and application numbers for the filing of compounds A–D.

On April 1, 2009, he synthesizes the molecule with the structure C. After establishing that the compound is active in well-established biological models, a provisional patent application is filed on April 15, 2009, describing the synthesis and biological activity of the compound C.

On July 11, 2009, he synthesizes what is to be the final compound in the series, compound D. The patent attorney, vaguely recalling that somehow there was a limit in terms of how far back a priority application could go back and still have priority claimed from it, decides he should write all the compounds up together (including compound D) and put them into a nonprovisional patent application and file it quickly; with considerable alacrity, the application is assembled and filed on July 29, 2009 (number 11/997,854). The nonprovisional patent application describes the claim to priority under the heading "Cross-Reference to Other Applications," where it is written, "This application hereby claims priority to U.S. provisional application 61/136,899 filed July 15, 2008, U.S. provisional application 63/147,234 filed December 29, 2008, and U.S. provisional application 63/257,424 filed April 15, 2009."[17] What date can be properly claimed as a filing date for compounds A, B, C, and D?

Let's start with the most recently filed patent application and then work backward. For compound D, there was no provisional patent application that was filed containing its structure. Rather, the compound made its first appearance in the nonprovisional patent application; therefore, it properly claims priority to a filing date of July 29, 2009. The third (last) provisional application that was filed contained compound C only, and it was filed on April 15, 2009, so compound C's effective priority date is April 15, 2009. The second provisional application that was filed contained compound B only, and it was filed on December 29, 2008, so compound B's effective priority date is December 29, 2008.

This isn't too bad, are you getting the hang of it? How about compound A? To properly claim priority to an earlier filed provisional application, that provisional must properly describe the subject matter the later-filed application is attempting to claim. Since the provisional patent application fully describing compound A was filed on July 15, 2008, the correct priority filing date for compound A is July 15, 2008. Right? Actually no, the earliest filing date that can be afforded to compound A is July 29, 2009. Why? Because for a regularly filed patent application to properly claim priority to an earlier-filed provisional application, the provisional must have been filed within 1 year of the regularly filed patent application.[18] In this case, the 1-year period had already lapsed, and as a result the patent application 61/136,899 containing compound A was abandoned on July 15, 2009, one year after its filing date. So the attempt to claim

---

[17]Notice the nonprovisional patent application needs only to refer to the provisional patent applications to which it claims earlier priority; it does not need to explain which provisional patent application supports what subject matter in the nonprovisional patent application.

[18]In some cases, an extra day or two beyond 1 year can be gained when the 1-year anniversary from the original filing falls on a weekend or regular USPTO holiday. In such a circumstance, the regular patent application can be filed on the next day that the USPTO is open for business. So, for example, if the 1-year anniversary of the original filing fell on Saturday, the regular application may be filed on the subsequent Monday and still properly claim priority from the provisional application, despite more than 1 year having passed.

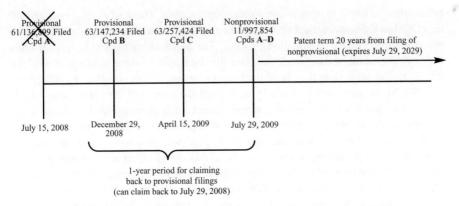

FIGURE 2.5   Timeline for patent applications from hypothetical 2.1.

priority to the application filed on July 15, 2008, will fail. Furthermore, the facts in this hypothetical scenario explained that each provisional case that was filed contained only the molecule described in that case—the first case contained compound A only, the second compound B only, etc. As a result, the regular (nonprovisional) patent application that described all of the molecules together is the first time that compound A was described in a patent application that was effective; it's as if the first provisional application had not been filed. A timeline covering the analysis for the three provisional and one nonprovisional application is shown in Figure 2.5.

One way to avoid the scenario just outlined for compound A, where effectively a whole year of priority filing for that compound has been lost, is to include the contents of each provisional patent application in the next provisional application. If that had been done in this example, the provisional application containing compound B that was filed on December 29, 2008, would also contain the subject matter from the earlier provisional application containing the compound A. This way, if for some reason the 1 year time frame was missed as in the example, the priority date for compound A would be December 29, 2008, rather than pushed all the way forward to July 29, 2009. This may not seem like much, but in a highly competitive area it can sometimes mean a lot.[19]

Can you appreciate the advantage of filing provisional patent applications? They offer ease of filing and reduced costs and thus facilitate the prompt filing of inventions as they occur in the course of a research program. Moreover, they allow for an entire year of effort to develop additional compounds with protection of the inventions as they develop without having to wait for the entire body of work to be finished

---

[19]As a matter of practical significance, the provisional patent applications will typically include not only the specific compounds that were made but also a Markush genus that more broadly describes the subject matter of the patent application. The Markush will cover both the specific examples of the invention and any related analog that the inventors are interested in making and believe would also be active. In this way, they still have the possibility in a provisional application of generically covering compounds that are not explicitly represented as specific examples.

before filing a single nonprovisional patent application. Think of a situation like that described in the hypothetical that we just reviewed, but without taking advantage of the provisional patent application process. The applicants might simply wait until they completed all of their work and then file the nonprovisional application on July 29, 2009, the same date they filed their nonprovisional in the original hypothetical. The risk in this approach is that they do not get the benefit of the earlier filing dates for the compounds that were discovered earlier, such as compound A, B, or C. If somebody else published on or filed a patent application containing the compounds A, B, and/or C between the time that A, B, or C was discovered and the time the nonprovisional patent application was filed, this could affect their ability to get a patent in countries outside the United States. Even in the United States, there are advantages to having the earlier filing date; the earlier filing still can act as prior art to another's earlier invention under some circumstances. Much more will be said about prior art and prior invention in Chapter 3, but for now let's appreciate that it is often preferable not to delay a patent filing. This is true especially given the low cost burden and flexibility inherent to the provisional patent filing process.

Another alternative to filing a series of provisional patent applications as just discussed would have been to file each of the compounds, as they were invented, into a separate nonprovisional patent application. For example, the first patent application could have been filed as a regular patent application containing compound A only. When compound B was discovered, the applicant could have filed another patent application either as a completely separate application or a continuation-in-part.[20] There are many aspects that make this alternative less desirable, including the increased filing costs and complexity of maintaining separate patent prosecution files for each case. More important, the option of filing a nonprovisional patent application instead of a provisional patent application would result in a loss of patent term. The reason for this is that the 20-year from patent filing lifetime for a nonprovisional patent application is tolled on the day that the patent application is filed. Therefore, if the applicant desired protecting the compounds as soon as they were invented by immediately filing a nonprovisional patent application, the applicant would sacrifice a longer patent life for immediate protection of the subject material. Filing a provisional patent application allows him to protect the subject material without starting the 20-year clock ticking.

Given the many advantages of filing a provisional patent application, it might be difficult to appreciate that there are circumstances in which it can be advantageous

---

[20]*Continuation-in-part applications* refer to the patent filing situation in which an earlier nonprovisional patent application is filed and then one or more subsequent patent applications containing at least some material not contained in the earlier patent application are filed and claim priority back to the earlier filing but only for the material that was in the original filing. A continuation-in-part cannot be filed for a provisional patent application. However, the nonprovisional patent application claiming priority from the provisional filing is functioning like a continuation-in-part application in that it too is claiming priority to an earlier filed patent application (in this case, the provisional patent application) and that the date of priority of the earlier patent application is good for only the material disclosed in that earlier filed patent application. A provisional patent application cannot claim priority to an earlier-filed patent application (provisional or otherwise).

to forgo a provisional patent filing and directly file a regular patent application. In particular, the primary reason for skipping a provisional filing and directly filing a regular patent application is that it gets things moving up to 1 year earlier at the USPTO. Remember that a provisional patent application is not formally examined; it acts more like the foot holding the door open. When moving things forward rapidly in the USPTO is of high importance, one may not want to wait the extra year for the provisional application to fully vest, but rather one may want to try to get the patent application published and issued as soon as possible. This might be of benefit when the applicant's business plan requires a more mature patent portfolio at an earlier time. In this regard, the sooner the patent publishes and issues, the sooner it can be presented to investors or licensees and/or be available for enforcement against infringers.

Because we wish to stay with the normal chronology of filing events as much as possible, let's consider the important third option for how a patent application may be filed in the United States: through the PCT. In this option, a provisional application may still be first filed (if desired) in the United States (or its equivalent elsewhere), and then an international patent application is filed in one of the receiving offices for the PCT. The application, through the initial stages, is processed according to the rules of the PCT.[21] In the same way that a regular nonprovisional patent application is filed in the United States within 1 year of the provisional patent application from which it claims priority, an applicant may instead opt to file a PCT patent application.

The *initial* effect of filing a PCT application can be *somewhat* akin to filing that patent application in every signatory country of the PCT, which is in excess of 140 countries and includes the United States, the European Union, China, and Australia. (The most notable nonmembers are Taiwan and Argentina.) Approximately 6 months after filing the PCT application that claims priority to the earlier U.S. provisional patent application (or earlier-filed application of another country), an international search is conducted and a brief opinion of patentability is presented. The international search and opinion of patentability will alternatively take place approximately 9 months after the filing of the PCT application if no priority application is relied on for an earlier filing date. After approximately 18 months after filing the priority application or the PCT application, if no priority application is claimed, the international patent application is published.[22]

Approximately 18 months after the international application is filed, the applicant will need to make the decision about whether and where among the various PCT participant countries he wishes to pursue his patent application. It is at this stage that the patent prosecution process gets expensive since separate filing and examination

---

[21]The first provisional application (if filed) may also be a national patent application filed in any of the other PCT countries. Thus, for example, the first application may be a German national patent application filed in Germany. While the particular procedures and effects given to such first filings can vary depending on the country that the application is filed in, for our purposes we can think of such a filing as a provisional application in terms of it reserving priority for the applicant to file the PCT application within 1 year of the filing. At least one of the applicants must be a resident or national of one of the PCT contracting countries.

[22]The PCT will not be automatically published if the United States alone is designated (Article 64(3) PCT).

fees will have to be paid for many of the individual countries, and translations of the patent application will have to be made for many of the patent offices. The costs of international patent prosecution (foreign patent agents and attorneys will need to be retained and paid) add up very quickly.

The convenience and advantages of the PCT filing process include that it allows a single application to be filed and it can be filed (and initially examined) in English.[23] Since it takes approximately 18 months from the time the regular PCT application is filed until applications into the separate countries are required (national stage entry) and the PCT application filing process also allows one to claim priority to a provisional or set of provisional applications filed within 12 months of the PCT application, the total time from the filing of the first provisional application describing the subject matter until entry into the separate countries is approximately 30 months. The process of first filing a provisional patent application followed by a PCT application within 1 year of the provisional patent application effectively buys the patent applicant 30 months to evaluate and/or develop the invention to better determine if what is described in the patent application is something worth committing significant resources to. In this vein, the PCT-initiated search for prior art and comment on patentability might help the applicant determine whether the invention is likely to be patentable, all before making the decision whether to enter patent applications in the many different countries where the applicant might eventually wish to enforce the invention.

So what are the disadvantages of filing by the PCT route? First, one needs to appreciate that despite the search and initial examination process, there is no patent that issues from a PCT application. The patents that will eventually issue must come from the separate patent offices, which often means a separate examination of the claimed subject matter will need to be conducted in the many different patent offices. In this regard, one should think of the PCT application primarily as a convenient filing mechanism. Previously, we discussed how the PCT filing buys the applicant precious time, a full 30 months from the time of a provisional filing to the entry into the national stage in the United States and any other national offices of interest. However, in cases in which a patent applicant desires a more rapid prosecution of the patent application, this extended timeframe can be a disadvantage.

If the convenience of the PCT application process is desired but there is also a concern about the 30-month delay, the applicant also has the option of taking a hybrid approach that blends the convenience of a PCT filing with rapid prosecution at the USPTO. The approach taken in such a situation is to file the PCT application but, at the same time, file a patent application containing the same content in the United States. This allows the prosecution of the application in the United States to begin while the PCT application is waiting. This can shave up to 18 months off the normal PCT timeframe for entry into the United States while still postponing costly translations and fees from multiple foreign filings. In effect, the prosecution in the United States will proceed as if the PCT were not filed but the door will still be left

---

[23]Other accepted languages include Arabic, Chinese, French, German, Japanese, Russian, and Spanish. Portugese and Korean were to be added in 2009.

open to the applicant should he decide to pursue prosecution in most of the rest of the world at a later stage.[24]

Now that we've reviewed some of the basic types of patent applications that can be filed in the United States (provisional, nonprovisional, nonprovisional entering through the PCT process) let's step in farther to see what happens once a nonprovisional patent application arrives at the USPTO. Whether an application arrives at the USPTO from the PCT as part of the national stage entry process or whether it arrives directly from an applicant to the USPTO, the first step taken by the office is to take a quick look at the patent application and make sure that it is complete.[25]

---

[24]The decision of which countries to enter at the 30-month timeframe is left to the applicant; it is not an all or nothing proposition. Thus, for example, one may decide that only a patent in the European Union, China, and the United States is desired. In that case, the applicant will instruct his attorney or agent to enter the application only into those patent offices and not others.

[25]As a preliminary matter, the nonprovisional patent application, whether filed in the United States directly or when it enters the U.S. national stage via the PCT, requires that all applicants must be inventors and all of the inventors names and addresses must be provided with the patent application. This is in contrast to many other patent jurisdictions, where the applicant can be a business entity, such as the company an inventor may have assigned his invention to. Additionally, all nonprovisional patent applications require an oath or declaration of inventorship to be included with the patent application or submitted within the prescribed time thereafter. The requirements of the oath or declaration of inventorship are explained in 37 CFR §§1.63–1.69. Initial entry of patent applications to the United States from the PCT are governed by the Code as outlined in 35 U.S.C. 371 National stage: Commencement.

(a) Receipt from the International Bureau of copies of international applications with any amendments to the claims, international search reports, and international preliminary examination reports including any annexes thereto may be required in the case of international applications designating or electing the United States.

(b) Subject to subsection (f) of this section, the national stage shall commence with the expiration of the applicable time limit under *article 22* (1) or (2), or under *article 39* (1)(a) of the treaty.

(c) The applicant shall file in the Patent and Trademark Office —

    (1) the national fee provided in *section 41*(a) of this title;

    (2) a copy of the international application, unless not required under subsection (a) of this section or already communicated by the International Bureau, and a translation into the English language of the international application, if it was filed in another language;

    (3) amendments, if any, to the claims in the international application, made under *article 19* of the treaty, unless such amendments have been communicated to the Patent and Trademark Office by the International Bureau, and a translation into the English language if such amendments were made in another language;

    (4) an oath or declaration of the inventor (or other person authorized under *Chapter* 11 of this title) complying with the requirements of *section 115* of this title and with regulations prescribed for oaths or declarations of applicants;

    (5) a translation into the English language of any annexes to the international preliminary examination report, if such annexes were made in another language.

The articles of the treaty referred to correspond to the PCT treaty and can be found in the MPEP. The basic requirements for completing a patent application are described in 37 CFR § 1.51. A more detailed explanation of the filing requirements for regular patent applications directly filed in the United States rather than through the PCT can be found at www.uspto.gov/web/offices/pac/utility/utility.html. The completeness of an application depends on the type of patent application. In some instances, an

This initial examination is done by the Office of Initial Patent Examination (OIPE), which will also assign an application number to the case. The next step at the USPTO will be to assign the patent application to an art unit that corresponds to the type of technology that the application deals with.[26] After being assigned to an art unit, the patent application will be assigned to an examiner working in that unit who will proceed to docket the case. The current backlog in the USPTO is such that one should not expect a very quick substantive response on the application. In fact, it would not be uncommon to take more than 2 years from the time a case is assigned to the examiner until the first substantive correspondence from the examiner to the applicant is made; substantive communication from the USPTO by the examiner to the applicant is typically in writing and is referred to as an *office action*.

Generally, as a parallel consideration, a regular patent application filed with the USPTO will eventually be published and thus made available to the public in much the same manner as that described previously for PCT applications.[27] The timeframe for publication is usually 18 months from the filing of the first application from which priority is claimed (as is the case for a PCT-filed application), except a patent application entering the United States from the PCT will already be 18 months from the nonprovisional filing date and up to 30 months from any earlier-filed provisional patent application from which it claims priority (but, of course, such an application will have already been published by the PCT as a WO publication at the 18-month time point).

So if an applicant files a provisional patent application on June 1, 2007, and files a nonprovisional patent application in the United States on May 28, 2008, that claims priority from the provisional patent application, then, under normal circumstances the patent application will be published around December 1, 2008. The application that is published is the nonprovisional patent application and not the provisional application. However, on the day the nonprovisional patent application is published, the provisional application together with the patent file history (including the formal correspondence between the applicant and the patent office) become available to

---

incomplete application (e.g., a nonprovisional application that does not include any claims) results in a failure to receive a filing date, and in others, the application is entered into the system and a notice of missing parts is sent to the applicant (e.g., if it does not contain a signed declaration of inventorship). If an applicant requests, notice of the receipt of the application by the USPTO can be obtained by sending a self-addressed stamped postcard with the application that the USPTO will mail back after receiving the application. This provides the applicant with confirmation that the application was received. If the applicants send the application in by US Express Mail Post Office to Addressee service via the U.S. Postal Service, the application filing date will be the date the application was deposited with the U.S. Postal Service, assuming the application is complete. See 37 CFR §1.10 for details.

[26]The art units dealing with chemistry-type professions (pharmaceuticals, coatings, perfumes, etc.) fall either between art units numbered 1614 and 1754 or 1764 and 1797.

[27]A published U.S. patent application can be readily distinguished from an issued U.S. patent by the publication code at the top right-hand side of the application. In contrast to an issued U.S. patent, which is indicated by a single unique patent number (e.g., 7,415,989), a published patent application will first list the year followed by a slash and a seven digit code. For example, a U.S. patent application published in 2009 might have a publication number of US2009/0123456.

the public.[28] There are situations in which an applicant can request that the patent application not be published (nonpublication request). Doing this requires that the applicant signs an agreement that the application has not and will not be submitted to another country or multilateral international agreement that requires publication of the application.[29]

Now patent prosecution begins in earnest. One of the first things the USPTO will do with respect to your patent application is prepare for a prior art search. Although we will learn much more about the specifics of patentability vis-à-vis the prior art, it's important to appreciate that the basic patentability of your invention will be determined in view of what is in the prior art. In the next chapter, we will go into great detail of what qualifies as prior art under U.S. patent law, so for now let's just consider that prior art as being equal to all of the things in the public domain that could affect the patentability of the invention that occurred before your patent application was filed (the USPTO will *initially* assume that the invention occurred on the date it was filed). In the United States, unlike most of the rest of the world, there is a burden on the applicant to provide any material to the patent office that the applicant *knows* about that may be material to the patentability of his invention.[30] This is what is known as the *duty of disclosure*, and it is a continuing duty up until the patent is issued. The disclosure of references that are material to patentability is submitted in the form of an information disclosure statement (IDS). In addition to noting appropriate references, the applicant is required to provide copies of the references as well as translated copies (if readily available) if the reference is in a language other than English. If a translated version of the foreign language document is not readily available, the applicant will need to provide a concise description or

---

[28]Public patent information can be accessed through the USPTO's website. If a patent application issues or is published and you know the application number, patent number, or U.S. publication number, you can look up the file history of the application in the Public PAIR section of the website (http://portal.uspto.gov/external/portal/pair). Most of the information is available as PDFs that can be viewed one page at a time. Before the advent of the Public PAIR database, file histories were available only as hard copies and tended to be very expensive to obtain. With Public PAIR, one can see much of the correspondence between a patent applicant and the USPTO. This is particularly useful when you are interested in learning more about another party's patent application status.

[29]Such a request might be made when the applicants are filing in the United States and do not want the competition to know what they are doing. The downside to this strategy is the obvious problem of not being able to file in other countries that require publication and also potentially limiting the enforceable lifespan of the patent application. In regard to this latter point, it is possible in some circumstances to trace royalties on a patent back to the date that the patent application was published if certain conditions are met. In addition, one may in some circumstances wish to give notice of his patent application through its publication to entice potential license partners. It is possible for the applicant who has made a nonpublication request to change his mind and elect to file a patent application in another country, provided the applicant notifies the USPTO within 45 days after filing the foreign patent. The USPTO will then publish the U.S. application as soon as practical. If the applicant does not notify the USPTO of the foreign filing within 45 days, the U.S. application will be considered abandoned (35 U.S.C. §122(b)(2)(B)).

[30]The duty of disclosure goes beyond just the applicant but includes anybody *substantively* associated with the filing and/or prosecution of the patent application, such as the inventors, applicants, assignees, and their representatives before the patent office. More will be said about inequitable conduct in section 2.3.

abstract of the document. Of course, a fully translated document can be provided as well, though translation fees can make this very expensive.

Copies of U.S. patents or published U.S. applications are not required to be submitted with the IDS (but you still need to cite them). If additional references come to the attention of the applicant or others associated with the filing during prosecution of the application, a supplemental IDS will need to be submitted to the USPTO disclosing the additional references. As a practical matter, one common way that additional references will come to the attention of the applicant during the patent prosecution is when that same application (or a closely related application) is being prosecuted in different countries where independent searches are also being conducted. In the course of such foreign patent prosecution, one or more of the foreign offices may cite one or more references against the application that the applicant was not aware of up to that point (or was aware of but did not think was material to patentability). Once the foreign office makes the applicant aware of the reference(s) and if the applicant believes the reference(s) to be material to patentability, he should submit a copy to the USPTO together with a supplementary IDS.[31]

Despite the duty of disclosure, the applicant does not have a duty to conduct an independent search.[32] The USPTO will conduct its own search of the claimed subject matter in the patent application, but quite often, before conducting the search, the USPTO will issue what is called a *restriction requirement*. A restriction requirement is an assertion by the patent office examiner that the claimed subject matter represents more than one patentably independent and distinct invention and that searching the subject matter in one application would be a serious burden. While the applicant has the right to file a petition further arguing against the restriction requirement, in reality most restriction requirements are accepted, unless the restriction requirement is deemed too excessive to continue separate prosecution of each desired group and when the separate prosecution of each independent group is necessary to the applicants. In such a case, a provisional selection still must be made while the petition makes its way through the patent examiner's chain of command.[33]

In the restriction requirement, the examiner will divide the subject matter up into two or more separate groups and ask the applicant to select the group that he wishes to pursue. The applicant will then select the group he wishes to pursue in that application and cancel the claims to the other groups. These other groups can be electively

---

[31] As a practical matter, the foreign cited references will almost always be included in a supplemental IDS even if the applicant does not think they are particularly material to patentability; better to play it safe early than be sorry later.

[32] However, a thorough and independent search is always a good idea before incurring the many legal expenses of patenting the invention and before incurring the costs of developing the invention in the hope of someday marketing it exclusively. Depending on the applicant's needs, the search should support one or both of two queries: Is the invention patentable in view of the prior art and do I have freedom to operate the claimed invention?

[33] A petition against a restriction requirement is granted as an administrative action as opposed to a right of appeal, which relates to substantive actions. The difference is that a petition will be reviewed as an administrative manner by a group director or supervisor. Applicants will typically argue the merits of the restriction requirement at the time they make their election. Even if they do not file a petition at that time, their objection is preserved in the record should they wish to pursue a petition at a later time in the prosecution.

pursued in one or more separate patent applications, termed *divisional applications*, that must be filed sometime during the pendency of the parent application (or one of its progeny) and will have the benefit of the filing date of the parent application. To better appreciate how a restriction requirement works, Figure 2.6 lists an abbreviated set of claims from an actual patent application (10/346,874) obtained from the USPTO Public PAIR. After the submission of the set of claims shown in Figure 2.6, the USPTO issued the five-way restriction requirement summarized in Figure 2.7.

In the applicant's response subsequent to the restriction requirement, an amended set of claims were filed where only one of the groups listed above was retained and the claims corresponding to the other groups canceled. The claims to the canceled subject matter were subsequently included in four separate divisional patent applications—the primary patent application eventually issued into US 6,939,578 and the four separately filed divisional applications issued into patents as well.

After the applicant has chosen a group to pursue after a restriction requirement (if they received a restriction requirement), the examiner will perform a prior art search. After reviewing the USPTO prior art search results together with any references that the applicant has submitted in the IDS, the examiner will issue an office action with a decision on whether all, some, or none of the claims is allowed. In some cases, the claims are rejected because the examiner believes that they are nonpatentable in view of one or more of the prior art references. Alternatively, the examiner may allege that the claims failed to meet the written description or enablement requirement. Or the examiner may not reject the claims outright but may have an objection to the form the claims are in. If there is an objection to form only, the claims can be rewritten according to the examiner's suggestion and the claims should then be allowed. Rarely are all claims in an application allowed in the first office action, and very often they are all rejected.[34] If some of the claims are allowed and others rejected, the applicant has the opportunity to cancel the rejected claims and have the allowed claims pass to issue in that application and, if desired, refile the remaining, rejected claims in a continuation application in which prosecution will be further pursued. In the case where all of the claims are rejected, the applicant can argue the grounds of the rejection, explaining to the examiner why his interpretation of the prior art and/or the law is not correct. Alternatively (or in addition) to the first option, the applicant may choose to amend the claims; often by narrowing them in the hopes of getting around the grounds (prior art, lack of enablement or written description) that were the basis of the rejection.[35] Figure 2.8 shows the overall procedure for handling patent applications at the USPTO.

---

[34]One way to view this is that the patent applicants are attempting to maximize the breadth of their invention and need to test the boundaries in the process. If all of the claims were allowed in the first pass, the applicant might wonder if he might have been able to claim his invention more broadly.

[35]Although the applicant is free to amend the claims in the patent application during the prosecution of the patent application, he cannot rewrite the claims in any way he pleases. Rather the claims must be fully supported by the text of the patent application as filed; any attempt to add matter in the claims that is not supported by the patent application as filed will draw what is often referred to as a *new matter rejection*, meaning that the applicant has attempted to introduce new matter to the patent application that was not present as originally filed. An attempt to introduce new matter into the claims during the patent prosecution process violates the written description requirement and will be discussed in more detail in Chapter 9.

What is claimed is

1.   A process for forming copper deposits on a substrate comprising
   a. Contacting a substrate with a copper complex (I), to form a deposit; and

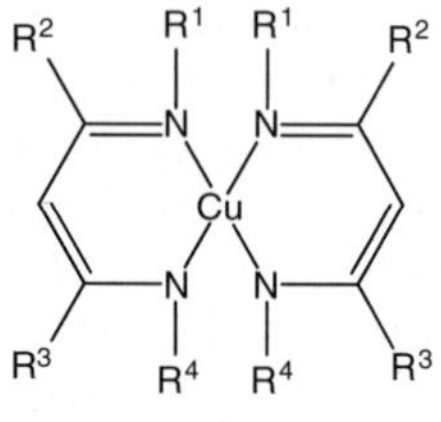

(I)

   b.   Contacting the deposited copper complex with a reducing agent.
   (Variable descriptors for R^1–R^4 taken out; claims 2–6 deleted.)

7.   A 1,3-diimine copper complex (II)

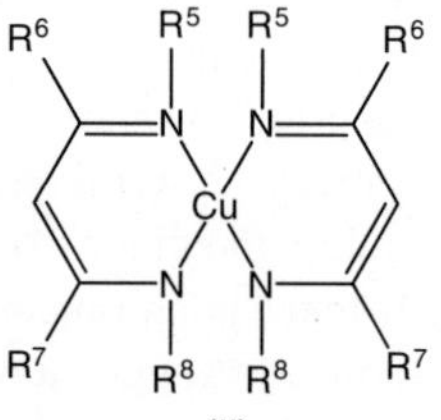

(II)

(Variable descriptors R6–R8 taken out; claim 8 deleted.)

9.   An article, comprising the 1,3-diimine copper complexes (II) of claim 7, deposited on a substrate.
   (Claims 10–11 deleted.)

12.   A process for the synthesis of diimines.
   (Subsequent steps left out; additional process claims 13–15 deleted.)

16.   A composition, comprising amino-imines (VII).

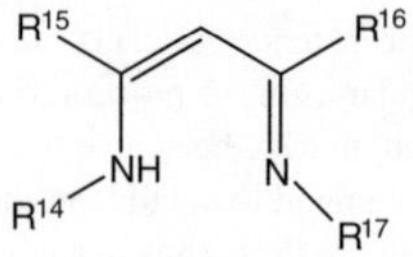

(VII)

(Variable descriptors and additional composition claim 17 taken out.)

FIGURE 2.6   Claims pending in application U.S. 10/346,874.

Application/control Number: 10/346,874

Art Unit: 1762

Detailed Action

Election/Restriction

Restriction to one of the following inventions is required under 35 U.S.C. 121:

I.   Claims 1–6, drawn to a deposition method, classified in class 427, subclass 252.

II. Claims 7 and 8, drawn to a copper complex, classified in class 556, subclass 1.

III. Claims 9–11, drawn to a coated substrate, classified in class 428, subclass 411.1.

IV. Claims 12–15, drawn to a method of making diimines, classified in class 548, subclass 100.

V.   Claims 16 and 17, drawn to a diimine, classified in class 548, subclass 100.

The inventions are distinct, each from the other, because of the following reasons. . . .

FIGURE 2.7   Restriction requirement by USPTO examiner.

Let's consider a hypothetical patent prosecution example representative of how applications flow through the patent process in the United States. In this hypothetical example, you will come to better appreciate the way in which an office action rejection by the USPTO is conveyed to the applicant as well as the way an applicant might attempt to amend the claim to overcome that rejection.

A chemist working for the Miracles in a Bottle pharmaceutical company (where the company motto is Discovering yesterday's drugs, tomorrow . . . maybe) has been working on compounds useful for the treatment of migraine headaches. During the course of his work, he discovers an indole-containing compound that appears to have a very salubrious effect in preclinical migraine efficacy models. He writes up a record of invention and submits it to the Miracles in a Bottle patent counsel, who drafts a provisional patent application. Even though provisional patent applications do not require claims, the patent counsel puts two in anyway, or as he says: "Just so I don't forget to put them in later."

The patent attorney drops off the patent application with the U.S. Postal Service (USPS) using Express Mail to Addressee Service and includes a stamped, self-addressed return postcard to make sure that the patent office gets the application (not that he doesn't trust the USPS). He deposited the provisional patent application at the post office on January 13, 2005, and later received the postcard confirming that the USPTO had received the provisional application. He dutifully placed the

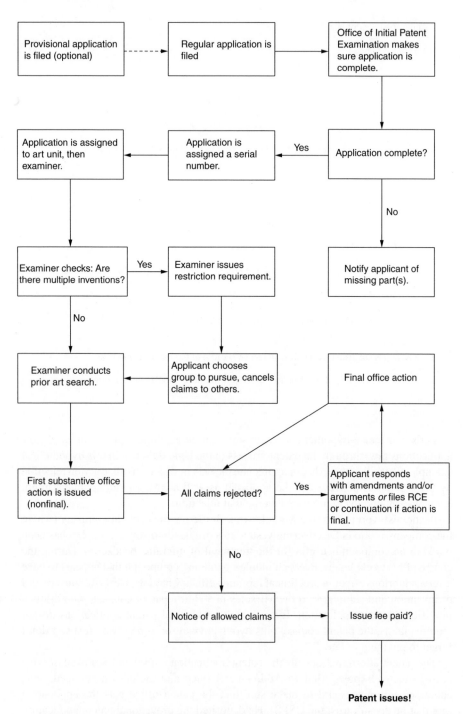

FIGURE 2.8   The patent process at the USPTO.

provisional application in the file folder corresponding to the company's internal docketing numbering system, where he also placed the filing receipt documenting the date he left the application off as well as the postcard. Several weeks later, he received the application number from the USPTO: 01/006,314. About 8 months after the provisional application was filed, the patent counsel got a tickle notice from his docketing system, letting him know that the 1 year time period for filing a regular application that could claim priority back to the first provisional (01/006,314) was 4 months away. "I'd better attend to that right away," he thought, "soon as I get back from lunch, I'll get right on it." Three hours later and just back from lunch, the patent attorney called the chemist on the telephone and asked, "Remember the indole compound you discovered for treating migraines last year? The regular application needs to be filed pretty soon, and I was wondering if you had made any additional analog that you wished to add in the regular filing?"

The chemist replied rather excitedly, "Well actually, I do. You see I had this funny idea to make benzothiophenes instead of indoles, and wouldn't you know it but one of them was even more active than the indole; of course, I jazzed it up further a little in the process!"

"I'll bet your boss is proud of you, eh?" the patent counsel replied.

"Nah, he didn't seem to happy . . . actually the indole was his idea, and I think he gets mad when my stuff works better than his," the chemist averred but then added, once again rather excitedly, "Hey, do you think you can mention that these compounds can prevent cancer, too?"

"Do we have any data to support that assertion?" the patent attorney asked somewhat skeptically.[36]

"Not yet, but trust me . . . I've got a good feeling about this one" the chemist replied.

"Well, all that aside, why don't you send me the experimental results for your new compound, and I will write it up and get a draft back to you and your boss in the next couple of weeks." "This will be easy," the patent attorney thought, "I'll just add the benzothiophene to the old case and file it all together as a regular application and list both guys as inventors." He also thought "What the heck, let's go for that cancer indication while we're at it. They don't call us Miracles in a Bottle for nothing." The claims from the first nonprovisional patent application are shown in Figure 2.9.

The patent attorney executed assignments of the invention from the chemist and his boss as well as signed declarations that they were in fact the inventors of the claimed subject material. In a procedure much like before, the patent attorney went to the post office and used Express Mail to send the patent application and a stamped self-addressed postcard; the date the application was mailed was January 11, 2006.

"Geez," the attorney thought, "I'm getting good at this, I made it with two days to spare!" Arriving back at his office, the patent attorney put his feet up on the desk and clasped his hands behind his head and leaned back in his chair. "Now all we gotta do is wait."

Approximately 3 months later, the first office action arrived. Seeing the envelope from the USPTO, the patent attorney exclaimed, "Man, I hope that's the patent!"

---

[36]More will be said about requisite demonstrations of utility in Chapter 6 and enablement in Chapter 7.

What we claim:

1. A compound of formula I:

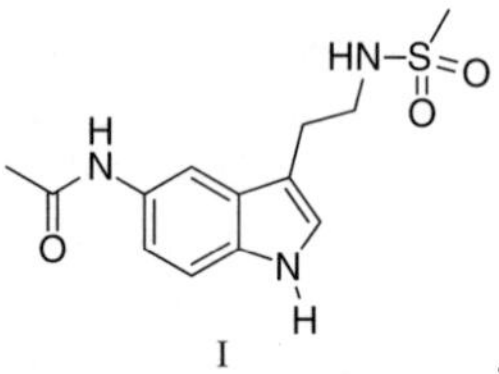

I .

2. A compound of formula II:

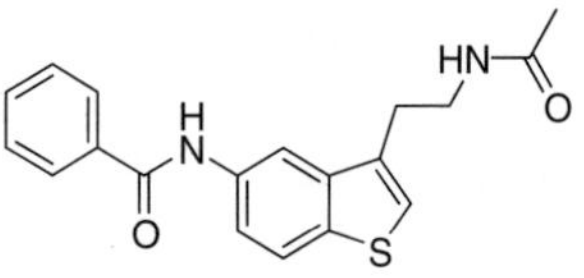

II .

3. A method of treating migraine headaches or preventing cancer comprising the administration of a compound of claim 1 or claim 2 to a patient in need thereof.

FIGURE 2.9   Claims from first nonprovisional patent application.

Eagerly tearing the envelope's contents open, he frowns as he sees it's a notice of a restriction requirement. The restriction requirement dated April 1, 2006, read as follows, in part:

Art Unit: 1672

Application Number 11/786,091

DETAILED ACTION

Election/Restriction

Restriction to one of the following inventions is required under 35 U.S.C. 121:

   I. Claim 1 is drawn to a heteroaryl group of class 548, subclass 490.[37]
  II. Claim 2 is drawn to a heteroaryl group of class 549, subclass 49+.
 III. Claim 3 is drawn to methods of treatment/prevention using compounds of class 548, subclass 490 and class 549, subclass 49+.

The inventions are distinct, each from the other because of the following reasons . . .

---

[37]The USPTO classifies claimed subject matter according to an internal classification system whose mysteries are not completely fathomable to the common man. The *Index to the United States* can be found at www.uspto.gov/web/patents/classification/uspcindex/indextouspc.html.

"Okay," our patent attorney thinks, "now instead of one patent we'll get three patents, and that should really help with my metrics this year." After conferring with the chemists over the phone, the patent attorney filed a response selecting group II and canceling the other two claims, which he will file independently in two divisional applications. He responds within the 1-month statutory time limit, selecting group II and at the same time files the two separate divisional applications, one containing the original claim 1, and the other containing the methods of claim 3. The amended claim set in the original application is shown in Figure 2.10, together with the claims from the separate divisional patent applications collectively claiming the canceled subject matter from the first patent application.

Within 3 months (shockingly quick for the USPTO, they apparently have their own metrics), the patent attorney received office actions for all three of the cases on the exact same day (now there is a strange coincidence).

"Geez, I feel like a kid opening up his birthday presents, which one do I open first?" he wondered. "I guess I'll close my eyes, have my administrative assistant hold them in front of me and I'll just pick one."

After cajoling his reluctant administrative assistant into holding the office actions up for him, the patent attorney closed his eyes and selected one. Gleefully opening it and then reading it, his face slowly lit up with a big, beaming smile. "The benzothiophene patent has been allowed!"

"That's one-for-one so far!" he shrieked, pumping his arm into the air in sort of a victory-type of celebration display.

"Now, let's try another, I'm on a roll now!" he giddily remarked, nervously jumping up and down in anticipation. He closed his eyes and picked another, all the while his administrative assistant looking at the clock, starting to worry that he wouldn't be getting too much work done that day.

"Okay, here goes number two," he blurted out, opening and then reading the contents from the second envelope, "The indole has been allowed! We got another allowance, that's two for two . . . that's close to 100%, right?"

By this time, some of the other workers in the office had started to migrate over to see what was going on, many of them with worried or puzzled looks on their faces. The patent attorney, aware now that he had begun to attract an admiring crowd decided in his sudden magnanimity to allow his faithful administrative assistant to open the last one.

"No really, it's your turn . . . I've already got to open two, now it's up to you to make the hat trick," the patent attorney generously offered. Quickly opening the letter, hoping to get this little ceremony over as quick as possible, the administrative assistant began to read the letter when his expectant boss, the patent attorney snatched it out of his hands saying,

"Darn it, you're going too slow," and turned to the apparently admiring crowd now surrounding him while he began to read,

"Claim 1 is pending in this application, claim 1 is . . . rejected," the let down was huge. Embarrassed to still be in front of the now disbanding crowd, the patent attorney turned to his assistant and said "I knew I shouldn't have let you open that one, you jinxed it." Noticing that the attorney was not really smiling when he said this, the

What we claim:

1. Canceled.

2. A compound of formula II:

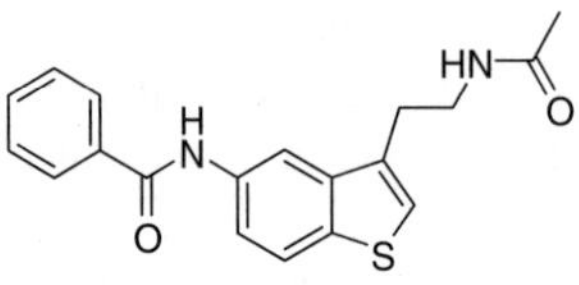

II

3. Canceled.

Claim from first divisional patent application

1. A compound of formula I:

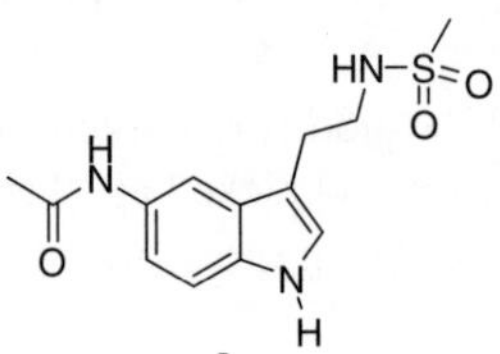

I

Claim from second divisional patent application.

1. A method of treating migraine headaches or preventing cancer comprising the administration of a compound of formula I or II to a patient in need thereof:

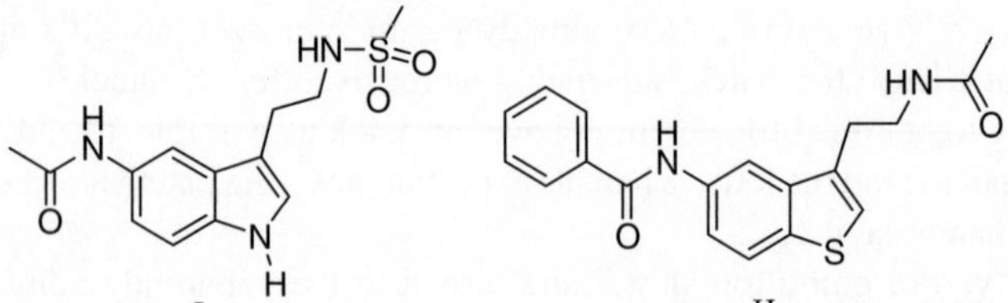

I   II

FIGURE 2.10   The first amended claim set in the nonprovisional patent application and the new claims to the canceled subject matter in two separate divisional application filings.

administrative assistant quickly ducked away, leaving his boss to carry on without him.

"Okay, let's see where the examiner went wrong," the attorney said, "but I'd better wait until after lunch, all of this hard work has made me hungry." Upon returning, the attorney unfolded the office action and began to read:

NON-FINAL REJECTION

The applicant has claimed that the indole and the benzothiophene can treat migraine *and prevent cancer*. However, the only data given in the application relate to the use of the compounds for the treatment of migraine headaches. There are no data to support the allegation that these same compounds will prevent cancer. In the absence of such data, the claimed method is simply not credible. For that reason, this claim is rejected for lack of enablement under section 112, paragraph 1.

"Not credible, not credible! Didn't he see the name of our company—maybe I should call him up and have a little interview just to let him know that he shouldn't be misunderestimating us," the patent attorney mumbled to himself.

Then on further reflection he thought, "Maybe it would be smarter just to amend the claim by removing the cancer aspect. He didn't seem to have any problem with the migraine stuff; I can leave out the cancer part of the claim and get the rest of the claim issued and just file a continuation and keep after the cancer indication in the continuation application—sooner or later I'll wear him down. That way, he gets what he wants and I get my patent before the end of the year review period. That's what I call a win–win situation!"

"Shoot, it's already four in the afternoon, I must as well head home now. First thing tomorrow, I'm gonna get right on it," he thought, heading out the door.

Upon arrival the next morning, the patent attorney drafted his amended claim and filed his response on September 1, 2006. The amended claim is shown in Figure 2.11.

What we claim:

1. Amended: a method of treating migraine headaches comprising
the administration of a compound of formula I or II to a patient in need thereof

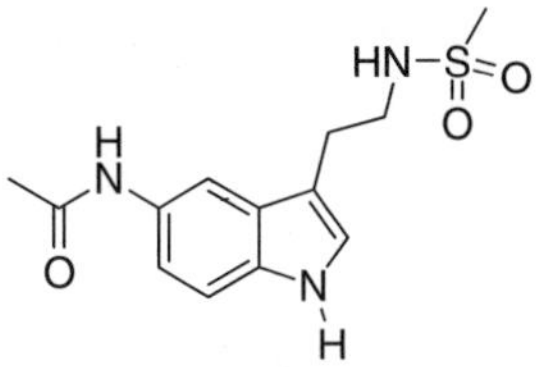

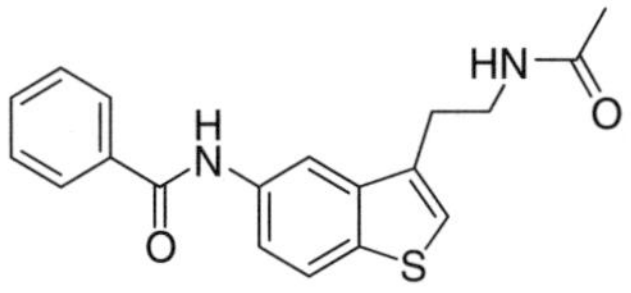

FIGURE 2.11   The amended claim in the rejected divisional method application.

Approximately 4 months after sending in the amended claim, the patent attorney received the letter from the USPTO. He opened the letter and read that his amended claim would be allowed as is. It is surprising that the attorney did not seem at all excited or even very happy about the fact that the claim was now allowed.

His administrative assistant, somewhat puzzled by the patent attorney's noticeable lack of excitement with this seemingly good news wandered over to the attorney's office and peeked his head in saying, "Why so glum? I noticed you got the case allowed."

The attorney responded, "Yes but it's one week too late, our performance reviews were last week, so this one won't even count."

The assistant thought for a minute and then responded, "But it will count for next year, won't it?"

"Who knows, I think they're changing the metrics again. And anyway, that's a long time from now," the attorney grumbled. "Hey, I think it's about lunchtime, you want to head out and grab a bite to eat?"

In the hypothetical just discussed, the applicant was allowed his claims in two out of the three cases in the first substantive office action. This is a little unusual in that very often at least some if not all of the claims in an application are rejected in the first office action. In this simple example, the applicants were trying to get single claims allowed, and the two cases containing compounds were not drawn to broad genera incorporating large numbers of compounds but, rather, were drawn to single species; perhaps that is the reason for the facile allowances.[38]

Once the first office action is received, the applicant is given a limited amount of time to respond. The amount of time given to respond depends on the nature of the office action. For a restriction requirement, an applicant is normally given 1 month. However, the applicant can extend the response time all the way out to 6 months if he is willing to pay a fee that increases for each additional month that the reply is extended. When the office action is on the merits, the response time is normally 3 months, with the possibility of extending the response time out to 6 months, once again by paying an extension fee that increases for each additional month. Failure to respond to an office action in the maximum amount of time will result in the abandonment of the application.[39]

After the first substantive office action is sent back to the USPTO, the examiner will review the arguments and/or amendments and send back what usually is a final office action. This final office action, which takes into account the applicant's previous amendments and/or arguments, may allow none, some, or all of the pending claims. In response to a final office action, the applicant's choices are much more limited. The applicant can no longer argue or make substantive amendments to the claim that would require additional argument or explanation; the only thing that can be done in response to this final office action is to make any nonsubstantive changes to the claim

---

[38] Or maybe their attorney was really just that good!

[39] It is possible in some circumstances to revive an unavoidable or unintentionally abandoned application. Such a revival depends on a showing that the abandonment was unavoidable or unintentional and, of course—you guessed it—the appropriate fee!

that the examiner has suggested (if he has made such suggestions) to put the claims in a condition for allowance.

If there are claims that have been allowed, the applicant can allow those to issue (by, what else, but paying an issue fee!) and file a continuation application to pursue the remaining and/or additional claims if so desired. If no claims are allowed, the applicant can give up and abandon the application (seldom done if the invention is worth pursuing), file a continuation application, or file a request for continued examination (RCE). In cases in which no claims are allowed in the parent application, most applicants choose to file an RCE because it is a simpler procedure. A continuing application is an entire new application that is assigned a new application number and is separately docketed. An RCE, in contrast, keeps the same patent application alive and basically allows the applicant a couple more shots at getting the desired claims issued.[40] However, if one or more claims are allowed in the parent application and the applicant wishes those claims to issue, then he will need to file a continuation application because the application containing the allowed claims that he wishes to issue will have matured into a patent with issued claims; no RCE on that application will be allowed.[41]

Whether the applicant is filing an RCE, a continuation, or a divisional application, an important aspect to keep in mind is that the filing date of the first regular patent application in the chain sets the patent term clock ticking. This means that any of the patents that claim priority (continuations or divisionals) from that original application will have their 20-year lifetime measured from the date on which the parent application was filed (absent special circumstances).[42]

To put a graphical representation on what just transpired in the hypothetical, a timeline is shown in Figure 2.12. A few extra details (issue dates and patent dates) have been added. Please note a couple of things about the figure. First, the two divisional applications, despite being separate patent application filings, still claim their priority back to their sister application, as a result, the patent life for the divisional applications is still 20 years from the date of the filing of the original application (January 11, 2006). As mentioned before, the same is true for continuation applications and continuation-in-part applications. The filing of the cancer claim has been added to Figure 2.12, it was not included in the narrative but was filed on January 14, 2007 as a continuation to the divisional application that claimed the

---

[40]If no claims issue from the RCE, the applicant may continue to file RCEs but he must make reasonable efforts at advancing some claims to issue; simply restating the arguments without amendment (or appeal) can result in a final rejection by the examiner without further comment. Additional RCEs without substantive changes in the claims or arguments will only nonproductively consume the applicant's resources.

[41]This means that the continuing application can be filed up to the time the issued claims actually publish in the issued patent. In practice, this means that the continuing application (if one is desired) will usually be filed at the same time the issue fee on the claims from the pending application is paid. It usually takes about 1 month from payment of an issue fee until the time an issued patent is published (and thus patented).

[42]Patent applications that issued before June 8, 1995, have patent terms that run 17 years from the date the patent issues. Patent applications filed before June 8, 1995, and issued after that date have terms of 17 years from the date of issue or 20 years from the date of filing; the applicant chooses. Applications filed after June 8, 1995, are subject to the 20-year from filing term.

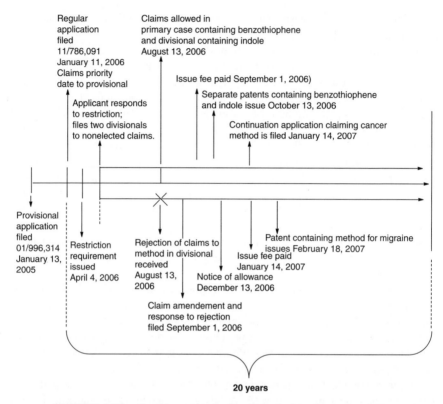

FIGURE 2.12   Graphic patent timeline from the hypothetical example.

cancer and migraine methods together (recall that the claim to the cancer method was canceled so the claim to the migraine could issue). The claim to the cancer would be appropriately pursued in the continuation application. Of course, this does not necessarily mean that the claim will issue because the applicants will still need to convince the examiner that the claim is enabled (which maybe a hard sell based on the facts!).

2.2   POST GRANT PROCEDURES AT THE USPTO

Once a patent is issued by the USPTO, the interactions between the patentees and the USPTO with respect to that patent will greatly diminish but will not cease. At a minimum, the patent assignees will need to keep current with the patent maintenance fees if they wish the patent to remain enforceable over its entire lifetime (discussed in section 2.2.a).

In addition to the mandatory maintenance fees, there are certain postgrant procedures that can take place under special circumstances. For example, patent interferences occur where a contest is waged in the USPTO to determine the first inventor of contested subject matter. These contests may occur between two or more applicants

TABLE 2.1   Patent Maintenance Fee Schedule for Fiscal Year 2009 (effective October 2, 2008)

Timepoint[a]	Small Entity[b]	Other Than Small Entity[b]
Due between 3 and 3.5 years	$ 490.00	$ 980.00
Due between 7 and 7.5 years	$ 1240.00	$ 2480.00
Due between 11 and 11.5 years	$ 2055.00	$ 4110.00

[a]Extra half-year grace period available with payment of appropriate fee.
[b]In U.S. dollars. See footnote xx for an explanation of the term *small entity*.

for patents or between one party that already holds an issued patent and another party that has an application pending.[43]

There are also postgrant procedures whereby a patentee can correct mistakes in a patent after it has already been issued. In general, for mistakes that are clerical, typographical, or minor in character and for which the correction does not introduce new matter or require reexamination, a certificate of correction is most likely the way to go.[44] Alternatively, under special circumstances, a patentee may wish to correct mistakes in the patent through a procedure involving the reissue of the subject patent. The claims in a reissue patent may be changed, and in some special cases, may even be broadened beyond those originally presented, making the reissue patent a sometimes very powerful post-grant procedure (see section 2.2.b).

Finally, a patent may be subjected to a reexamination procedure in which the issued patent is reexamined in view of additional prior art not of record in the original prosecution file for that patent application. The reexamination procedure can proceed by different routes, depending on the circumstances, and it can be a very useful tool for patentees but more likely for those who wish to challenge one or more claims in an issued patent but who do not wish to engage in full scale litigation to do so (see section 2.2.c).

2.2.a Patent Maintenance Fees

After a patent issues in the United States, a maintenance fee schedule is established using the *issue date* of the patent application as the starting time point. The general payment timelines and amounts for U.S. patents for the fiscal year 2009 are shown in Table 2.1.

These amounts are adjusted each year in concert with the consumer price index; as a result, the amount of a maintenance fee that is due for an issued patent needs to be checked before its payment. The first patent maintenance fee must be paid in

---

[43]Interferences will be explained more in Chapter 3 under the discussion of §102(g).
[44]See MPEP sections 1480–1485 for a description of the processes involved in the issuance of a certificate of correction. As a practical matter, you may have encountered certificates of correction in your normal patent experience. The corrections made to patent specifications are not made in the patent specifications per se but rather are included as auxiliary sheets placed at the end of the issued patent and provide instructions to the amendments that have been entered into the record. The patent as corrected will have the same legal effect as if the patent had been correctly issued in the first place (35 U.S.C. §254).

the window between 3 and 3.5 years from the patent issue date, the second patent maintenance fee must be paid in the window between 7 and 7.5 years from the patent issue date, and the third patent maintenance fee must be paid in the window between 11 and 11.5 years from the patent issue date; the maintenance fees cannot be paid in advance of the payment window. If the patent assignee misses the window period through nonpayment or insufficient payment, there is a half-year grace period during which the fee can be made up by paying the regular fee plus a surcharge.[45] A failure to make the required payment within the prescribed timeframe, including the half-year grace period (available with the surcharge) constitutes abandonment of the patent.

Although the best advice is to be 100% certain that the patent maintenance deadlines are met with the required payment, mistakes can be made and deadlines missed. For example, a patent assignee may rely on another party to make the payments for her and that other party (e.g., agent, attorney or payment agency) may fail to fulfill his duty. In such an unfortunate circumstance, there still may be hope for resurrecting what has become a prematurely expired patent, depending on the circumstances of the nonpayment. A patent that has expired for failure to make payment within the statutory grace period may be revived if the assignee can show by petition that the failure to make timely payment was either unintentional or unavoidable.[46]

---

[45]The surcharge for a payment falling within a half year of the normal payment deadline is $130 for a non-small entity and $65 for a small entity; these were the fees for the year beginning October 2, 2008. These fees are adjusted up on an annual basis and a current fee table should be consulted immediately before making payment.

[46]37 CFR 1.378 Acceptance of delayed payment of maintenance fee in expired patent to reinstate patent.

(a) The Director may accept the payment of any maintenance fee due on a patent after expiration of the patent if, upon petition, the delay in payment of the maintenance fee is shown to the satisfaction of the Director to have been unavoidable (paragraph (b) of this section) or unintentional (paragraph (c) of this section) and if the surcharge required by § *1.20*(i) is paid as a condition of accepting payment of the maintenance fee. If the Director accepts payment of the maintenance fee upon petition, the patent shall be considered as not having expired, but will be subject to the conditions set forth in *35 U.S.C. 41*(c)(2).

(b) Any petition to accept an unavoidably delayed payment of a maintenance fee filed under paragraph (a) of this section must include:

(1) the required maintenance fee set forth in §1.20 (e)–(g);

(2) the surcharge set forth in § 1.20(i)(1); and

(3) a showing that the delay was unavoidable since reasonable care was taken to ensure that the maintenance fee would be paid timely and that the petition was filed promptly after the patentee was notified of, or otherwise became aware of, the expiration of the patent. The showing must enumerate the steps taken to ensure timely payment of the maintenance fee, the date and the manner in which patentee became aware of the expiration of the patent, and the steps taken to file the petition promptly.

(c) Any petition to accept an unintentionally delayed payment of a maintenance fee filed under paragraph (a) of this section must be filed within twenty-four months after the six-month grace period provided in § 1.362(e) and must include:

(1) the required maintenance fee set forth in § 1.20 (e)–(g);

(2) the surcharge set forth in § 1.20(i)(2); and

(3) a statement that the delay in payment of the maintenance fee was unintentional.

2.2.b Reissue Applications and Patents

Under certain circumstances after a patent has already issued, the patentee can withdraw the issued patent and subject it to a reissue process, which can result in a new patent: the *reissue patent*. Reissue patents are identified by the prefix *RE* followed by the number of the patent (e.g., RE 39,424). It is not surprising that a reissue patent originates from a reissue application. The reissue application process occurs when a patent is withdrawn from issue by its patentees and submitted as an application to the patent office for further prosecution. In order to do so, the patentee must assert that the patent is:

> through error without any deceptive intention, deemed wholly or partly inoperative or invalid, by reason of a defective specification or drawing, or by reason of the patentee claiming more or less than he had a right to claim in the patent, the Commissioner shall, on the surrender of such patent and the payment of the fee required by law, reissue the patent for the invention disclosed in the original patent, and in accordance with a new and amended application, for the unexpired part of the term of the original patent. No new matter shall be introduced into the application for reissue (35U.S.C. §251).

One possibility for seeking a reissue patent includes the situation in which the patentees discover prior art that they were not aware of during the prosecution of the patent application, but that could affect the validity of one or more of their patent claims such that they might like to narrow the claim(s) to avoid the prior art and thus provide them with the maximum patentable scope for their invention.

Conversely, patentees might wish to broaden one or more claims in their patent. This might occur, for example, when they become aware that they could have more broadly claimed their invention then they actually did and wish to broaden one or more of the claims after the original patent has issued. It might seem a little bit odd or perhaps even unfair to allow patentees have a patent issued and then go back at a later date and reissue the patent but with broader claims than what were originally present. When we as practitioners of the chemical arts look at an issued patent, our presumed belief is that the claims determine the scope of the invention; therefore, we might be surprised to learn that those claims can actually be broadened under certain circumstances. Perhaps you might be (at least partially) relieved to learn that the reach of a broadening reissue is limited by three important factors.

> (d) Any petition under this section must be signed by an attorney or agent registered to practice before the Patent and Trademark Office, or by the patentee, the assignee, or other party in interest.
>
> (e) Reconsideration of a decision refusing to accept a maintenance fee upon petition filed pursuant to paragraph (a) of this section may be obtained by filing a petition for reconsideration within two months of, or such other time as set in the decision refusing to accept the delayed payment of the maintenance fee. Any such petition for reconsideration must be accompanied by the petition fee set forth in § *1.17*(f). After the decision on the petition for reconsideration, no further reconsideration or review of the matter will be undertaken by the Director. If the delayed payment of the maintenance fee is not accepted, the maintenance fee and the surcharge set forth in § *1.20*(i) will be refunded following the decision on the petition for reconsideration, or after the expiration of the time for filing such a petition for reconsideration, if none is filed. Any petition fee under this section will not be refunded unless the refusal to accept and record the maintenance fee is determined to result from an error by the Patent and Trademark Office.

First, a reissue application that seeks to broaden any of the claims (in any way) of the original issued patent can do so only if the reissue application is submitted within 2 years of the original patent's issuance. However, the reissue application needs to be only *filed* within the 2-year period from the original patent's issue date; the actual broadened claims in the reissue application may be added after a period of greater than 2 years from the original patent's issue date, provided they are added as part of the reissue patent application that was filed within the 2-year period.

Second, the patentee cannot claim anything in the reissue application that is not supported in the original application. In other words, they cannot introduce new matter in the reissue application; they are bound in substantive content by the disclosure of their original patent.[47] For example, if in their patent specification the patentee(s) had adequately described *only* a single embodiment of a certain polymer with an average molecular weight of "between 18,000 and 25,000 daltons," they probably would not be able to claim that polymer in a narrower or broader range during the reissue process. If, however, the specification adequately described the additional narrower and/or broader molecular weight ranges for that polymer, then one or more of these additional ranges possibly could be claimed in a broadening reissue.[48]

Finally (and perhaps most important), under the doctrine of intervening rights (35 U.S.C. §252) relief is provided for defendants who did not infringe an original patent but do infringe the patent as reissued. Paragraph 2 of 35 U.S.C. §252 provides the following:

> A reissued patent *shall not* abridge or affect the right of any person or that person's successors in business who, prior to the grant of a reissue, made, purchased, offered to sale, or used within the United States, or imported into the United States, anything patented by the reissued patent, to continue the use of, to offer to sale, or to sell to others

---

[47]The prohibition against placing new matter into an application after it's already been filed is governed by the written description requirement as stated in 35 U.S.C. §132(a). The prohibition against introducing new matter into a patent application after filing is discussed in detail in Chapter 9.

[48]From a freedom-to-operate standpoint it is important for the person reviewing issued patents to review the entire patent and not just the issued claims. Why? As we see in this section, it is possible for patentees under some circumstances to reissue patents with broader claims than what were in the originally issued patent. Considering that the patentees have 2 years from the issue of the original patent to file the reissue patent application and that it might take several additional years until the final reissued claims are published as part of the reissued patent, it ultimately could take several years to really know what claims might issue from a single patent application. However, since the claims in any issued patent must be adequately supported by the disclosure in the patent application that was originally filed, a more complete understanding of what *might* ultimately issue from a single patent application is the specification of the patent application as filed (including the original claims). As we read later in this section, there are defenses to persons who did not infringe the original patent but do infringe the reissued patent; these remedies however, are not always easy or inexpensive to invoke. Further, when one is reviewing issued patent claims one also needs to consider the possibility that additional continuation (or divisional) patent applications could still be pending that may claim subject matter within the scope of the patent application as filed but different from any claims that have already issued. For this reason also, one should review the entire patent document and not just the issued claims or at least have a full appreciation for the possibility that claims will issue at a later date that are different from the claims in the issued patent that you are reviewing. Patent continuity data for U.S. patents and patent applications can be obtained from the USPTO's website under the subheading Public PAIR. There you can see what, if any, continuation applications have been filed as of the time you do the checking.

to be used, offered for sale, or sold, the specific thing so made, purchased, offered for sale, used or imported unless the making, using, offering for sale, or selling of such thing infringes a valid claim of the reissued patent which was in the original patent. The Court before which such matter is in question *may* provide for the continued manufacture, use, offer for sale, or sale of the thing made, purchased, offered for sale, used or imported as specified, or for the manufacture, use, offer for sale, or sale in the United States of which substantial preparation was made, before the grant of the reissue, to the extent and under such terms as the Court deems equitable for the protection of investments made or business commenced before the grant of the reissue (emphasis added).

In effect, the doctrine of intervening rights provides that a noninfringing act will continue to be noninfringing even where the reissue process changes at least some of the claim scope such that the activity technically infringes the reissue patent; paragraph 2 of 35 U.S.C. §252 directly addresses the issue of basic fairness to those who may have relied on the issued patent and planned their activities accordingly. What is especially noteworthy about the doctrine of intervening rights is that it contains two separate categories of rights. The first category of rights describes an absolute intervening right because the reissue patent *shall not* abridge the right of the party whose rights would be affected by the reissue patent (absent the doctrine). The absolute intervening right occurs when the party involved is already engaged in the activity that was not claimed in the first patent but was covered by the reissue patent.

In contrast, an equitable assignment of intervening right *may* occur even when the subject party has not yet engaged in the infringing activity covered by the reissue patent. This latter situation is considerably more fluid in that it requires a showing of substantial preparation and allows for the tailoring of a remedy to meet the variable demands of the particular situation. In effect, this latter scenario can allow a party to infringe the reissue patent even when she had not established the actual infringing activity before the issuance of that patent. This equitable doctrine of intervening rights depends on the notion that it would not be fair to allow a party to invest resources in a reasonable reliance on her intended activity being noninfringing only to have the patent landscape change when an earlier patent is reissued with new claims that her intended activity will infringe. The equitable judgment needs to be carefully tailored since an overly liberal construction of the doctrine could effectively eviscerate the value of the category of reissue patents.

Besides the legal and equitable limitations on reissue patent enforcement, there are direct limits on what a patentee can attempt to capture in a broadened, reissue claim. In particular, the recapture doctrine prevents the patentee from claiming territory in reissue patents that she has given up during the prosecution of the original patent application. If a patentee amends elements or features of claims to make that claim patentable (e.g., in view of prior art) or argues that certain elements or features of the claim help to make that claim patentable over the prior art during the original application process, then the patentee cannot later try to reissue the claims of that invention by broadening out the relevant elements or features that were previously narrowed to attain patentability. This is fair in principle, but as a practical matter it is not always easy to discern whether the particular amendment or argument in question relates specifically to the relevant aspects of the claim that are being broadened. The

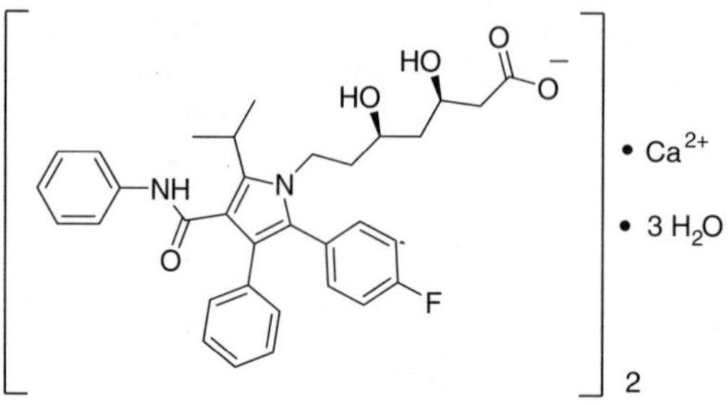

FIGURE 2.13 Chemical structure of atorvastatin (Lipitor®).

USPTO describes a number of general situations related to recapture that serve as guidelines.[49]

To help clarify how a reissue application might be effectively used, let's consider a recent high-stakes litigation contest, *Pfizer v. Ranbaxy Laboratories*,[50] over the number one selling drug in the world, atorvastatin (Lipitor®) and how the patent holder (Pfizer) used the reissue process to add and/or amend claims identified in litigation with one of their patents covering the compound. Atorvastatin is a member of the statin class of drugs primarily prescribed for the lowering of low-density lipoprotein (LDL) cholesterol (the bad cholesterol) and has been demonstrated to reduce heart attacks and strokes in at-risk individuals. The compound is manufactured and sold by the Pfizer Pharmaceutical Corporation as a single enantiomer with the structure shown in Figure 2.13.

Pfizer's patent US 5,273,995 contained several claims of which the following three were relevant to the issue at hand.[51] Claims 1, 2, and 6 read as follows:

> 1. [R-(R*, R*)]-2-(4-fluorophenyl)-β, δ-dihydroxy-5-(1-methylethyl)-3-phenyl-4-[(phenylamino)-carbonyl]-1H-pyrrole-1-heptanoic acid or (2R-trans)-5-(4-fluorophenyl)-2-(1-methylethyl)-*N*,4-diphenyl-1-[2-(tetrahydro-4-hydroxy-6-oxo-2H-pyran-2-yl)ethyl]-1H-pyrrole-3-carboxamide; or pharmaceutically acceptable salts thereof.

> 2. A compound of claim 1 which is [R-(R*R*)]-2-(4-fluorophenyl)-β-δ-dihydroxy-5-(1-methylethyl)-3-phenyl-4-[(phenylamino)carbonyl]-1H-pyrrole-1-heptanoic acid.

> 6. The hemicalcium salt of the compound of claim 2.

Pfizer asserted claim 6 against Ranbaxy, alleging that Ranbaxy's proposed launch of generic Lipitor® would infringe claim 6 of their '995 patent. Among the many counterclaims Ranbaxy raised in its defense against infringement was that claim 6 did not actually cover Lipitor®, due to the language in Pfizer's claims. In particular,

---

[49] See MPEP 1412.02 for the patent office's treatment of recapture issues.

[50] *Pfizer, Inc. v. Ranbaxy Laboratories Ltd.*, F.3d 1284 (CAFC 2006).

[51] There were actually two patents at issue in the litigation. The other patent, US 4,681,893 covered the compound generically (essentially in any form), whereas US 5,273,995 specifically identified and claimed the single enantiomer and the single enantiomer as the hemicalcium salt that is Lipitor®. The significance of this single enantiomer calcium salt patent to Pfizer's economic interests cannot be easily overstated. The US 4,681,893 generically covering the compound is set to expire (including regulatory patent extension

Ranbaxy alleged that because the compound of claim 2 was specifically described only as the carboxylic acid, it did not allow for the possibility of a hemicalcium salt meaning that the compound of claim 6 could not be the salt either; claim 6 depends on claim 2 and, therefore, cannot expand the definition of claim 2.[52] The CAFC agreed

under the Hatch-Waxman legislation) on March 24, 2010 (see FDA Orange Book—this can be readily viewed online by searching for "electronic Orange Book" or going to www.fda.gov/cder/ob/default.html). Patent US 5,273,995 (including a pediatric market exclusivity extension) expires June 28, 2011. At the 2007 sales rate of approximately $6 billion in the United States, the ability to keep generic competition from the marketplace for an additional 15 months (the difference in expirations dates between the two patents) would likely mean several billions of dollars of additional revenue for Pfizer. In the United States, a generic drug maker may file an abbreviated new drug application (ANDA) with the FDA, and the applicant needs to show that their product is bioequivalent (meaning that certain pharmacokinetic parameters fall within established ranges relative to the branded drug product; this is in contrast to the originator company which must conduct multiple clinical trials to establish both safety and efficacy). They must also file a certification for each patent that the innovator (drug originator) has listed in the FDA's Orange Book. If there is no listed patent covering the pharmaceutical product, than the generic company may file a certification to that effect (paragraph I certification). Alternatively, if there is one or more listed patents, the certification must assert that for each patent listed in the Orange Book that the patent is expired (paragraph II certification), the patent will be expired when the generic comes to market (paragraph III certification), or the patent is invalid or will not be infringed by the generic product or that the product will be used for an indication not covered by the listed patent (asserting noninfringement and/or invalidity is known as a paragraph IV certification). Where a generic company has filed their ANDA with a paragraph IV certification (asserting noninfringement and/or invalidity of the listed patent), the innovator company has 45 days to respond with an infringement suit that will in turn invoke an automatic 30-month stay (or until the litigation is resolved, whichever comes sooner) during which time the litigation can commence. The first generic to file an ANDA under paragraph IV that successfully comes to market is granted 180 days exclusivity; a period of time when no other generic marketer can come to market (unless they filed on the same day). These provisions are part of a comprehensive legislation known as the Drug Price Competition and Patent Term Restoration Act of 1984, or commonly as the Hatch-Waxman Act, that strove to balance the need for affordable drugs with the inducement necessary for funding expensive, high-risk pharmaceutical research.
[52]Claiming chemical inventions will be discussed in more detail in Chapter 5, but for now please appreciate that there can be independent and dependent claims. Independent claims stand alone and do not incorporate limitations from another claim. In contrast, dependent claims are used to narrow the independent claim that they are dependent from. A dependent claim can usually be recognized by language such as: "The process according to claim x wherein..." or "The compound according to claim y...." The dependent claim incorporates all of the limitations that the claim it is dependent from and then adds one or more additional limitations, removes one or more possibilities or in some other way narrows the claim that it is dependent from. This is a requirement of the patent statute, 35 U.S.C. §112 paragraph 4 "Subject to the following paragraph, a claim in dependent form shall contain a reference to a claim previously set forth and then specify a further limitation of the subject matter claimed." In this example, claim 2 limits the compound claimed to the specific structure specified; there is no inclusion in this claim of the possibility of salt forms (unlike claim 1). Since claim 6 is dependent from claim 2 and not claim 1, it must bring all of the limitations of claim 2 into it, including the fact that it is a carboxylic acid and not its calcium salt. Since claim 2 did not contemplate the possibility of a calcium salt, the attempt to make a dependent claim from claim 2 that claims a calcium salt is an impermissible expansion of the definition of the claim 2. It is interesting that, whereas dependent claims always refer to another claim, the mere reference to another claim does not always make a claim dependent; a claim may borrow a definition from another claim without automatically incorporating all of its limitations. In the instant example, Pfizer conceivably could have argued that claim 6 was actually independent and was only borrowing the definition from claim 2 and not all of the limitations of claim 2. However, Pfizer had stipulated in the context of the litigation that claim 6 was a dependent claim.
    Making claim 6 dependent from claim 1 rather than claim 2 during the prosecution of the original patent application probably would have avoided the problem for Pfizer. Alternatively, the addition of the language

with Ranbaxy's argument while still recognizing that Pfizer had improperly claimed "what might otherwise have been patentable subject matter."

Having thus witnessed what could eventually amount to potentially billions of dollars of lost revenue due to a drafting error, Pfizer pursued the reissue of US 5,273,995. In the prosecution of their reissue application, the USPTO initially rejected Pfizer's attempt to claim the single enantiomer but eventually allowed the patentability of the single enantiomer subject matter in the form of a pharmaceutical composition as well as methods of using the pharmaceutical composition (see RE40667, issued on March 17, 2009). Presumably, Pfizer will reassert the new patent against any generic litigants and the reissue patent application could ultimately be subject to a new round of litigation.[53] Of course, patent protection for Lipitor® under the reissued patent RE40667 will still end on the same date as the original patent 5,273,995, which is June 28, 2011.

2.2.c   Ex Parte Procedures

Cause, Baby it ain't over 'til it's over

—Lenny Kravitz

Patent prosecution sits at the busy intersection where complex legal and technological issues intersect. In the crowded dockets of USPTO examiners as well as the

---

"or pharmaceutically acceptable salts thereof" to claim 2 should also have done the trick. Likewise, simply writing claim 6 as an independent claim meaning writing the name of the claimed calcium salt without referring to any other claim would have achieved Pfizer's goals as well; nevertheless, these things were not done during the original prosecution most likely due to a drafting oversight, thus the need for the reissue patent application. Finally, claim 1 generically covers Lipitor® as well but it's not clear from the written CAFC opinion why Pfizer did not assert claim 1 in the litigation.

[53] In a protest by Ranbaxy, the reissue application was challenged on many grounds, including the grounds that the invalidated claims were not appropriate subject matter for reissue under 35 U.S.C. §251. In particular, Ranbaxy argued that the patent did not claim more or less than it should have but rather the scope of claim 6 in the '995 patent was the hemicalcium salt of atorvastatin; in other words, the patent did not claim too much or too little but merely claimed it improperly, which they argued was not an appropriate ground for reissue. If you are interested, the file history of the reissue patent application can be found on the USPTO's website at http://portal.uspto.gov/external/portal/pair. Click on the "Public PAIR" link. After a verification protocol, you will be directed to the page that will allow you to review electronic versions of the documents submitted to and from the USPTO. The entry input can take several forms, including a patent application number, patent publication number, or patent number. For the US 5,273,995 patent reissue, you can select the choice "patent" and then enter 5,273,995 as the patent number. You should be directed to a screen with many choices of information to be selected. For the reissue patent application, click the tab marked "Continuity Data" and select the reissue patent application 11/653,830 (filed on 01-16-2007). To view the correspondence between the USPTO and the applicant, click on "Image File Wrapper" from the tabs near the top and then you will be able to select and read various documents. The Public PAIR system can be an extremely useful and powerful learning and research tool since it allows you to explore a lot of patent information without having to order and pay for it. In general, nonconfidential correspondence between the USPTO and patent applicants can be viewed as part of the patent prosecution file history where the patent application or patent issuing from it has published. If the file history is not available electronically on Public PAIR, it can be obtained by various patent services on a per page charge.

patent practitioners who prosecute their client's cases before the USPTO, it can be very easy to miss the existence or the significance of one or more pieces of prior art. As much as patent examiners work to find and understand the prior art that might affect the claims of the patent application before them, it may not always be possible to adequately evaluate the complete prior art in a manner that bears up well on further scrutiny. In fact, patent examiners may not have all of the relevant prior art in front of them when they perform the patent examination, or depending on the particular technology being examined, the breadth and depth of the prior art may simply be too much to completely sound out in a normal search.

When one considers the powerful exclusionary powers an issued patent confers on its owner, the importance of issuing valid patents is paramount. Of course, all efforts toward getting the patent examination done correctly in the first place must be made, but nevertheless it has come to be recognized that additional mechanisms should be available, short of full-scale litigation, for testing the validity of issued patents. In Europe and elsewhere, systems of public opposition to recently granted patents have been in place for some time. The idea behind a postgrant public opposition period is that parties who might be adversely affected by the granting of a patent are afforded a chance to contest it without being first subjected to a suit for patent infringement. In many instances, the parties contesting the issuance of the patent are highly motivated and especially knowledgeable in the area of the patent and can thereby subject the patent to a degree of scrutiny that otherwise might not be possible. In the United States, postissuance reexamination procedures provide opportunity for patent challenges as do the postgrant opposition procedures available in Europe and elsewhere.[54]

The most commonly used patent reexamination procedure before the USPTO was instituted in 1980–1981. This procedure is referred to as the ex parte reexamination procedure, and its basic framework is outlined in the patent laws 35 U.S.C. §301–307.[55] Some key features of the patent laws regarding ex parte patent reexamination are as follows:

1. *Any* person at *any time* (during the enforceable life of the patent) may file a request for the reexamination of a U.S. patent.[56] This includes other parties, the patentee or even the USPTO itself, although most ex parte reexamination requests are filed by *other parties*. Reexamination requests filed by parties other

---

[54]This is more true for the inter partes reexamination procedure, which will be discussed shortly.

[55]The term ex parte is Latin legalese for "on one side only" (*Black's Law Dictionary*, 6th ed.). In the context of this discussion, it means that the actual reexamination process, once begun, is conducted between the patent holder and the USPTO only. This is in contrast to the inter partes proceeding that to a much greater extent involves the party requesting the reexamination as well as the holder of the patent and the USPTO. While an ex parte reexamination can be initiated by the patent holder, an inter partes action cannot since there would be no other party to occupy the other side. The ex parte reexamination procedure is described in detail in Chapter 2200 of the MPEP.

[56]Any person includes corporations and/or governmental entities. Very important: the *any time* aspect of the reexamination proceeding before the USPTO means that a party does not have to wait until she is threatened with an infringement allegation or action before she requests the reexamination of a patent.

than the patentee must be accompanied by a certification that the patent owner has been served with a copy of the reexamination request; if the patent owner cannot be located, then an extra copy of the request must be provided to the USPTO.

2. The real party of interest behind the request for the reexamination may remain anonymous throughout the ex parte reexamination process.[57]

3. The request for reexamination must be accompanied in writing and must be accompanied by the reexamination fee ($2,520 in 2009).

4. The request for reexamination can be based on prior art only and must explain the relevancy and manner of applying that prior art to each claim for which reexamination is requested.

5. The cited prior art must consist of *printed* publications or patents.[58]

6. Within 3 months of the filing of a request for reexamination, the USPTO will determine whether a *substantial new question of patentability* affecting any claim of the patent has been raised by the request. Questions regarding patentability in reexamination are limited in scope to questions regarding the novelty and/or nonobviousness of the claimed invention. However, a substantial new question of patentability does not necessarily require that *new* prior art be identified; the prior art relied on in the reexamination request may have already been part of the original patent examination. The pertinent question is not whether the references were previously considered or cited but whether they raise a substantial new question of patentability in "the appropriate context of a new light as it bears on the question of the validity of the patent."[59] If a substantial new question of patentability has been raised with regard to one claim, then the entire patent will be reexamined during the ex parte process.

---

[57]The name of the person who actually files the request for reexamination will not be held in confidence but the actual party in interest need not be revealed. MPEP 2212 provides: "Some of the persons likely to file requests for reexamination include the patentees themselves, licensees, potential licensees, *attorneys without identification of their real client in interest*, infringers, potential exporters, patent litigants, interference applicants, and International Trade Commission respondents."

[58]See MPEP 2128 for discussion of what constitutes a written publication.

[59]House report No 107-120 accompanying legislation amending the previous version of 35 U.S.C. 303(a). The U.S. Congress disagreed with federal court interpretation (*In re Portola Packaging*, 110 F.3d 786 (CAFC 1997)) of what is meant by "substantial new question of patentability," so the law was modified in 2002 to make it explicitly clear that the standard could still be met even though a patent or printed publication was previously cited by or to the USPTO during the prosecution of the patent. 35 U.S.C. 303(a) provides the following:

> Within three months following the filing of a request for reexamination under the provisions of section 302 of this title, the Director will determine whether a substantial new question of patentability affecting any claim of the patent concerned is raised by the request, with or without consideration of other patents or printed publications. On his own initiative, and at any time, the Director may determine whether a substantial new question of patentability is raised by patents and publications discovered by him or cited under the provisions of section 301 of this title. *The existence of a substantial new question of patentability is not precluded by the fact that a patent or printed publication was previously cited by or to the Office or considered by the Office.* (emphasis added.)

7. A finding by the USPTO that no new issue of patentability has been raised will be final and nonappealable.

8. If the USPTO finds that a substantial new question of patentability has been raised, a reexamination will be ordered, and the applicant will have at least 2 months to file a reply including any claims that she wishes to cancel or amend. A copy of the patentee's response must be served to the party requesting the reexamination procedure and that party has 2 months to file a response to the patent holder's statement. Any claim amendment cannot broaden any of the claims of the patent.

9. The patent reexamination proceeding is then conducted between the patent holder and the USPTO examiner handling the reexamination.

10. Any final action unfavorable to the patentee by the patent examiner may be appealed by the patent holder to the Board. Further appeal by the patent holder to the Court of Appeals of the Federal Circuit is possible (35 U.S.C. §§141–147). The right of appeal in the ex parte reexamination proceeding is limited to the patent holder and not the reexamination requestor.[60]

11. The proceedings of reexaminations can be viewed on the USPTO's Public Patent Application Information Retrieval (PAIR) web page.

12. Once the reexamination has concluded, a certificate of reexamination is issued that makes any changes to the claims that result from the reexamination.

Since 1981, there have been >9000 requests for ex parte patent reexaminations with >500 per year in recent years. The patent office has granted over 90% of the reexamination requests made to it, with approximately 75% of the reexamined patents being at least partially modified if not completely revoked. While this cannot be taken as statistical reflection of all patents issued by the USPTO, since the challenged patents are a selected group to begin with; it does provide those wishing to challenge issued patents by an ex parte reexamination at least some reason to be optimistic.[61]

An important aspect of the ex parte reexamination proceeding is the ease by which any person may request reexamination of an issued patent, especially when one considers the costs and burdens associated with one of the alternatives, an infringement suit with potential damages coupled with litigation in the federal courts. Furthermore, the presumption of patent validity (35 USC §282) means that a challenge to a patent's validity in the federal courts requires a "clear and convincing" showing that the patent is invalid. In contrast, patent reexamination is conducted like an initial examination, meaning no such assumption of validity is made. In fact, a significant amount of reexamination requests take place in the context of patent litigation where one of the

---

[60]The reexamination requestor in the inter partes proceeding (discussed in the next section) has the right to appeal adverse decisions.

[61]Presumably, the patents that have been subjected to reexamination have apparent flaws in view of the prior art that make the person bringing the challenge believe that they will be successful. Moreover, the degree to which the enforceability of the modified patents (short of those that have been completely revoked) has been affected by the reexamination proceeding is not conveyed in this gross statistic—a minor change to the language in one claim might still qualify as a modified patent for statistical purposes.

parties (or even the USPTO itself) may request a reexamination of one or more of the claims asserted in the litigation. This procedure is particularly useful for the defendant in a patent action who finds the reexamination route presents strategic advantages.[62] The popularity of ex parte requests for reexaminations is likely to increase with an ever-growing recognition for its ease of initiation as well as a fairly high success rate for those requesting patent reexaminations.[63]

2.2.d   Inter Partes Procedures

During the ex parte proceedings, the party filing the reexamination request makes their arguments against patentability when they file the request and in their rebuttal, but from that point on, the process is conducted between the patent holder and the patent examiner. Although the reexamination requestors can monitor the progress of the reexamination by reading the documents generated between the patent holder and the USPTO, they are sealed in silence behind the glass with nothing to do but watch the process unfold. Even though their initial arguments accompanying the reexamination request may have been very persuasive, the vibrant resonance of their logic may fade to only a soft echo with the examiner as the process wears on, especially in the presence of a very motivated and persistent patent holder. Perhaps in recognition of the limited oppositional nature of the ex parte proceeding, there is a second reexamination procedure available that allows the party bringing the action to make her arguments heard throughout the course of the patent reexamination. In contrast to the ex parte procedure, the inter partes procedure allows the reexamination

---

[62]It is interesting that the right to request patent reexamination can endure even *after* the same patent (and even same issue) have been litigated in federal court. In the appeal from reexamination (*In re Swanson* (CAFC 2008)), the CAFC denied the appellants assertion that because their patent had already been held not invalid in view of certain prior art in federal court litigation that the same prior art could not raise a substantial new question of patentability in a subsequent request for reexamination. In this instant, a reference cited in the reexamination application had already been considered by the federal courts, and the patent's validity upheld. Somewhat counterintuitively the CAFC held: "As properly interpreted a 'substantial new question of patentability' refers to a question that has never been considered by the PTO; thus, a substantial new question can exist even if a federal court previously considered the question." Given this holding, it appears that even after a drubbing in federal court the game is not necessarily over for the defendant, though one wonders how often the USPTO would be willing to seemingly override a federal court decision on the same point of law, especially where the loser on reexamination can appeal once again to the federal court. However, it is important to remember that the requirement for overturning a valid patent in the federal court typically requires a showing of clear and convincing evidence that the patent is invalid, whereas the USPTO performs the reexamination without the same burden—thus providing one possible explanation for the disparate results. If this seems to you like a recipe for endless litigation, then you are probably not alone. A more compact scheme would require that, if a request for reexamination were to take place, any federal court litigation be stayed, pending the result of the patent litigation and in fact this is often done.

[63]Moreover, Internet advertising may further attract more participants. For example, at www. patentassassins.com, the following assertion is made: "Eliminate the Patent Restrictions Standing in Your Company's Way. You can easily infiltrate an existing patent while greatly reducing your company's patent infringement risk. Use the Patent Assassins to raise *substantial new questions of patentability* through reexamination" (emphasis in original.)

requestor to file her own responses to adverse positions taken by either the patent holder or the patent office.[64] Moreover, the reexamination requestor may even submit additional prior art beyond the initial submission (under certain circumstances) and may even appeal adverse decisions to the Board and from there to the CAFC.

Sounds like a pretty good deal for the reexamination requestor, eh? Well not completely since there are significant downsides to the inter partes procedure that have, at least to date, limited its popularity relative to the ex parte procedure. First, one can imagine that potential infringers might be reluctant to identify themselves to the patent holder, but 35 USC §311 (b)(1) requires the identity of the real party in interest be made known. If the requesting party had not yet been threatened with a patent infringement suit before the reexamination request, she might be more closely scrutinized after identifying herself by initiating the inter partes request. Second, the cost and time invested in the procedure will be higher than for the ex parte procedure. Besides a significantly more expensive filing fee, properly conducting the procedure is likely to require a lot of legal advocacy time.[65] The final and perhaps most important consideration for many contemplating patent reexamination is the effect that the inter partes proceeding will have on future litigation. Due to the oppositional nature of the inter partes proceeding and the judicial policy that does not allow the same litigants to repeatedly litigate the same issues, the inter partes proceeding can prevent the requestor from raising the same issues in any subsequent patent infringement litigation. Furthermore, the reexamination requestor is also barred from challenging any claims in later litigation on any ground that *could have been raised* during the inter partes reexamination proceeding.[66] So unlike the anonymous ex parte reexamination proceeding in which the requestor essentially gets a free bite at the apple, the inter partes requestor needs to be considerably more cautious since she might irrevocably limit certain defenses she might otherwise adopt in subsequent litigation. As a result, the patent holder whose patent survives the inter partes proceeding could be in a stronger position in any subsequent infringement litigation, particularly in any infringement litigation against the reexamination requestor. Finally, one must also consider the "halo effect" that the inter partes reexamined patent claims may have during subsequent patent litigation. The trier of fact could be especially influenced by the fact that the patent being litigated had already been contested before the USPTO and upheld. Even though different issues will be litigated at the federal court level, the *penumbras and emanations* cast from a patent revalidated by the USPTO may carry weight with a fact finder (judge or jury) beyond the topics actually litigated.[67]

---

[64] Inter partes reexamination proceedings are covered in Chapter 2600 of the MPEP.

[65] The cost to file an inter partes reexamination request is $8,800 (2009) but some of the money can be refunded if the request is denied. Despite the higher costs of filing and advocacy expense, the costs will still pale significantly compared to federal court litigation, where the final tally can easily run into millions of dollars before all is said and done—not to mention that most patent litigation takes a very long time.

[66] 35 U.S.C. §315(c) and 35 U.S.C. §317(b).

[67] Apparently penumbras and their emanations can be very powerful, especially in constitutional law. In *Griswold v. Connecticut*, 381 US 479 (1965), the majority of the Supreme Court justices found that a zone of privacy was created from the "penumbras and emanations" of the first, third, fourth, and ninth amendments to the U.S. Constitution. The constitutional right to privacy thus cobbled together has been

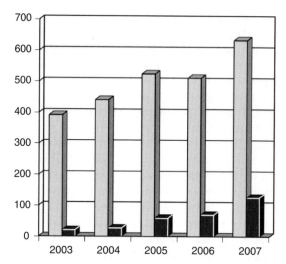

FIGURE 2.14 Ex partes and inter partes reexamination requests by year.

Inter partes patent reexamination proceedings were added to the federal patent laws in 1999, and although significantly lagging the number ex parte reexamination requests, the growth rate in applications is impressive (Figure 2.14).

2.3 INEQUITABLE CONDUCT IN PATENT PROSECUTION

As you've no doubt learned by now, a patent grants its holder potentially very valuable patent rights. The holder of a patent can bring enforceable actions in the federal court system and seek both monetary as well as injunctive relief.[68] It seems that practically every other day you can read in the newspaper about some ongoing, high-stakes patent litigation. Businesses can rise or fall on the validity of a single patent claim. For these reasons alone, the importance of the USPTO's ability to issue meaningful, valid patents cannot be understated. Against this backdrop, however, is the inconsolable fact that the patent office must process hundreds of thousands of patent applications every year (Figure 2.2) and examine millions of claims. In the active give and take between the patent applicants and the USPTO, a very significant

subsequently found to provide constitutional protection, in the zones created by these penumbras and their emanations, to rights related to contraception, abortion, individual possession of pornography, and consensual homosexual sex.

[68] If money damages are deemed an inadequate remedy for making the plaintiff whole, the federal courts also have the power to issue injunctions that force the defendant to perform or cease to perform certain actions. In patent law, injunctions normally take the form of a command to the defendant to cease the infringing activity. In certain instances, injunctions are preliminary in nature, meaning they are issued before a final judicial determination on the merits. Preliminary injunctions are applied only under circumstances in which the plaintiff is deemed to be likely to prevail on the merits and the type of harm caused by the defendant's continued actions is irreparable.

amount of correspondence will be generated. In most instances, the examiner will need to rely on the veracity of the statements of the applicants because the patent office does not have the means or the mandate to carry out independent tests to verify the applicant's characterization of the prior art and/or their invention. Very often the decision of whether a patent is granted distills down to the statements that applicants make in their patent application and/or their communications with the office during the prosecution of the application. The very credibility of the U.S. patent system depends on the credibility of those persons involved in trying to get U.S. patents issued. For these reasons, the patent laws impose a duty of candor and good faith on those individuals associated with the filing or prosecution of patent applications.

Beyond the honesty in communication (candor) required by the U.S. patent system to issue strong patents, there is an affirmative duty on applicants for patents in the United States to disclose any material information that they are aware of that might affect the patentability of their invention. Although we have already learned that examiners at the USPTO will independently search the prior art during patent examination, they often do not have the familiarity with the subject matter that the patent applicants and their representatives do. This collective duty of disclosure, candor, and good faith are critical to the mission of not only the USPTO but to the applicant as well, for at least three reasons.

The first reason is that patentees should want the strongest possible system of patent procurement and enforcement possible. To have such a system, all parties involved in the patent process need to work collectively to uphold the integrity of the system. No jury or judge will view your patent favorably if the entire system itself is a sham.

Second, the holder of the patent wants the strongest patent possible. If a patent is to have value, it must survive the rigors of litigation, and patents that have been procured through less than fully honest means are going to work against the patentees should they try to enforce the patent in federal court.

If the first two reasons were too tangential to convince you then may be the third will:

A patent that is obtained through *inequitable conduct* can be found nonenforceable, and therefore essentially useless, even if the inequitable conduct applies only to a single claim of that patent. This means the entire patent is contaminated despite the possibility that the inequitable conduct was more limited in nature.[69] This sanction has teeth and is enforced vigorously in federal court where patent infringement actions are heard.[70]

---

[69] A finding of inequitable conduct in the prosecution of one patent can even extend beyond a finding of nonenforceability of the claims of that patent but also may render the applicant's closely related patents unenforceable as well.

[70] The USPTO has neither the resources nor the mandate to enforce actions amounting to inequitable conduct. The finding of inequitable conduct requires at least some level of intent and presumably the best way to determine whether there was the requisite intent is through an adversarial process, including the power to take depositions under oath-the USPTO is not in that position. So even though the USPTO has the authority to deny a patent grant where "fraud on the office" is committed, the primary investigators in actions asserting inequitable conduct are the federal courts, where the issue is usually raised by the defendant as an affirmative defense to escape an allegation of patent infringement.

You may wonder, How does the issue of inequitable conduct typically arise since the USPTO is not likely to raise the matter? The answer is as follows. Somebody is infringing your patent and you don't like him doing that, at least without his paying you royalties for all of your hard work, expenses, and sheer inventive genius. You write him a letter explaining that you believe your patent covers what he is doing and either he needs to quit doing what he is doing or negotiate a license with you. He responds: "Patent, you call that a patent? Sue me!" So you take him up on his challenge and hire a lawyer (more like a team of lawyers) to bring a cause of action against the infringer. The accused infringer then hires a team of lawyers, and you take your case to the federal district court having the appropriate jurisdiction. You sue him for infringement and allege incredible damages, he counterclaims, first arguing that he *never* infringed your patented claims to begin with. Second, *even if* he did infringe your claims, those claims are not valid because the subject matter is not novel. Next, he'll say *even if* he does infringe your claims and *even if* they are novel, they are obvious and therefore invalid. Fourth, he'll say *even if* he did infringe your aims and *even if* they are novel and nonobvious, that they are invalid because you failed to properly enable the invention. Finally (and he has probably been saying this all along), he will say *even if* he did infringe your patent and *even if* it was novel and nonobvious, and *even if* the patent was fully enabled, that you used inequitable conduct to obtain at least one of the claims in that patent and thus the entire patent is invalid.

Sound far-fetched? It might seem hard to believe that very many patent practitioners or inventors would be guilty of such ignoble conduct, after all we are mostly honest and hardworking people, right? Well, somewhat surprisingly, you might find that the inequitable conduct defense results in quite a few invalid patents, including some pretty important and high-profile cases. "Why is this?" you might ask, and the answer will to a large part depend on who you ask. If you ask the defendants (e.g., generic drug companies seeking to invalidate innovator patents), they will say because the greedy individuals and companies overstepped their bounds and were so hell-bent on monopolizing the market that they would say anything or do anything to get their patent issued and then, on top of all that, have the sheer audacity to try to enforce it. On the other side of the line, you will have the patent practitioners and inventors who will point to the incredibly complex nature of patent prosecution, the heavy burden on the applicant to decide what references might be material to patentability, an ever-changing and ever-growing list of sins that amount to inequitable conduct resulting in a way-too-low bar being used to invalidate patents, the sometimes preternatural dislike for companies trying to enforce their patents when the industries represented by those companies are not in high favor, and, finally, the occasional possession and use of amazingly powerful "retrospectroscopes" by the federal courts.[71]

For at least some of these latter reasons, the federal courts have at least espoused a reluctance for too easily finding inequitable conduct, but a quick glance at some of

---

[71] Retrospectroscopes are amazingly powerful cognitive devices that lend one perfect insight and judgment, especially when analyzing another's actions. Unfortunately, they work only in hindsight.

the recent case law may give one pause. In any event, whatever reason you choose for your motivation to act equitably in patent procurement, it is important that at a minimum you come to understand exactly what that burden is, as an inventor (or patent practitioner) in the patent procurement process.[72]

Because it is such an important topic and because the inventor is a party whose actions and statements can directly affect the outcome of the patent prosecution, the entire substance of the duty of disclosure, candor and good faith is produced here for you to read in its entirety before further explanation and some real examples from actual court cases.

37 CFR 1.56 Duty to disclose information material to patentability

(a) A patent by its very nature is affected with a public interest. The public interest is best served, and the most effective patent examination occurs when, at the time an application is being examined, the Office is aware of and evaluates the teachings of all information material to patentability. Each individual associated with the filing and prosecution of a patent[73] has a duty of candor and good faith in dealing with the Office, which includes a *duty to disclose to the Office all information known to that individual to be material to the patentability* as defined in this section. The duty to disclose information exists with respect to each pending claim until the claim is canceled or withdrawn from consideration, or the application becomes abandoned. Information material to the patentability of a claim that is canceled or withdrawn from consideration need not be submitted if the information is not material to the patentability of any claim remaining under consideration in the application. There is no duty to submit information which is not material to the patentability of any existing claim. The duty to disclose all information known to be material to patentability is deemed to be satisfied if all information known to be material to patentability was cited by the Office or submitted to the office in the manner prescribed by §§ 1.97(b)–(d) and 1.98.[74] However, no patent will be granted on an application in connection with which fraud on the office was practiced or attempted or the duty of disclosure was violated through bad faith or intentional misconduct. The Office encourages applicants to carefully examine:

(1) Prior art cited in search reports of a foreign patent office in a counterpart application, and

(2) The closest information over which individuals associated with the filing or prosecution of a patent application believe any pending claim patentably defines, to make sure that any material contained therein is disclosed to the Office.

---

[72] It is interesting that the burden ends after the actual issuing of the patent. So for example, if you realize after your patent issues that you forgot to submit what was an important reference, it is too late to undo the wrong committed—inequitable conduct cannot be undone once the patent is issued.

[73] This may include you.

[74] §§1.97(b)–(d) and 1.98 describe the manner of submission of information disclosure statements.

(b) Under this section, information is material to patentability when it is not cumulative to information already of record or being made of record in the application, and

    (1) It establishes by itself or in combination with other information, a prima facie case of non-patentability of a claim; or

    (2) It refutes, or is inconsistent with, a position the applicant takes in:

        (i) Opposing an argument of non-patentability relied on by the Office, or

        (ii) Asserting an argument of patentability.

            A *prima facie* case of non-patentability is established when the information compels a conclusion that a claim is non-patentable under the preponderance of evidence, burden of proof standard, giving each term in a claim its broadest reasonable construction consistent with the specification, and before any consideration is given to evidence which may be submitted in an attempt to establish a contrary conclusion of patentability.[75]

(c) Individuals associated with the filing or prosecution of a patent application within the meaning of this section are:

    (1) Each inventor named in the application;

    (2) Each attorney or agent who prepares or prosecutes the application; and

    (3) Every other person who is substantively involved in the preparation or prosecution of the application and who is associated with the inventor, with the assignee or with anyone to whom there is an obligation to assign the application.

(d) Individuals other than the attorney, agent or inventor may comply with this section by disclosing information to the attorney, agent, or inventor.

(e) In any continuation-in-part application, the duty under this section includes the duty to disclose to the Office all information known to the person to be material to the patentability, as defined in paragraph (b) of this section, which became available between the filing date of the prior application and the national or PCT international filing date of the continuation-in-part application.

Okay, you got all that? If not everything, we hope you at least gathered up a few pertinent items. The first being that the applicant for patent as well as other parties involved with the application process (including you, the inventor) have an affirmative duty to submit information that is material to patentability. You might reasonably ask, "But how do I know what is material to patentability and even if I did, who would I report it to?" Let's take the second part first because that one is easier. If you are an inventor or are otherwise substantively involved with the case, perhaps you are the assignee with knowledge and interest in the case and you have a patent agent or attorney of record who is handling the case, then you need to give the information to your representative so he can submit it to the patent office. However, this does not

---

[75]Preponderance of evidence is the typical civil burden of standard meaning the evidence establishes that the thing being asserted is more likely than not.

by itself alleviate you of your duty as an inventor because as you might glean from section (d) of 37 CFR 1.56, the duty on the inventor cannot end on reporting the information to the attorney or agent.[76]

If you are the inventor and are representing yourself (pro se), then you will need to supply the information yourself to the USPTO. The information submitted must be listed on an information disclosure statement, and the forms can be found on the USPTO website. Briefly, there is a section for U.S. patent and published patent applications, foreign patent applications, and patents as well as nonpatent references. For U.S. patents and patent publications, only a citation is necessary. For non–U.S. patent references or any other reference, the citation must be indicated on the form and a hardcopy of the reference must be provided. If the reference is written in a language other than English, a summary of the relevancy of the document in English or an English translation must be provided.

Having settled the how to file the information let's turn our attention to the sometimes thornier problem of deciding *what* references need to be cited and/or sent to the USPTO. First the good news: There is no affirmative duty on the applicant, inventors, or any other individuals associated with the relevant parties and the filing of the patent application to search for information that might be material to patentability. Of course, a thorough knowledge of the prior art is often critical to getting and staying competitive in your field. Furthermore, understanding how the claims in your patent application relate to (and are different from) the prior art is good to know before you go through all the trouble and expense of developing your invention and filing the patent application. A lone inventor working in his basement, entirely cut off from the outside world since birth might write a patent application for his mousetrap invention and submit the application with no information disclosure sheet, despite the fact that numerous embodiments of his claimed invention have been known in the art for quite some time. He may not be guilty of inequitable conduct, though he may not get a patent on his mousetrap either (unless, perhaps, it truly is a better mousetrap).

If you fail to submit a reference that you are aware of that is deemed to be material to patentability in patent litigation, then you risk losing the entire patent to inequitable conduct. However, many technical areas are so broad that an attempt to include all potentially relevant reference would require renting a truck for the weekend to haul all of the references to the USPTO for delivery. If you are an inventor or somebody most involved in the patent process from the technical side of things, it is best to provide your agent or attorney with any and all references (or citations) that you think could be relevant so that they can make the ultimate determination. You can assist in the process by highlighting to your representative any references you feel are especially relevant. In that manner, if there are a large number of references, the agent or attorney will be less likely to miss those that are most crucial. As a practical matter, you should be particularly alert to anything you or your colleagues have published on

---

[76]This means you should verify with the attorney or agent that any information or references that you believe are material to patentability have been included in the information disclosure statement sent to the USPTO. If not, it is reasonable to ask why not, to make sure that there has not been an oversight or misunderstanding in regard to the information you provided.

any subject matter related to what it is you now intend on patenting and also any prior art that you have taken into account when conceptualizing and/or reducing to practice your own invention. In this regard, you may have to fight any instinctual tendencies toward mitigating the contribution of others in your field toward the genesis of your own ideas—invention often does not involve, nor does it require, a spontaneous flash of genius. Most ideas are somehow related to or are in some way derived from other ideas, and it is important that you also consider any of your source materials when making disclosures to the patent office. Conversely, it is common to err in the opposite manner as well. Just because you did not use a particular piece of prior art in the development of your invention does not mean it is not material to patentability. For example, you might have independently conceived of a compound and then later discovered that somebody else published very similar compounds. The mere fact that you did not know about his work or did not take his work into account when you made your discovery does not alleviate you from your duty of making that citation or reference available to the USPTO—it may not have been material to your invention but could still be material to the patentability of your invention.

Beyond submitting any known prior art, parties of interest must confirm that the patent application is an accurate reflection of their invention. If statements or data characterizing the invention in the specification are relied on to support the patentability of one or more of the claims in the patent application then those statements or that data need to be accurate and also should be representative of the claimed invention. This also applies to data and statements in declarations and related documents submitted to the patent office to support the patentability of the claimed invention; selective disclosure of only favorable data while omitting contradicting data is frequently not a good idea ("when in doubt, send it out").[77]

To better appreciate the standard by which inequitable conduct is measured, let's review an actual recent court opinion, *Bristol-Myers Squibb Company v. Rhone-Poulenc Rorer, Inc.*,[78] where this issue was heard. This case serves to explain how materiality and intent are inextricably related. Perhaps less abstractly, it will also show why it is important for the scientists to pay close attention to what they publish as well as to documents that they are asked to review.[79] This decision arose from an appeal from the U.S. District Court for the Southern District of New York (D-SDNY), which held the Rhone-Poulenc Rorer (RPR) patent relating to methods of preparing Taxol® (hereinafter referred to by its generic name paclitaxel) as well as claiming key intermediates in the paclitaxel synthesis to be unenforceable due to inequitable conduct. A brief review of the facts of the case is presented next.

The patent at issue claimed a semisynthesis of paclitaxel as well as certain intermediates useful in that synthesis. Paclitaxel is a very popular drug for use in the

---

[77]Of course, some bad data are bad data because a measurement or experiment is demonstrably wrong and clearly not representative of the claimed invention, in which case the patent applicants need to use their sound judgment. The applicant doesn't want to drown the examiner in meaningless data points if it obfuscates what could or should be a clear scientific or legal conclusion.

[78]*Bristol-Myers Squibb Company v. Rhone-Poulenc Rorer, Inc et al.* 326 F.3d 1226 (CAFC 2003).

[79]It's also for a caution to lawyers to consider any documents that could be material to patentability and not just those that are prior art.

treatment of certain cancers, including lung, ovarian, and breast cancers. The drug was originally discovered by the National Cancer Institute through screening samples from natural sources for cytotoxicity. It is interesting that paclitaxel was initially isolated from the bark of the Pacific yew tree in very low yields. For example, in one of the original isolations, 1200 kg Pacific yew bark had to be harvested to isolate just 10 g of paclitaxel. The naturally isolated material was tested and showed very promising activity in early clinical studies, but due to logistical and environmental concerns, it quickly became evident that an alternative method to produce the compound would be needed. The government looked to the pharmaceutical industry to form a partnership to provide the material and run the clinical trials necessary for FDA approval. In 1989, Bristol-Myer Squibb (BMS) was awarded the contract, and they were able to get the drug approved by the end of 1992. Although the drug is currently produced directly by a fermentation process and then isolated, there was an intermediate period where the drug was efficiently produced by semisynthesis from a metabolic precursor, 10-deacetylbaccatin (10-DAB), a compound that could be isolated from a renewable source, the needles of the European yew.

The inventors of the patents-in-suit claimed an efficient method of producing paclitaxel from 10-deacetylbaccatin in the eventually issued patent US 4,924,011. The broadest claim of the patent (claim 1), together with the first dependent claim (claim 2) are reproduced for your reference in Figure 2.15.

When organic chemists use the term *protecting groups*, they are referring to groups of atoms that can be used to mask reactive functionalities when reactions at those functionalities are not desired. Once the desired reactive transformations are completed elsewhere on the molecule, the protecting groups can then be removed. The keys to successful protecting group strategy are that the groups must be able to be placed selectively onto the desired functionalities, that they remain where they need to be for the period when they need to be present, and they can be removed selectively at the end without causing unwanted changes elsewhere on the molecule. The more different types of functionalities a molecule has, the more complex the protecting group scheme can become. It is seldom easy to predict ahead of time with great certainty what combination of protecting groups will result in a successful synthesis. This is part of the art and challenge of organic synthesis. In this vein, it is important to appreciate the contextual nexus between the complexity of the taxol molecule with its densely functionalized skeleton and the very huge demand for the chemical semisynthesis of the molecule. For some time, this was a very hot area of research in both industrial and academic labs. The work described in the patent application summarized the culmination of some very successful chemistry of potentially high commercial and public health significance.[80]

---

[80] At the time of the patent suit, it appears that BMS used a semisynthetic process that was licensed from Robert Holton's group at Florida State University. Holton's group not only discovered a highly efficient semisynthesis of paclitaxel from 10-DAB, but also were one of the first two groups (KC Nicolau's group was the other) to complete the total synthesis of paclitaxel, which means they assembled the molecule from commonly available precursors—essentially from scratch. The total synthesis, however, is not and probably never will be a commercially viable way to make this very complex organic molecule.

1. A process for preparing taxol of formula:

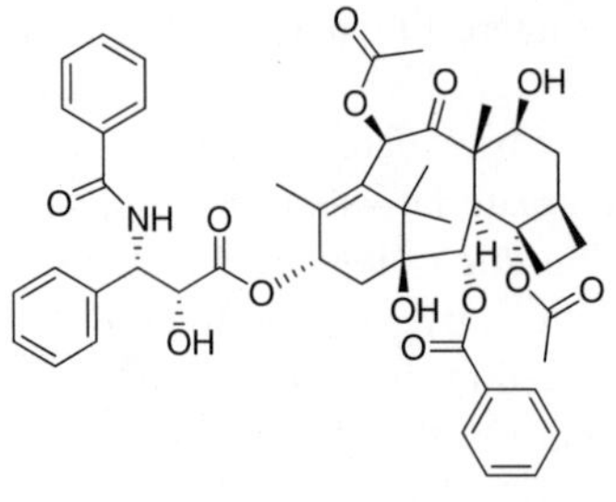

in which a (2R, 3S) 3-phenylisoserine derivative of general formula:

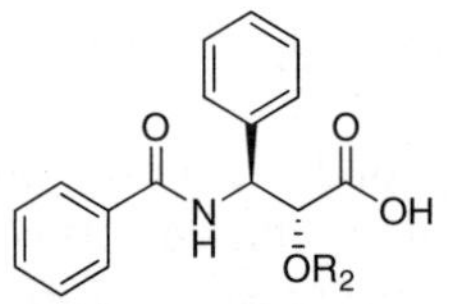

in which R_2 is a hydroxyl-protecting group and is esterified with a taxan derivative of general formula:

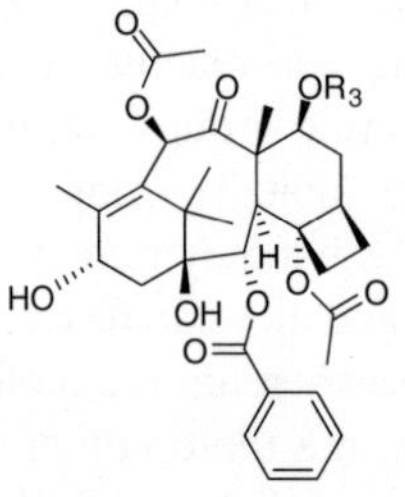

in which R_3 is a hydroxyl-protecting group, and the protecting groups R_2 and R_3 are then both replaced by hydrogen.

2. A process according to claim **1**, in which R_2 *is chosen from methoxymethyl,* 1-ethoxyethyl, benzyloxymethyl, (β-trimethylsilylethoxy)methyl, tetrahydropyranyl and 2,2,2-trichloroethoxycarbonyl, and R_3 *is chosen from trialkylsilyl groups* in which each alkyl portion contains from one to three carbon atoms.

FIGURE 2.15   Process claims for taxol semisynthesis.

The decision by RPR to patent the method came about when one of the discoverer's of the method sent RPR's patent counsel a draft publication that described the synthesis of paclitaxel from 10-DAB that they intended to submit for publication to the *Journal of the American Chemical Society* (JACS). Of particular note, the draft publication explained that the conversion of 10-DAB to taxol "could be successfully achieved only with specific protecting groups and under unique reaction conditions." It also discussed some of the limitations of certain protecting groups as follows:

- "A methoxymethyl [MOM] protecting group at C-2 could not be removed following esterification"
- "The trimethylsilyl [TMS] group could also be selectively introduced at C-7, but it proved unstable to the subsequent esterification conditions"
- "The *tert*-butyldimethylsilyl [TBDMS] group could not be cleanly introduced" at C-7.

The RPR patent agent working in France recommended that a patent application be filed before submission of the manuscript. The French patent agent drafted a patent application covering the subject matter and claim 1, which refers to a protecting group at R_2, and R_3 contained no limitation as to what type of protecting groups might be used. In claim 2, which depends from claim 1, the description is given that R_2 is chosen from groups including the MOM-group, and R_3 is chosen from groups including trimethylsilyl (this occurs where the alkyls are all C-1 (see the bolded portion of claim 2). This means that claim 2 explicitly included some of the protecting groups that the inventors stated in their JACS publication would not work.[81]

Well, this oversight should have been no problem because the patent agent then sent a draft copy of the application to two of the inventors, asking them to *carefully* read it over and make any comments or changes as necessary. Of course one should probably only assume that they read the document over very carefully (everybody loves to read patent applications, right?) but it appears they still included more than they should have in their claims because the claimed protecting groups appeared to contradict the statements in their own JACS publication. Further compounding the error, the French patent agent sent the patent application for filing in the United States but *did not* send the then-published JACS paper to the U.S. attorney for submission or consideration as a possibly material reference. The patent attorney in the United States prosecuted the patent application without sending the reference, not least because the reference had never been brought directly to his attention. The patent eventually issued as 4,924,011.

In a subsequent action, RPR instructed a U.S. attorney to file a reissue application on their behalf.[82] The reissue application included the original claims plus three additional claims to actual intermediates useful in the synthesis; the reissue application issued in 1993 and was assigned the number RE 34,277. It is interesting (but ultimately not dispositive to the CAFC's decision) that the JACS reference was submitted during the reissue application proceeding, and the examiner signed off that he had seen the reference.

---

[81]This also means that claim 1 included those groups as well since claim 2 is dependent from claim 1.
[82]Reissue applications were discussed above in Section 2.2.b.

In 1995, BMS filed suit in the D-SDNY seeking a declaratory judgment that they did not infringe the RE 34,277 reissue patent and that in any event the patent was invalid and unenforceable.[83] RPR then filed suit against BMS asserting infringement in a different court, and the case was transferred for consolidation in the D-SDNY. During the initial court proceedings, the D-SDNY issued an opinion that *by clear and convincing evidence*, RPR had attained the RE 34,277 patent through inequitable

---

[83]Federal courts have limited jurisdiction. While the subject matter of a patent suit is certainly a federal question and appropriate subject matter for a federal district court, there are additional requirements put to those seeking redress in the Court. One of the requirements is that there must be a "case or controversy" before the federal court, which means that a party seeking a judgment on a patent question must have much more than just a passing interest in the subject. For example, a patent holder who believes that another party is infringing his patent may bring a suit against that party because there is a case or controversy to be settled because he is seeking to assert rights conferred on him by his patent. On the other hand, a potential infringer cannot simply petition a federal court to decide if the patent he might be infringing is, in fact, being infringed, or even if the patent is being infringed by his activities, whether the patent is, in fact, valid. Upon those facts without more, the question of any consequences is simply a moot point where there is no threat of a lawsuit; the inquiry is too theoretical or abstract in nature (no "case or controversy" before the Court). However, where the potentially infringing party has a reasonable apprehension that he is going to be sued for infringement then there may be a "case or controversy" even though he has not yet been sued. If the risk of a suit is real enough, the federal courts are allowed to intervene to solve the dispute on the merits. It is not clear from the CAFC decision what triggered BMS's reasonable apprehension of a suit such that they moved for declaratory relief but often such an apprehension is triggered by a notice from an entity holding a patent or patents sent to an entity who they believe might be infringing one or more of their patents (commonly referred to as a "cease and desist" letter). Depending on the specifics of the letter, a federal court could find that the reasonable apprehension of a suit exists. To some extent, the standard of what is required for a declaratory judgment plaintiff to be heard in a federal court has been lowered since the BMS Taxol case was heard. While the reasonable apprehension of a suit by the patentee would still do the trick, the trigger for a declaratory judgment jurisdiction to be established has been lowered by a recent decision by the court, *MedImmune, Inc. v. Genentech, Inc.*,127 S. Ct. 764 (2007). The *Medimmune* case dealt with a specific issue created when licensees to a patent wish to get a declaratory judgment in a federal court on whether the patent(s) they have licensed are valid. As long as a party has a license, they technically do not risk a suit because they hold the license so they would fail to have a case or controversy in the federal court under the old standard. Before *Medimmune*, the licensee would need to renege on the contract and expose themselves to a suit of infringement in order to challenge the validity in federal court of the patent they had licensed. Although *Medimmune* dealt with a fairly narrow set of facts, the language has been interpreted as a repudiation of the previous standard. While the new standard certainly lowers the bar to get a federal court to grant declaratory judgment jurisdiction on a patent validity challenge, it may take time to glean the exact parameters. In a post-*Medimmune* decision heard by the CAFC, the Court had this to say with regard to their partial interpretation of what the new standard is:

> We need not define the outer boundaries of declaratory judgment jurisdiction, which will depend on the application of the principles of declaratory judgment jurisdiction to the facts and circumstances of each case. We hold only that where a patentee asserts rights under a patent based on certain identified ongoing or planned activity of another party, and where that party contends that it has the right to engage in the accused activity without license, an Article III case or controversy will arise and the party need not risk a suit for infringement by engaging in the identified activity before seeking a declaration of its legal rights.

See *SanDisk Corp. v. STMicroelectronics, Inc.* 480 F.3d 1372, (CAFC 2007).

conduct; therefore, the patent was unenforceable.[84] RPR appealed the D-SDNY's opinion to the CAFC, which upheld the D-SDNY's opinion.[85]

Before getting into the CAFC's reasoning, let's briefly recapture what the D-SDNY found that RPR did wrong. Their primary sin, it seems, is that they *did not* submit the JACS reference to the examiner during the prosecution of the application that resulted in the issuance of 4,924,011—there is no doubt from the facts given that the French agent knew about the publication, but from these same facts, it does not appear that the U.S. attorney knew about it. Notice that RPR *did submit* the reference during the prosecution of the reissue application which was examined by the same examiner and is the actual patent that RPR is trying to enforce.

Here are some additional facts we should keep in mind as we review the basis for the D-SDNY's decision as well as the CAFC's upholding of that decision:

1. The primary mistake for RPR was that the *French* patent agent did not send the U.S. attorney prosecuting the case the JACS reference. Since intent to deceive is a basic requirement of a finding of inequitable conduct, then we need to take into account the entire context of the French agent's actions to help glean that

---

[84]Once issued, a U.S. patent whether a regular U.S. patent or a reissue patent has a presumption of validity. In theory at least, the presumption of validity requires a higher evidentiary burden for overturning than the normal civil standard where a finding is on the basis of a preponderance of evidence (more likely than not). The clear and convincing standard means that something is "substantially more likely to be true than not." This standard is still lower than "beyond a reasonable doubt," which is the standard for proving criminal guilt. Obviously all of these standards are qualitative and subjective, but they should help guide the fact finder, whether judge or jury, in generally the right direction.

[85]When the CAFC reviews a finding on inequitable conduct in a federal district court, they are obligated to do so with great deference. To overturn the D-SDNY's finding of inequitable conduct, the CAFC would need to find that the District Court "abused its discretion." Deferential review by the CAFC of district court decisions on matters related to findings of fact are the rule since the district court is in the best position to review matters concerning the facts of the case (as opposed to the interpretation of laws). The district court judge directly observes the witness' testimony and is in the best position to judge the credibility and the weight to give each piece of evidence. Intent to deceive and the materiality of the deception are the touchstones of inequitable conduct, and both are considered to be fact-based inquiries, best evaluated by the district court judge at the trial because the weight granted to any witness testimony is largely based on the credibility of that witness in view of the entire gestalt of the trial. You may wonder, then, why much of the reasoning of the CAFC panel deciding this case is presented as their reasoning rather than a simple affirmation of deference to the District Court. Without reviewing the District Court opinion in its entirety, it may be difficult to understand how much deference has actually been lent to that court's decision since the CAFC may be mostly reiterating the lower court's reasoning. For this reason, one should appreciate that even though much of the reasoning in the discussion is being attributed to the CAFC panel hearing this appeal, it might more accurately be viewed as a district court decision that the CAFC is upholding; however, the CAFC may not have come to the same result if they were hearing the issue without deference. In contrast to findings of fact, findings of law by the district court are reviewed de novo (a new review without taking into account the lower court's reasoning) on appeal to the CAFC. In this regard, the CAFC may engage in more extensive discussion of their own because it is their job to at least make sure the district court gets the law right. While this all sounds clear and logical in theory, the line between what is a matter of law and what is a matter of fact at times can get very blurry since the facts must ultimately be interpreted after being projected through the prism of some legal standard. Since the CAFC will review any legal determinations de novo, it is particularly important for the district court to make sure that they are at least articulating the correct legal standard.

intent. In this regard it is helpful to understand that European patent law does not place an affirmative burden on the applicant to submit references to the European patent office, either now or at the time this case took place. It would be hard to expect a French agent to have the same sensitivity and appreciation of this requirement as the U.S. practitioner. While this does not relieve the French agent of his responsibility to comply with U.S. patent law, we still have to ask whether the failure to give the reference to the U.S. attorney was done with the specific intent of deceiving the USPTO or could have been an oversight.

2. The reference was not a typical prior art reference in that its disclosure did not affect the novelty of the invention. Patent attorneys and agents are naturally more attuned to looking and thinking about references in terms of how they affect the novelty and nonobviousness of an invention, and this requires that the references be prior art. The JACS paper was not only not prior art (the paper was submitted after the patent application) but it disclosed information about how the inventors performed the synthesis, not whether the synthesis was novel and nonobviousness in view of a prior art reference. Was the failure to think about the reference as a "material reference" despite it not being prior art and not relating to the novelty or nonobviousness of the invention an intentional deception by the French agent or was it an oversight?

3. The USPTO examiner conducted an independent search of the subject matter, and in the search report, the JACS paper turned up—in other words, the examiner's search identified the paper even though it was not submitted. Nevertheless, there was no evidence that he actually reviewed the reference but the search report was entered into the file. Remember that the U.S. attorney who understood the U.S. law and the affirmative burden it places on applicants *did not know* about the reference, the French agent did. The French agent would have seen the search report in the file and would have known that the U.S. examiner saw at least the abstract to the reference. If you were the French agent after having seen the JACS reference turn up in the examiner's search report, might you have assumed the examiner looked at his own search report and if he had a question about any reference, asked the applicants?

4. In the reissue application, the U.S. attorney (this time a different attorney) prosecuting the case was informed by the French agent about the JACS reference, but even the U.S. attorney did not submit it *initially*. In fact, he informed the French agent by letter that the JACS reference would not affect the previously issued patent (4,924,011) because in his view it was not prior art (which is actual evidence of what was argued in point 2 above). In any event, he later submitted the reference, and the *exact same examiner* from the original patent who was now examining the reissue application, signed off on the reference and issued the reissue application with the same claims (plus three new ones). So in fact, how material did the reference turn out to be?

In its decision upholding the D-SDNY, the CAFC panel noted that its own precedent required both materiality and intent, and that if there was more evidence of one, there could be less evidence of another. In particular, the federal courts will more easily find intent where there is a high degree of materiality. In regard to whether the failure to submit the JACS reference was material, the CAFC noted that the reference was material because it called into question whether the applicants had provided adequate written description to get the first claim issued. In particular, they found the reference contradicted the scope of what they were claiming in the patent. Since it was a publication by some of the inventors themselves, one would expect the reference would be given extra weight in terms of the credibility of the assertions in the publication. In order for a patent to be granted, it must satisfy the written description requirement: The patent application must describe the claimed subject matter in such a way that one of ordinary skill in the art can make and use the claimed invention without requiring undue experimentation. In this case, the JACS publication calls into question at least some of the claimed scope of the invention. Clearly, the invention wasn't completely inoperable but some nonperforming embodiments had been included in the claims; there were specifically claimed protecting groups that probably would not work for their intended purpose. Is this material to patentability and if so, how material? The answer from the D-SDNY and the CAFC was yes and very much.

But why did the failure to cite the reference in the first patent application count against the reissue application where the reference was submitted and the reissue patent still issued in view of the JACS reference? This is a good question that implicitly contains two separate questions. The first of the two questions might be, Isn't the reissue application a fresh start? The quick answer is no, once a patent issues and is contaminated by inequitable conduct, the damage cannot be undone through a reissue application. The second question, which requires more analysis, goes to the fact that the reference was reviewed under an actual examination, with the same examiner against the exact same claims, and the claims were found to be patentable in view of that reference. Does it matter that in the real-world examination that took place in the reissue application, the reference did not change the outcome of the patentability of the claims? In addition, in the real-world litigation, the D-SDNY judge (the same judge who found inequitable conduct) refused to grant BMS's request for summary judgment of invalidity of the patent for nonenablement.[86] This means that

---

[86]Many issues or controversies never make it to the fact finder of the case (jury or judge sitting as jury). It is possible for issues in many situations to be decided by a judge, determining that as a matter of law, there is no disputed issue of material fact, even if all of the facts alleged by one side to the dispute were true, they would still lose. This is a summary judgment, and you can think of it as the sometimes outcome of a little mini-pretrial used to ferret out nonsupportable or nonmeritorious positions. Once both sides have staked out their assertions but before conducting the actual trial, they often ask the judge for a summary judgment verdict in their favor. They are essentially asking the judge to decide the case in their favor before it goes to a trial. When a judge issues a summary judgment at the district court level for patent matters, the party that received the unfavorable decision can appeal that decision to the CAFC where the review will be complete, with no deference to the lower court's decision. If the CAFC upholds the verdict of summary judgment, there will be no trial on that issue (assuming there is no successful appeal to the U.S. Supreme Court—a good assumption to make). If the CAFC issues a reversal of summary judgment it is usually accompanied by a remand to the lower court (e.g., federal district court) with some additional instruction. It doesn't mean that the winner of the appeal will necessarily win the case or even that issue. In a typical scenario, an appeals court reversal of a summary judgment will be followed by the lower court's trial of

the District Court did not believe that the sum of the evidence justified a conclusion, as a matter of law, that the claims were not enabled. Together, these facts might seem to be strong evidence of an actual low level of materiality for the failure to cite the JACS reference. However, in regard to materiality, there is a catch: The materiality of the reference must be judged in a perceived sense from the vantage point of a hypothetical examiner—not the particular examiner who actually examined the patent in question[87]—as the panel in this opinion quoted from a previous decision from a similar case:

> The standard to be applied in determining whether a reference is "material" is not whether the particular examiner of the application at issue considered the reference to be important; rather, it is that of a "reasonable examiner."[88] Nor is a reference immaterial simply because the claims are eventually deemed by an examiner to be patentable thereover. (*Molins PLC v. Textron, Inc.*, 48 F.3d 1172, 1179 (CAFC 1995)).

The CAFC further explained that their earlier decision in *Molins* also pertained to an affirmation of inequitable conduct, "notwithstanding that the withheld reference was later cited in a reexamination and the claims were allowed to issue."

> We recognize that [the withheld references] were cited eventually to the PTO and that the examiner initialed them and passed the reexamination application to issue thereafter. However, the references were not cited when they should have been.[89] (*Molins*, Id at 1182).

The CAFC dealt with the second argument regarding the District Court's rejecting the defendant's summary judgment motion for invalidity for nonenablement by stating:

> The issue of whether a reference would be considered important by a reasonable examiner in determining whether a patent application is allowable, including whether the invention is enabled, is a separate issue from whether the invention is actually enabled.

In other words, considering whether a reference is material to enablement is not the same as the final determination of whether the invention is actually enabled. The legal determination of enablement versus nonenablement takes into account much more than any single reference but that does not mean that any single reference is notmaterial.

that issue or combination of issues that make up the case, especially including a consideration of all of the relevant and admissible facts introduced by the parties to the action. In the instant case, the apellees urged the CAFC to find the fact that the District Court itself did not feel that the JACS paper was sufficient as a matter-of-law to prove nonenablement in a summary judgment was sufficient to show the nonmateriality of the JACS paper.

[87] Who needs the opinion of the actual examiner when hypothetical examiners are so much more reliable?

[88] Are they saying the examiner in this case was "not a reasonable examiner"?

[89] Does this sound more like an argument in support of their ultimate conclusion or the conclusion itself?

The conclusion of inequitable conduct requires a finding of intent and materiality such that the overall formula might be represented by materiality × intent = inequitable conduct so that a higher value for materiality means a lower required intent. In effect, a higher level of materiality allows one to more easily conclude that the requisite intent was present. The putative reason for this is because intent can seldom be proven directly (absent a confession), but rather intent is inferred from the surrounding circumstances. As a practical matter, in terms of how the federal courts appear to apply the finding of intent, one of those surrounding circumstances is the materiality of the omission or commission in question.[90] If a highly material item is omitted, one is more likely to assume that the omission was intentional because people are presumably less likely to unintentionally leave out important references and moreover, the motivation for making a material omission or commission is present because there is a reason to do it; the person is trying to get a patent, and the act in question is deemed significant enough to affect that process. In contrast, if the omission or commission is of something inconsequential, then one is less likely to infer intent because it is natural to pay less attention to insignificant details and the requisite motivation is less apparent. So from that sense, one can appreciate the CAFC's logic by which they are reading materiality not so much from the standpoint of whether the omission of the reference was actually material but rather if it would have been judged material from the standpoint of the applicants at the time of their omission–it is their intent we are concerned with. Where the apparent materiality is high (even though the real materiality was not so high), then the Court is willing to infer the requisite intent to deceive; in this decision, the CAFC emphasized the hypothetical materiality as it *should have been* perceived by the applicants. The CAFC, in effect, appeared to attempt to place themselves into the applicant's representative's shoes and tried to approximate what the applicant's representatives should or would have thought to do under those circumstances.

Until now, we have primarily focused on the omission by the French agent to submit the JACS reference but we also need to consider the role of the inventors as well, since we have already discussed that the affirmative duty to submit material information extends to them as well. We do know that the French agent was already aware of the JACS reference, but it probably would be too much to expect that inventors check to make sure their agent or attorney actually submitted the references to the USPTO. However one should expect that the inventors will review the draft of the patent application before it is submitted, to make sure that what is placed in the patent application is an honest appraisal of what the invention actually is and does. So this brings us to the view from the inventor's perch. The inventors were the ones most familiar with the technology that was being patented. They knew, better than anybody, whether the patent application was an accurate description of the method that they

---

[90]If you find the reasoning process a little circuitous, you are not alone. Many have alleged that the requisite intent is often bootstrapped by the materiality itself. If something is material, the intent is essentially inferred. The danger of this linkage is that what really should be an extraordinary remedy with a high standard devolves from an intent-based standard to a simple negligence standard, which is something that the CAFC has professed to not want to do.

intended to patent. So why didn't they correct the patent application? If the patent application had been corrected, the likelihood of the reference being material would have decreased because they would not have been attempting to claim embodiments that their own reference specifically taught against trying.

This begs many questions regarding the context of the inventors, but that notwithstanding, you can now better appreciate that inequitable conduct is a real concern for inventors, agents, and attorneys. We cannot control the federal court jurisprudence, but we can try to make sure the inequitable conduct vulture never nests in any of our patents. As inventors or people involved in that end of the process, it is important that all references we have that might be relevant in any way to our invention are submitted to our attorney or agent. If we know that some of these references are particularly relevant, we need to explain very carefully how they are relevant. From the case we just discussed, we learned that relevant documents goes beyond prior art documents; any reference relating to the potential patentability of our invention needs to be handed over to our agent and explained. We also learned that as inventors we have a duty to carefully read over everything our representative passes on to us related to the patent application. For example, patent application drafts are not something to simply be glanced at and then filed away. We inventors know the technical material in those patent applications much better than our patent representative, the draft in our hand is often just his best guess at what our invention is; we need to make sure that it is right. If we're not sure about something, then we need to ask and not assume that something has been written a certain way for a special legal reason (please don't give your attorney too much credit!).[91]

---

[91]One area of patent application drafting that is almost entirely in the control of the inventors is the experimental description sections. Most typically, the inventors will be requested to draft a description of the preparation and possibly testing of the examples of the invention. Since the person drafting the actual patent application is often not the inventor but rather is the patent agent or attorney, he will often not be familiar enough with the experimental details of the invention to provide meaningful feedback with respect to the accuracy of the inventor's experimental description. This places an extra duty of care on the inventors because the duty of candor and disclosure apply equally to them. If experimental examples do not accurately describe what has been done, then the inventors have taken the first ramp onto the inequitable conduct superhighway. Of course, the errors will have to be material and accompanied by the requisite intent to deceive but as we've already seen, the reach of the test is not always easy to avoid. One line of cases where inequitable conduct and inaccurate example presentation has arisen is in the context of prophetic examples. Prophetic (or constructive) examples refer to examples that are hypothetical in nature but provide the applicants a mechanism to explain how they think the example could be made or performed. It is sometimes true that an applicant will have sufficient knowledge of the invention and technical area to provide explicit instruction without actually having carried out the representative example. Often times, such examples are used to augment the presence of other actual examples to demonstrate how the invention can be made or used beyond the scope of the actual work that has actually been done. The use of prophetic examples must be approached cautiously. In particular, it is very important that an applicant clearly identify examples or procedures that have not actually been performed. Failure to clearly identify work that has actually been done from work that has not actually been done has been used to support findings of inequitable conduct in more than one case. In *Hofmann-La Roche v Promega Corp.* 323 F.3d 1354 (CAFC 2003), a patent was invalidated because the use of the past tense instead of the present tense implied that the experiments described had actually been performed as they were described—in patent parlance, examples written in the present tense indicate a prophetic example, whereas those written in

If you think you have the idea of inequitable conduct by now and prefer to move onto the next chapter then please feel free to do so. If however, you enjoy reading about the misfortunes of others who have seen their hard-won patents evaporate during litigation due to a finding of inequitable conduct then by all means, please read on. Our next case, *Aventis Pharma S.A. et al. V. Amphastar et al.*[92], relates to recent generic drug litigation that did not go well for the company that discovered and built the market for their blood-thinning product. This opinion from the CAFC resulted from an appeal of a District Court's finding of inequitable conduct against Aventis Pharmaceuticals for their prosecution of the patent covering their product marketed as Lovenox® in the United States. Lovenox® is a mixture of low molecular weight heparins that has been a highly successful antithrombotic agent. The defendant in this case, Amphastar Pharmaceuticals, is a generic drug manufacturer that wished to market a generic version of Lovenox®. Amphastar filed a paragraph IV certification (under Hatch-Waxman rules) alleging (among other things) that Aventis US reissue patent RE 38,743 was obtained through inequitable conduct and was thus unenforceable (sound familiar?). During the prosecution of the RE 38,743 reissue, the patent examiner cited a prior art reference that he alleged disclosed the same composition that Aventis was trying to claim (thus alleging that it was not novel and thus not patentable), and even if it were not the same composition as what was in the prior art, it was close enough to be considered obvious (and thus not patentable). To try to overcome the examiner's rejection, Aventis relied on declarations from one of its research directors (who incidentally was not one of the listed inventors).[93] To establish that their claimed material defined a different product, the research director explained that their composition had a longer half-life, but apparently what he did not say was that the half-life comparison was done at a different dose.[94] The Aventis mixture used for the comparison submitted to the USPTO was tested at a dose of 40 mg, and the product prepared by the prior art process was tested at 60 mg. The examiner still rejected the application asserting, in part, that the differences in half-life were not statistically significant. In response, a second declaration was submitted by the same Aventis research director. In this second declaration, the researcher submitted a

---

the past tense indicate that the experiments were actually carried out. It's a thin but deep line. Clearly separating what actually has been done from what could be done or what has sort of been done is a good start. The biggest challenge might be in describing experimental procedures in which most of it has occurred as written, but perhaps different experiments have been combined since multistep procedures are often not done in a linear fashion. There is no one size fits all way to address this situation other than to clearly delineate the preparation and testing of examples and to not imply something has been done that has not.

[92]*Aventis Pharma S.A. et al. V. Amphastar et al.*, 525 F.3d 1334 (CAFC 2008).

[93]Declarations are often submitted to the patent office to help explain to the examiner one or more aspects of the invention. In particular, these declarations are usually submitted by people with technical knowledge of the subject matter. The usual goal is to convince the examiner that the claimed invention for which a patent is being sought is patentably distinct from prior art being cited against the invention. The declarations might be submitted for other purposes related to patentability as well, such as for example, establishing inventorship or inventorship dates or that an invention is enabled.

[94]The half-life of a drug is generally defined as the time it takes for a drug concentration in a subject's plasma to be reduced by 50%.

number of tables containing the raw data of the mean half-life comparisons together with statistics showing that they were statistically significant. The dose of the Aventis compound could be found in the table, but the dose of the prior art mixture was not given. The District Court found inequitable conduct because the half-life was material to patentability, that there was a strong inference of intent to deceive the USPTO because it found no credible reason for comparing the two materials at different doses and because comparisons at the same dose actually showed very little difference (and the applicants knew this). On appeal to the CAFC, Aventis argued (among some other things) that the District Court erred because:

1. If the dosing was material, the examiner would have asked what the dose of the prior art composition was.
2. Half-life comparisons are often done at different doses.
3. Half-life is typically independent of dose.
4. Aventis also argued that they had, in fact, informed the examiner that the half-lives were determined at different doses because in the first declaration, the Aventis declarant stated: "this represents an increase of 250% in the half-life and is very significant because *it enables the same effect to be achieved at a lower dose.*

Unfortunately for Aventis, the CAFC upheld the District Court and also decided against Aventis; they simply were not convinced that leaving out the dosage of the composition they were testing against was inadvertent. The problem for Aventis was that even if it were inadvertent, the whole thing looked suspicious because every other dose of their compound compared had the same half-life as the prior art compound. As the District Court had stated, "[the half-life justification] was incredible because: (1) there was no statistical difference in half-lives when the 60 mg dose of [the prior art composition] was compared to the patented composition at a 20 mg, 60 mg or 80 mg dose, i.e., there was a statistical difference *only* when a 40 mg dose of the patented compound was compared" (emphasis added). If Aventis was trying to prove their claimed composition was different than the prior art composition, they needed to do so at the same dose. If the half-life was different between the two compositions at the same dose, then that would establish that they were different. Further, the CAFC pointed out that "Even if we acknowledge that half-life data at other doses for the patented compound were provided to the examiner, the date were provided in a very misleading way."

The CAFC decision was not unanimous. The dissenting judge opined that inequitable conduct should be reserved for only the "most extreme cases of fraud and deception" and that the threat of inequitable conduct was an "atom bomb" remedy, presumably more useful as a threat then actually used. The dissenting judge argued that inequitable conduct has taken on a "new life as a litigation tactic." In supporting his policy views, the dissenting judge harkened back to a previous case in which the CAFC had referred to the use of the inequitable conduct as a "plague."[95] To alleviate

---

[95]Citing *Burlington Indus. V. Dayco Corp.*, 849 F.2d 1418, 1422 (CAFC 1988).

the plague of inequitable conduct litigation, the dissenting judge highlighted the fact that the CAFC took the case *Kingsdown Med. Consultants, Ltd. V. Hollister, Inc.*[96] en banc, specifically to deal with that issue. In revisiting *Kingsdown*, the dissenting judge averred that "inequitable conduct was not a remedy for every mistake, blunder or fault" in patent prosecution. Particularizing his criticism, the dissenting judge pointed out that the CAFC (and presumably the district courts) were allowing the materiality of the offense to completely swallow the intent requirement. What might be simple mistakes in the patent prosecution process have resulted in findings of inequitable conduct because the mistake allegedly was material to the patentability even when the mistake was clearly not intentional, and even if intentional, it was hard to justify the "atom bomb" of inequitable conduct. Turning his attention to the facts of the instant case, the judge highlighted the fact that the data the declarant submitted were from a clinical study chart that the declarant did not prepare himself. Rather he explained the chart as it appeared to him; he did not intentionally conceal data that were already on the chart, the data were not there and he did not add to them. Acknowledging that such an omission might be negligent, the dissenting judge opined that it did not rise to the standard of intent.

Not mentioned by the dissenting judge but something worth highlighting given the gist of this text is that the declarant whose conduct resulted in the nonenforceability of this very important patent was not a lawyer. He most likely was not aware of U.S. federal court law as it relates to inequitable conduct. We do not know if his agent or attorney explained to him his duty in light of the declaration he was signing. Perhaps if the attorney had gone carefully over the declaration, he might have caught the omission, but he did not. As scientists, we may not know the law of inequitable conduct (but we hope you know a little more now), but that does not alleviate us from the duty of candor and disclosure with the USPTO. Being clever in our day-to-day life is normally a good thing but is not always a good thing in the patent procurement process. Because of the ever-shifting quicksand nature of inequitable conduct, the only safe advice is to be completely open and forthcoming in your dealings with both your agent or attorney and the USPTO.

---

[96] *Kingsdown Med. Consultants, Ltd. V. Hollister, Inc.* 849 F.2d 1418, 1422 (CAFC 1988).

CHAPTER 3

Prior Art and the Chemical Invention

> Bernard of Chartres used to say that we are like dwarfs on the shoulders of giants, so that we can see more than they, and things at a greater distance, not by the virtue of any sharpness of sight on our part, or any physical distinction, but because we are carried high and raised up by their giant size.
>
> —John of Salisbury, 1159[1]

3.1 WHAT IS PRIOR ART?

The invention process can take place in many ways. In some cases, serendipitous discoveries are made where the lucky but observant person happened to be the right person at the right place at the right time. As attractive and well established as this notion is in our common mythology, the truth is usually more mundane—invention is often more evolutionary than revolutionary. Many if not most areas of chemical research or technology are in relatively mature areas with many years or decades of prior work leading to the present moment, and so new inventions are often the latest step in a long line of persistent research, representing the collective effort of many. A big challenge in patent law and a recurrent theme in our wanderings therein will be determining where the prospective inventor's work begins and the previous efforts of others end. We have already developed a fond appreciation for the special exclusionary privilege a patent grants to its owner, and so we also appreciate that the inventor needs to give something back in exchange for that privilege. The bargain works where the inventor has brought something to the public that was not already available. But therein lies the rub, for what do we mean when we say that something is already available to the public? How publicly available does something need to be?

As we will learn shortly, not all "public" information is available for purposes of determining whether an invention is worthy of a patent. For example, some information may be considered in the examination of a patent application by one party but not another party, depending on who made the prior public disclosure. The types of public information that qualify for determining whether an invention is worthy of a

---

[1] Wikipedia, "Standing on the Shoulders of Giants."

The Chemist's Companion Guide to Patent Law, by Chris P. Miller and Mark J. Evans
Copyright © 2010 John Wiley & Sons, Inc.

patent are usually referred to collectively as *the prior art*. In this chapter, we will delve into what categories of information are available as prior art and what types are not; some of the distinctions might surprise you.

In several subsequent chapters, prior art will play the starring role in determining whether patents are valid and allowable, for it is the prior art that the patent application will be examined against. Whether determining if an invention is novel (Chapter 7), nonobvious (Chapter 8), or even adequately described (Chapter 9), the prior art will take its rightful place on center stage. As such, getting to know what is and isn't prior art must precede any analysis of how a patent application is examined. Mirroring this order, any erstwhile chemical inventor should be familiar with the prior art in her area before she makes costly research investments, particularly if her goal is to eventually patent her inventions.[2]

The categories of prior art have been defined collectively under the aegis of 35 U.S.C. §102 and are reproduced here. Taken in sum, the seven sections of §102 define

---

[2]There is no duty for applicants to search for prior art (although there is a duty to disclose prior art that is known (see Section 2.3), since the USPTO will conduct an independent search as part of the patent application examination process. However, it is highly recommended that a chemist be apprised of the relevant prior art before committing significant resources to a particular research endeavor. This can save significant time and money by ensuring that what she proposes to do is in fact patentable, assuming that is the end goal. Even if obtaining a patent is not the goal, it will help ensure that her research efforts will not unknowingly infringe another party's patent (keep in mind, however, that freedom to operate and patentability are different things, and it is not surprising that the searches for the prior art and appraisal of the prior art may be similar but are not identical; see section 1.4). Even though the USPTO will conduct its own search, it is the applicants who must have confidence in the validity of their own patents. There is a common notion among chemists that it is the patent attorney who should be aware of the prior art and that the practicing chemist need not be concerned with such "legal" matters. In fact, typically it is the chemist who should be most aware of the prior art in her respective area of research while it is the patent attorney who can best understand the legal implications of that prior art. Thus any prior art is best understood through the combined efforts of a chemist and a patent attorney. As a general matter, the chemist should conduct a thorough search of the prior art in a prospective area of research before beginning a significant program of research in that area. This search can be configured according to the chemist's expectation of the scope of the research to be conducted. Ideally, the chemist involved would then meet with her patent attorney to discuss the prior art before embarking on an extensive research effort in the proposed area. After embarking on a program of research, the chemist should discuss with the patent attorney any additional prior art of which she becomes aware that might impact the course of the research. Normally, chemists should keep current with developments in their area through any number of mechanisms including the use of various chemistry search tools such as the Internet, STN, SciFinder, Beilstein, and Derwent, or a combination of these. Please note that the conducting of a thorough search of the literature often requires the help of an information scientist because some of the search tools are either not available to the practicing chemist or require significant training and practice (and may be quite costly, especially if one is not proficient in their use). It is also important that chemists understand that the some of the search tools commonly available to them (such as Beilstein and SciFinder), while being extremely useful in their own right, are often not solely capable of conducting a full patentability search. For example, if a particular reference describes a genus of compounds with no actual exemplification of a compound within that genus, it is likely that reference will not show up as a hit within either Beilstein or SciFinder since these databases encode actually made compounds and not Markush structures. While this distinction is of particular concern in the freedom-to-operate arena, in some instances it might be relevant to the patentability inquiry as well. This is why the conscientious chemist (such as yourself) will want to work closely with both a patent attorney/agent and an information scientist if possible.

the categories of information or activities that the invention under consideration will be examined against.

35 U.S.C. 102 Conditions for patentability; novelty and loss of right to patent.

A person shall be entitled to a patent unless—

(a) the invention was known or used *by others* in this country, or patented or described in a printed publication in this or a foreign country, *before the invention thereof by the applicant* for a patent, or

(b) the invention was patented or described in a printed publication in this or a foreign country or in public use or on sale in this country, more than *one year prior to the date of application for patent in the United States;* or

(c) he has abandoned the invention, or

(d) the invention was first patented or caused to be patented, or was the subject of an inventor's certificate, by the applicant or his legal representatives or assigns in a foreign country prior to the date of the application for patent in this country on an application for patent or inventor's certificate filed more than twelve months before the filing of the application in the United States, or

(e) the invention was described in—(1) an *application* for patent, *published* under section 122(b), *by another* filed in the United States *before the invention by the applicant for patent* or (2) a patent granted on an *application* for patent *by another* filed in the United States before the *invention* by the applicant for patent, except that an international application filed under the treaty defined in section 351(a) shall have the effects for the purposes of this subsection of an application filed in the United States only if the *international application designated the United States* and was published under Article 21(2) of such treaty *in the English language;* or

(f) he did not himself invent the subject matter sought to be patented, or

(g) (1) during the course of an interference conducted under section 135 or section 291, *another inventor involved therein establishes*, to the extent permitted in section 104, *that before such person's invention thereof the invention was made by such other inventor* and not abandoned, suppressed, or concealed, or (2) before such person's invention thereof, the invention was made in this country by another inventor who had not abandoned, suppressed, or concealed it. In determining priority of invention under this subsection, there shall be considered not only the respective dates of conception and reduction to practice of the invention, but also the reasonable diligence of one who was first to conceive and last to reduce to practice, from a time prior to conception (emphasis added).

Please note the keyword *unless* in the first sentence of §102. This means that during examination of the patent at the USPTO, it will be the responsibility of the examiner to provide a basis for rejection of the patent application. If the patent examiner cannot affirmatively reject a claim in the patent application, then that claim must be allowed. To do this, the patent examiner will rely on the "best available art" when making a rejection, meaning the examiner is instructed to avoid making rejections that are cumulative in nature. A cumulative rejection is one that merely stands on the same grounds as one or more other rejections. In the case in which the strongest references or set of references is not able to sustain a rejection, additional weaker references standing on the same grounds will not add anything to the rejection and thus are to be avoided.

Before getting started in earnest, it is important to understand the one thing that most distinguishes U.S. patent law from the rest of the world is that under certain circumstances a patent is awarded in the United States to the first person to *invent* the claimed subject matter rather than the first to *file* the patent application. As one consequence of the first to invent system, it is possible that one can receive a patent in the United States arising from a patent application that was filed *after* another party filed their patent application, even when both patent applications claimed the same invention. In the rest of the world, the first applicant to file the application is the one who will be awarded the invention (assuming she has met all of the requirements of patentability in that country). As another consequence of the first to invent system, we will see that certain types of prior art may not apply to an applicant who is able to successfully assert that she invented the subject matter before the effective date of that prior art. The process of proving invention before the date of a piece of prior art is often referred to as antedating (or *swearing behind*) a prior art reference, and this feature flows from the first to invent characteristic of the American patent system. As a result, one might be able to obtain a patent in the United States but not the rest of the world (and vice versa). However, there are limitations to the first to invent rule because there are certain events that can absolutely deny patentability to the first inventor—these are referred to as *statutory bars*. U.S. patent law reflects a balance between rewarding the first inventor with a patent and yet encouraging that same inventor not to sit on the invention too long before filing the patent application.

The seven sections of §102 may be grouped according to whether they apply to the patent applicant's invention date or the patent application filing date. Accordingly, §§102(a), 102(e), 102(f), and 102(g) are grouped together as they relate to events that occur before the patent applicant's *date of invention*. As mentioned earlier, it is sometimes possible to remove those references by antedating them by showing invention before the effective date of the prior art reference (or prior art activity). The remaining §§102(b), 102(c), and 102(d) cover the statutory bars and are directed to types of prior art occurring before the patent applicant's *date of filing* and that cannot be removed, sworn behind, or antedated by demonstrating invention before the reference date. We will cover each of the types of prior art separately.

3.2 PRIOR ART THAT CAN BE ANTEDATED

3.2.a §102(a)

Prior art under §102(a) includes only the works of "others" and thus one's "own" work does not apply.[3] This follows because for §102(a) to apply as prior art, the

---

[3]One's own work can serve as prior art, just not under the circumstances of §102(a). Sections 102(b), 102(c), and 102(d) are all examples where one's own work can serve to function as prior art, but only under the conditions that each describes.

invention that one is attempting to patent must be "known or used *by others* in this country, or patented or described in a printed publication in this or a foreign country, *before the invention thereof by the applicant.*" Why does §102(a) apply only to the works of others? The primary reason is that it is unlikely that an inventor or group of inventors could know or use, patent or describe *their own* invention in a printed publication, in this or a foreign country before their actually inventing the subject matter (unless their invention was for some sort of time machine that they then used on themselves). Section 102(a) of the code simply reflects the fact that the United States is a limited first-to-invent country by setting the defining time point before the invention and as a result, the basis for a patent application rejection under §102(a) must be effectively dated before the *invention* of the applicant.

The types of prior art in §102(a) include publications and patents as well as knowledge (known) and use. These publications, patents, knowledge, or uses might have multiple parties listed or involved, just like the patent application, which may list one or more inventors. For purposes of §102 (a), the word *other* is narrowly construed. If the person or group of people who are the collective authors of the prior art publication (or patent), or the collective users or knowers of the prior art invention are *in anyway different* from the listed inventor or joint inventors on the patent application under consideration, then that prior art can be considered to be that of "another," and that art applied under §102(a) against the invention. The list of inventors on a patent application are collectively referred to as the *inventive entity*, and so another way of thinking about this requirement is to understand that the inventive entity must be different from the source of the prior art in order for section §102(a) to apply. For example, a publication with the authors Traney, Collino, and Miller could still be cited as prior art under §102(a) against a U.S. patent application naming only two of the authors (e.g., Collino and Miller) as inventors. Since Collino and Miller together constitute the inventive entity, the original publication is considered to be by "others" for purposes of §102(a). Presumably the reason for this rule is that it is *possible* that the author that was *not* listed as an inventor on the patent application (Traney in this instant) actually "invented" some of the subject matter that was in the publication (and later claimed by the other two authors in their patent application). In such a case, the listed inventors might not have invented the subject matter of all of the claimed invention before the material in the publication since some of the claimed subject matter could have been invented by the author who was not listed as an applicant—the third author (Traney) *may* have invented some of the subject matter first.[4]

Assuming that the patent applicants believe they are the correct and complete list of applicants for the invention and that the additional co-author(s) of the §102(a) reference did not invent any of the claimed subject matter of the patent application, the publication can be removed as a citable §102(a) reference in a couple of different

---

[4]Such a situation might also present an inventorship issue; the claimed invention must list all of the inventors and if one of the authors of the paper was also an inventor, then that person needs to be listed on the patent application as well. Note that the criteria for authorship are usually different from the criteria for inventorship; much more detail will be provided on this topic in Chapter 4.

ways. One way of removing the reference is for the author of the paper who was not listed as an applicant on the patent application to write an affidavit stating that the pertinent portions of the cited publication originated with the patent applicants rather than the nonapplicant co-author.[5] As an alternative, the applicants for patent may themselves assert by declaration that the relevant portion of the cited reference is the applicants' own contribution/invention.[6] However, if it is determined that the non-listed inventor was actually a co-inventor who should have been listed as an inventor but was mistakenly left off, the patent application may be amended to correct the mistake in inventorship provided the correction is in accord with the requirements for amending inventorship.[7] The addition of the nonlisted author to the patent application results in the list of the publication authors and the inventive entity being the same. The publication is no longer by "others" and thus no longer available as prior art under §102(a).

The definition of "printed publication" includes works that are disseminated or otherwise available such that one of ordinary skill in the art, exercising reasonable diligence can locate it.[8] As applied, this definition has been interpreted very broadly to even include a thesis dissertation that was cataloged into a German university's library system, distribution of six copies of an oral presentation delivered to a congress, an Australian patent application kept on microfilm at the Australian Patent Office, and even a poster that was presented for 3 days at a conference since that poster did not explicitly require confidentiality or explicitly restrict the making of copies of the poster.[9] There still remain some gray areas, such as whether a transient display of slides accompanying an oral presentation can be considered a "printed publication." In

---

[5] See MPEP 2132.01, citing *Ex parte Hirschler*, 110 USPQ 384 (Bd. App. 1952).

[6] See MPEP 2132.01, citing *In re Katz*, 687 F.2d 450, 215 USPQ 14 (CCPA 1982) and explaining: "Katz stated in a declaration that the coauthors of the publication, Chiorazzi and Eshhar, 'were students working under the direction and supervision of the inventor, Dr. David H. Katz.' The Court held that this declaration, in combination with the fact that the publication was a research paper, was enough to establish Katz as the sole inventor and that the work described in the publication was his own. In research papers, students involved only with assay and testing are normally listed as coauthors but are not considered co-inventors." The *CCPA* as referred to in this case cite was the predecessor to the Court of Appeals of the Federal Circuit (CAFC).

[7] See MPEP 2132.01 citing 35 U.S.C. 116 third paragraph: "Whenever through error a person is named in an application for patent as the inventor, or through an error an inventor is not named in an application, and such error arose without any deceptive intention on his part, the Director may permit the application to be amended accordingly, under such terms as he prescribes."

[8] *In re Wyer*, 655 F.2d 221 (CCPA 1981).

[9] *In re Hall*, 781 F.2d 897 (CAFC 1986), *Massachusetts Institute of Technology v. AB Fortia*, 774 F.2d 1104 (Fed. Cir. 1985) and *In re Klopfenstein*, 380 F.3d 1345 (Fed. Cir. 2004). In the latter of these three cases, the CAFC identified four factors for consideration in the determination of whether a poster at a conference was a publication: (1) the length of time the material was displayed, (2) the expertise of the audience, (3) reasonable expectations that the material will not be copied, and (4) simplicity or ease with which the displayed poster could be copied. In contrast to the result in *In re Hall*, in which the thesis in question was deemed to be prior art since that thesis had been cataloged in the library's card catalog system, the Court in the case of *In re Cronyn*, 890 F.2d 1158 (CAFC 1989), came to a different conclusion because the thesis in question was present in the college's library and thus available to the public but the thesis was not cataloged according to subject matter but rather alphabetically by the author's names. Close calls such as these can turn very narrowly on the particular circumstances of the disclosure.

a 1981 district court decision, this was not considered to be a printed publication, but modern technology allows easy photography of slides (a common occurrence even at meetings that ban photography) and this may change.[10] Finally, information available on the Internet can be a printed publication.[11] A common theme in determining whether a disclosure is a publication so as to qualify as prior art focuses on whether the publication was accessible to the public in a "meaningful way."

You may have noticed that the "known or used" provision of §102(a), unlike printed publications, applies only within the United States. Thus, for example, the use of an herbal treatment for a particular malady as practiced by a native Andean tribe in South America might not prevent a patent to an *independent* but later discovered invention of the same thing in the United States. However, if the tribal medicine man had published on the treatment in the *Today's Medicine Man* journal, this could then serve as prior art under §102(a). Probably more than anything else, this distinction is rooted in the historical concerns regarding the ability to obtain evidence of activities occurring outside of the United States versus within the United States. While this concern is still real, the world has shrunk very significantly in recent years, and the wisdom of making this distinction has become increasingly questioned.

It is interesting that §102(a) requires literally that the invention only be "known or used" in this country. What if somebody (or a limited group of people) knew about or used the invention and never disseminated information about their use? Should that be considered prior art under §102(a)? In a strict construction one could say literally that the invention was "known" or "used" by a person or persons in this country, but would denying the granting of a patent on such a previous but secretive use make sense? In keeping with the patent system's priority on encouraging public disclosure of inventions in exchange for a period of exclusivity to the first inventor, denying somebody that exclusivity because of prior secret knowledge would not lead to a very sensible result. The courts have agreed, the "known or used" verbiage requires that the knowledge or use must be accessible to the public. If a person or group of people knew about or used the invention but kept that knowledge secret, then it would not benefit the public at large. Since the patent system rewards and encourages the public disclosure of inventions, the proper balance is still in place where the prior art has not met that burden but the applicant's invention has.

As you were reading this section, you may have at some point come to wonder what is meant by the *effective date* of the prior art. In keeping with the general purpose and intent of the prior art, we can appreciate that its effective date should relate to the date that it begins to fulfill its purpose of informing the public. In so doing, prior art becomes effective when it actually becomes generally accessible by the public. Thus, for example, a magazine or journal article becomes prior art not on the day it is printed or on the day it is mailed but rather on the day it is first received by an

---

[10]*Regents of the University of California v. Howmedica Inc.*, 210 USPQ 727 (D. NJ 1981).

[11]In *SRI International, Inc. v. Internet Security Systems, Inc.*, 511 F.3d 1186 (CAFC 2007), the CAFC reversed the District's Court's finding on summary judgment that a paper placed on an ftp file server (and left on for 7 days) was a printed publication, reasoning that the server on which the file was located did not have an "index or catalog." The case was remanded to the District Court for further determinations of fact.

addressee. Many publications have online versions and the effective publication date could very well be the day it was posted on the Internet.

Since §102(a) refers to prior art events that have an effective date before the invention by the applicant, it is critical to know how to determine when an invention occurs. For this, it is first necessary to appreciate that an invention is conveniently thought of as consisting of two parts: conception and reduction to practice. The conception portion relates to the mental aspect of the invention, *the idea of the invention before it is put together in tangible form.*[12] The reduction to practice can be actual or constructive. An actual reduction to practice is the putting together of a working example, model, or prototype of the invention and *demonstrating that the invention works to fulfill its intended purpose.* A constructive reduction to practice is the filing of the patent application itself. The linchpin connecting conception and reduction to practice is the concept of *due diligence.* The due diligence in regard to an actual reduction to practice means *reasonable diligence* on the part of the inventor.[13] The establishment of reasonable diligence is necessarily fact specific but will require proof that the invention was worked on during the entire required diligence period, or acceptable excuses must be provided for lapses in time; mere statements will not suffice, and the work relied on to show diligence must relate directly to the reduction to practice of the invention (attempts to commercialize or finance the reduction to practice have been held to be not sufficient).[14] In a constructive reduction to practice, reasonable diligence must be exercised both by the inventor and the attorney or agent who might be assisting in preparing the patent application.[15] An applicant for a patent can establish an invention date before a cited §102(a) reference by demonstrating that they either (1) reduced the invention to practice before the effective date of the §102(a) reference, (2) conceived the invention before the §102(a) reference date and then exercised due diligence from before the §102(a) reference date to the actual reduction to practice, or (3) conceived the invention before the reference §102(a) reference date and exercised due diligence from before the §102(a) reference date to the filing date of the patent application (constructive reduction to practice). From the latter two examples, one can see that the invention need not actually be completed before the effective date of the §102(a) reference date as long as it was conceived before the reference date and the inventor(s) exercised due diligence from some time before the reference date to the reduction to practice (actual or constructive). Proving

---

[12]*Conception* has been defined in various ways. A synthesis of many of these thoughts and perhaps a reasonable way to think of conception is that it is the complete idea of the invention such that the reduction to practice does not require additional inventive input or excessive effort.

[13]MPEP 2138.06 citing *Emery v. Ronden*, 188 USPQ 264, 268 (Bd. Pat. Inter. 1974).

[14]MPEP2138.06 citing *Scott v. Koyama*, 281 F.3d 1243, 1248–49 (CAFC 2002).

[15]If a long time occurs between an actual reduction to practice (meaning the completion of the invention) and the filing of a patent application, then the concept of due diligence is not in play but rather the question becomes one of whether the inventor abandoned, suppressed, or concealed the invention. This possibility is covered separately by §102(c), which pertains to the abandonment of the invention and, more important, by §102(g) where, in a contest to prove who had invented the subject matter first (patent interference), a party found to have abandoned, suppressed, or concealed the invention can lose to a later-inventing party who did not commit one of those sins. More on these two sections later in this chapter.

prior invention in order to antedate a §102(a) reference for the purposes of obtaining a patent requires that the inventor or owner of the patent application submit an oath or declaration establishing the facts necessary to show prior invention. This is known as *antedating* or *swearing behind* the cited §102(a) reference. To antedate or swear behind the cited reference, the inventor(s) will need to have invented the claimed section matter in question before the effective date of the reference, and the invention will need to have been made in the United States, a NAFTA country, or a WTO member country (this probably does not leave too much out). The actual showing of proof will normally be in the form of an affidavit, which includes drawings or records (or copies thereof) unless there is a satisfactory explanation for not having such record of proof. If the applicant's proof is accepted, the §102(a) reference can properly be removed as prior art.[16]

Perhaps as an inventor or as a prospective future inventor, you have been reminded of the importance of keeping accurate records documenting and dating your research efforts. You also might have been asked to have a co-worker witness and sign your lab notebook pages. These requirements are primarily due to the critical importance of having proof of the actual invention date. As we have seen in this section, §102(a) prior art references (as well as §102(e) and §102(f) references to be discussed later) can be antedated to show invention before the reference's effective date. In the case of §§102(a), 102(e), and 102(f), the proof of prior invention will be used to demonstrate invention before the effective date of the reference itself. In the case of §102(g), two or more different inventors will be competing to demonstrate who made the invention first. In both cases, the success or the failure can well turn on the quality of evidence that the inventor can produce. If (and until) the United States adopts a first-to-file system, the accurate documentation of one's ideas and efforts will continue to be of significant legal importance.[17]

3.2.b   §102 (e)

A second 102 section relating to the work of "another" is §102(e), which is concerned with *patent applications:* (1) that are filed in the United States and published by the USPTO or (2) that are filed in the United States and issued into patents by the USPTO or (3) that are filed as international PCT patent applications and are published (e.g., WO published patent applications) in English and designate the United States.[18] The

---

[16]Persons beside the inventor(s) may make the showing under certain circumstances when the inventor is otherwise unavailable, such as where the inventor is dead, not legally sane (truly a mad inventor), refuses to cooperate, or cannot otherwise be reached. The guidelines for the antedating affidavit are described in 37 CFR §1.131.

[17]In *Procter & Gamble (P&G) v. Teva Pharmaceuticals* (CAFC 2009), the inventor of the osteoporosis drug risedronate attempted to swear behind a prior art reference; however, the court rejected the evidence because the notebook "was unwitnessed and was not corroborated by any other evidence." P&G won the case on other bases, but nevertheless you get the idea.

[18]One aspect of §102(e) worth also noting is that the word *describes* is used and not the word *claims.* If a prior art patent claims the invention, then §102(g) may be relevant and an interference may be declared. Section 102(g) will be covered subsequently.

effective prior art date for a §102(e) reference is the date that the patent application is *filed* and not the date it was *published* (or patented). Moreover, if the published U.S. patent application or patent claims priority to a U.S. provisional patent application, then the filing date of that provisional patent application might become the effective date of the §102(e) reference.[19] International PCT applications can also be effective prior art as of their filing date (rather than their publication date) under §102(e) when the international application was filed on or after November 29, 2000, designated the United States, and was published in English. As a practical matter, this means that when you see a published US or PCT application (designated with publication numbers preceded by *WO*), its effective prior art date could very well be before the publication date of the patent application.[20] These rules have the interesting effect of "pretending" that the contents of a patent application becomes available to the public on the filing date of the application, even though the patent application will not be published and actually available to the public for some period of time, typically about 6 months from the nonprovisional patent application filing or 18 months from the earliest provisional filing.

In a similar fashion to §102(a) prior art, §102(e) art can be sworn behind or antedated, except that it is necessary to prove invention before the filing date of the application or even an earlier provisional application. However, §102(e) prior art can also function as §102(a) prior art and, depending on whether it's been published for over a year, it can serve as §102(b) art as well (we will cover §102(b) shortly). You might wonder how many times can you hang somebody with the same piece of prior art and so does it really matter how many different ways the same reference is cited? Interestingly enough, it can. Sometimes it is possible to avoid or remove a reference

---

[19] A U.S. provisional patent application is not published or becomes publicly available until a regularly filed application that claims priority to it is published. At the time the regular application is published, the file history then becomes available to the public, including any provisional applications from which the regular application claims priority. Electronic file histories for recent U.S. patent applications are available for free at the USPTO's website under the "Patent Application Information Retrieval" (PAIR) menu.

In order for the provisional application to be relevant §102(e) prior art as of the date it is filed, the subject matter of that provisional application must provide written support (enablement—i.e., teach one to make and use the invention) for the portion of the published patent application or issued patent that is being relied on as §102(e) prior art.

For the provisional application to have the potential of being prior art as of its filing date, it needs to be a provisional application filed in the United States; applications filed in other countries and relied on for priority for their U.S. filing cannot be §102(e) references against other patent applications as of their filing date. For the owners of the foreign filed applications, their foreign filing date can be used to antedate other references (they can claim the earlier filed provisional patent application for a filing date to demonstrate constructive reduction to practice for establishing an invention date) but cannot be used as a §102(e) reference against other U.S. patent applications.

[20] A quick review of some of the material presented in Chapter 2: In most cases, patent applications filed after November 29, 2000, in the United States (and the rest of the world) are published approximately 18 months after the earliest application in the chain was filed. However, if the application is no longer pending (e.g., abandoned) or subject to a secrecy order, it will not be published. Also, an applicant for patent in the United States can request that the application not be published but the applicant must certify that the application "has not or will not" be filed in another country that requires publication 18 months after application.

cited under one section of 35 USC 102 but not another. For example, an applicant's own patent application cannot be prior art against him under §102(e) but, if it has been published for more than 1 year, that same patent application can be nonremovable prior art under §102(b).

3.2.c §102 (f)

Section 102(f) is not a common source of prior art since it is concerned with the situation in which one party conceives an invention and that invention is learned of by another who then attempts to claim the invention himself in a patent application.[21] Section 102(f) is useful in circumstances in which a citation under one of the other 102 sections is not appropriate. For example, under §102(a), we learned that public use outside of the United States would not be a basis for rejection under that section, but deriving one's invention from witnessing that public use and filing a patent application on that derivation could still be citable under section §102(f) since §102(f) does not have any geographic restraints. Perhaps more generically, one might best think of §102(f) as describing a situation in which the patent application fails to properly list the inventors. Under many circumstances, this type of error can be corrected (see Chapter 4).

3.2.d §102(g)

Section 102(g) describes another's prior invention as prior art. Section 102(g) is broken up into two sections, the first and most important section deals with the situation where two or more parties are trying to establish who invented something first. Since the United States is a first-to-invent country and the same invention cannot be patented twice, it stands to reason that when two separate inventive entities independently derived their inventions (if one copied from the other then we have a section §102(f) problem), only one of those entities will be awarded the patent—the party that first invented the subject matter. As we learned before, the first to invent is the one who reduces the invention to practice first, unless there is an earlier conception date followed by due diligence up to the reduction to practice. The contest to prove inventorship between two separate inventive entities is referred to as an interference. The actual interference proceeding itself is somewhat akin to mini-patent litigation in which two or more parties are pitted against each other in an evidentiary proceeding either before the USPTO's Board of Patent Appeals and Interferences or in the federal courts.[22] The party that loses the interference will have the other party's invention cited against him as prior art under section §102(g), which will serve to prevent them from getting a patent to the contested subject matter.

---

[21]MPEP 2137 citing *Kilbey v. Thiele*, 199 USPQ 290, 294 (Bd. Pat. Appeals and Interferences 1976).
[22]The Board has jurisdiction when at least one patent application is involved. If the interference occurs between two issued patents, jurisdiction is conferred on the federal courts. MPEP 2300.01.

Procedurally, an interference occurs when one or more claims in a patent application is substantially the same as one or more claims in another patent application with certain time bar limitations when the action can occur. These time bars are a consequence of §102(b), which sets a time limit on the first-to-invent aspect of U.S. patent law. A later filed patent application that claims or could be drawn to claim the same subject matter as an earlier filed patent application or patent must be filed within 1 year of the time the earlier patent application was published or patented. If the later applicant files after the 1-year period, then the earlier art becomes §102(b) prior art that cannot be removed by proof of prior invention. As a practical matter, §102(g) prior art can be especially problematic because it can take the longest time to discover, and moreover, an interference can take significant time (and money) to resolve.

One way that interferences occur is when inventors sees a published patent application or patent claiming the same invention that they have already conceived or made. Within a year, they file their patent application claiming substantially the same invention and bring the conflicting subject matter to the attention of the patent examiner. In another, perhaps more common scenario, two or more separate inventors or inventive entities are working in a competitive area, and they file applications claiming or that could claim the same invention in a similar timeframe. During the course of the examination of the two applications, an examiner discovers the common subject matter and alerts the applicants to the situation and suggests a claim for one or both of the applicants that will provoke the interference.

The second prong of §102(g) deals with prior inventions as prior art but not in the context of a patent interference. Section 102(g)(2) might be relevant, for example, when one party is trying to invalidate another party's patent but he is not trying himself to get a patent issued to the same subject matter. A common way to try to invalidate somebody's patent is to allege that it is not novel in view of prior art under one or more 102 sections. One possible category is §102(g)(2), if it can be established that a third party invented the subject matter first (in the United States). The reason why this is not expected to be a highly used provision is that the prior invention by another party might be hard to establish since to know that another party invented the subject matter first typically would require some sort of public performance of the invention. This prior public performance of the invention could also qualify under §102(a) or §102(b), which might be cited instead. If the invention by another party was being cited under §102(g)(2) but not under §102(a) or §102(b) due to the fact that there was no public disclosure, then one has to deal with the simple evidentiary problem of how one would discover the prior invention in the first place. Last, you might have noticed that the invention cannot have been concealed, abandoned, or suppressed to qualify under §102(g)(2). If the prior art qualifies under section §102(g)(2) and not under §102(a) or §102(b), the implication is that there was not a public use but at the same time there was also no concealment, suppression, or abandonment—not a sure contradiction but a thin slice to stand on nevertheless.

To better appreciate §102(g) and the possibilities presented by some different scenarios, let's consider a hypothetical example. Imagine that two chemists working at different companies separately synthesize a novel pigment for use in a new kind of floor coating and paint. Chemist A got the idea for the pigment composition on June 1, 2007. He thought that by combining a certain metal salt with a particular oxidizer, he could make a metal oxide with just the right stoichiometry to make a great pigment for use in different coatings. On that same day he recorded the idea and had it witnessed in his notebook. He shelved the idea for a few months because he had some other projects that needed to be finished up first, and he did not have the time to get to it right then. So on September 1, 2007 when he was finally finished with his other work, he decided to test his idea by first making the pigment that he had earlier written down in his notebook. He worked on the idea diligently for the next several weeks, optimizing the conditions for making the metal oxide pigment. Once he had several batches of the pigment prepared, he combined them with a base material and demonstrated that the pigment could indeed function as good ingredient for the paints and coatings as he had originally envisioned; this all occurred on and around September 26, 2007. The company's patent attorney drafted and filed the patent application in record time (his own record, that is), and so the patent application covering the pigment was filed on December 9, 2007.

Unbeknownst to Chemist A, over a similar period of time, Chemist B stumbled upon the exact same pigment. For Chemist B however, the discovery process took a more serendipitous yet expeditious route. Here is the honest truth as to how it happened. On August 27, 2007, Chemist B happened to be in the lab when her arm bumped into a flask containing an unknown powder that spilled out onto the bench. "Oh darn" she said and grabbed a roll of paper towels but when going to wipe the powder up, she then accidentally knocked over a flask containing a clear solution that spilled all over the powder. "Double darn!" the unlucky chemist exclaimed but then glancing down at the mess thought "Ooooh, that's really pretty. I'd better save that stuff." She grabbed a chamois cloth and soaked up the pigment and squeezed it out into what she thought was an empty bucket but then realized the bucket was partially filled with some sort of liquid. "Triple darn" she cried, "Now I've got a real mess!" As she bent over to pick up the bucket, she accidentally kicked it over onto the floor. She was so mad that the words she attempted to sputter were choked off in her throat. She was never going to get home in time for dinner now. Resigned to her fate she let out a deep, relaxing breath as she had been taught so often to do, grabbed a mop and began to swab the lab floor. As she mopped the floor, she wasn't sure if what she was seeing was really true because it was hard to visualize clearly through her tear-glazed eyes. It appeared that whatever was in that bucket made one of the most beautiful, fast-drying paints she had ever seen—it had a pearlescent, almost acrylic glow that was growing even more beautiful by the minute, right before her very eyes! Each push of the mop increased the shine. "I've got to find out what all that stuff was," she excitedly thought. She called the chemist whose lab it was and asked him what each of the ingredients was. As she learned that day, the powder was a particular salt the chemist had been working on, and the flask contained an oxidizing agent that he figured must have converted the metal salt into its metal oxide. The metal oxide was then inadvertently dumped into the bucket, which contained a polyacrylic-containing base material that, when combined with the freshly synthesized pigment, formed the beautiful coating that now permanently adorned their lab floor. "I've got to document this stuff right now before I forget," Chemist B hurriedly thought, abandoning all ideas of dinner and getting home on time. "This could be the big break I've been waiting for!" She wrote everything down in her notebook and had it witnessed by the chemist whose lab she was using and who understood the nature of the invention when explained to him. Over the next several months, she performed some additional controlled experiments to verify the early result and wrapped up the work on February 17, 2008. She contacted the company's attorney who drafted their application for patent and filed it on April 1, 2008. The patent application covered the same metal oxide as Chemist's A.

During the course of his normal review, the patent examiner at the USPTO discovered that two pending applications assigned to two different organizations contained claims to the same subject matter—the metal oxide useful as a pigment in paints and/or coatings. An interference was declared, and the parties set out to establish who invented the subject matter first. In this case, Chemist A will be considered as the senior party and the presumption will be made that Chemist A is the first to invent. This is because Chemist A was the first to file a patent application, on December 9, 2007. Chemist B will be considered the junior party since her patent application was not until April 1, 2008, and it will be up to Chemist B to overcome the presumption that Chemist A was the first to invent. Let's go through the arguments from each side.

During the course of the proceedings, Chemist A was able to prove a conception date of June 1, 2007, due to his meticulous notebooks. He was able to successfully argue that he had fully conceived of the pigment and its preparation; he needed only to actually synthesize the material and test it out to prove it work for its intended purpose (reduction to practice). The actual reduction to practice did not require anything beyond ordinary testing to prove that the pigment could work to form a good coating material. In view of this information, Chemist A's patent litigation counsel has adopted the position that the date of invention was the same as the date of conception, June 1, 2007. Since it occurred before Chemist B had any idea about any such pigment (her accident did not occur until August 27, 2007), Chemist A should be awarded the patent. The patent attorney relaxes and prepares for his annual four-week summer Caribbean vacation, confident in the strength of his client's position.

Let's review the law to see if the patent attorney should be so confident. In our discussion of section 102(a), we learned that prior invention could be proved by a reduction to practice before the reference date or a conception before a reference date followed by a period of due diligence until the actual or constructive reduction to practice. In the case of Chemist A, there was an actual reduction to practice that occurred when Chemist A synthesized the pigment and showed that it worked for its intended purpose, which occurred on September 26, 2007. The filing of the patent application is a constructive reduction to practice and would serve as a reduction to practice had an actual reduction to practice not occurred first. To use the conception date, however, Chemist A will need to show that he was diligent from the conception date to the date of the reduction to practice. If he is not able to show due diligence from the conception date to the reduction to practice, he will not be able to mark the conception date as the date of the invention. Instead, the date of the invention is the time period from when the due diligence began and ran without stop up to the reduction to practice. In our instant case, we can readily see that Chemist A did not practice due diligence from the conception of the invention up to the actual reduction to practice. Rather, he shelved the idea until he finished some unrelated work and then started working on it in earnest on September 1, 2007. We thus have the timeline for Chemist A: June 1, 2007, conception; September 1, diligence starts; September 26, actual reduction to practice occurs; and December 9, constructive reduction to practice occurs. So, the attorney's argument that the date of invention was June 1, 2007, is not correct because he glossed over the diligence requirement—the effective date of invention is actually September 1, 2007.

Now that we have the priority date for Chemist A, let's see what we can come up with for Chemist B. First, unlike the case for Chemist A, Chemist B never conceived the invention in the usual sense; rather, she recognized and appreciated the invention after the physical act of combining the chemicals had already occurred. Since an invention requires both a mental part and a physical act of some sort (even if the physical part is only the filing of a patent application), the invention will not have occurred until both parts have come together. Normally the way this is explained is that the reduction to practice is not complete until the results are recognized or appreciated. In the present example, Chemist B recognized the results and appreciated them on the day that the invention was reduced to practice. First, she saw the results of the inadvertent mixture after it had formed, thus establishing a utility for the invention, and then she contacted the chemist whose lab it was to understand what the invention actually was. In this situation it is likely that Chemist B's invention date is the day the accident happened because that is the same day that she learned what actually made up the invented subject matter; it appears she fully understood the invention at that point, and the subsequent efforts were directed at further routine testing of the pigment. In this case, due diligence in the normal patent sense of the word was not required, we used that term to link an earlier conception to a later reduction to practice to establish the earlier date of conception as the invention date. In the present hypothetical, there is no issue of an earlier conception date with respect to Chemist B's invention since Chemist B did not conceive the invention in the usual sense of the word. The timeline for Chemist B is thus: August 27, 2007, actual reduction to practice and recognition of invention; and April 1, 2008, constructive reduction to practice. Accordingly, her invention date was August 27, 2007.

Not to be deterred, Chemist A's legal counsel convinces his client that he should not concede yet. Chemist A's counsel informs his client that he has carefully reread the text of the §102(g) statute and he has cleverly discovered in the fine print that it gives priority to the earlier inventor, but only in cases in which that inventor has not "abandoned, suppressed or concealed" the invention. Chemist A's side needs only to establish that Chemist B committed one or more of these three sins. Since §102(g) is concerned with priority contests between parties, the guilty party will be put behind the other party and lose the priority contest together with any patent to the contested invention. This begs the question of what it takes for an invention to be abandoned suppressed, or concealed, and, not surprisingly, this is a question of the particular facts of the case.[23] In the facts of this case, it was approximately 7 months from the date of Chemist B's discovery and the filing of the patent application, and it appeared the time was used to further test and evaluate the discovery as well as file the patent application. Under these facts, it is unlikely that Chemist B will be found to have concealed or suppressed the invention; Chemist A is unlikely to prevail with this argument.

Not ready to give up until he has exhausted all of his client's legal arguments (and finances) Chemist A's patent counsel argues that Chemist B never conceived the invention at all. Without a conception, there can be no invention and therefore Chemist B's accident does not qualify as an invention in the first place. Do you agree? *Conception* in the patent law sense of the word has been defined as "the formation in the mind of the inventor of a definite and permanent idea of the complete and operative invention as it is thereafter to be applied in practice."[24] In the instant case, Chemist B did not conceive of the invention prior to her serious of serendipitous stumbles. In effect, the invention was conceived after the physical happening occurred; no immaculate conception here but perhaps a type of conception nevertheless. Unlike Chemist A's discovery, Chemist B's invention was essentially complete when she conceived it, since the physical proof that the invention would work for its intended purpose had already occurred. In that sense, Chemist B's conception of the invention was as much a recognition or appreciation of the invention as anything else – once it was recognized and appreciated, it was complete. Despite the unusual etiology of Chemist B's invention, it was an invention nevertheless, and it can be dated to August 27, 2007,when both the reduction to practice and recognition occurred.

3.3   PRIOR ART THAT IS AN ABSOLUTE BAR

Sections 102(b), 102(c), and 102(d) represent absolute bars to a patent applicant's invention in the sense that references cited under any one of these sections *cannot* be

---

[23]MPEP 2138.03 Abandonment, suppression, or concealment were explained in *Correge v. Murphy*, 705 F.2d 1326, 1330 (CAFC 1983) (quoting *International Glass Co. v. United States*, 408 F.2d 395, 403 (Ct. Cl. 1968)), as "The Courts have consistently held that an invention, though completed, is deemed abandoned, suppressed or concealed if, within a reasonable amount of time after completion, no steps are taken to make the invention publicly known. Thus failure to file a patent application; to describe the invention in a publicly disseminated document; or to use the invention publicly, have been held to constitute abandonment, suppression, or concealment." In the facts of *Correge*, 7 months after an actual reduction to practice, the invention was publicly disclosed and a patent application was filed 8 months after the public disclosure. The Court found that this was not an unreasonable delay. The Court also held that waiting up to 1 year after the public disclosure to file the patent application is not unreasonable because this is what is allowed under §102(b). Since waiting longer than 1 year after a public disclosure to file a patent application results in a §102(b) bar, the unreasonable delay can occur only during the period after the reduction to practice and before the public use or the filing of the patent application. In this regard, and more appropriate to the facts of our present hypothetical, the actual time between the reduction of practice and filing an application is not as important as the perceived intent or purpose of the delay in filing.
[24]*Townsend v. Smith*, 36 F.2d 292, 295 (CCPA 1930).

antedated by proving prior inventorship—the patent application in question must have been filed before the events listed under these sections. Furthermore, these absolute bars apply both to an applicant's own work as well as the work of others.

3.3.a §102 (b)

A key element of §102(a) prior art is that it relates to the date of the invention by the applicant. For a §102(a) reference date to effectively block the applicant's work, the §102(a) event must have occurred before the date of the invention by the applicant. However, the provisions of §102(b) provide an important limit to the first-to-invent rule. The section of prior art defined in §102(b) is referred to as the *absolute bar* because it limits how far one can go in antedating a prior art reference to prove earlier invention in the United States. Specifically, a patent for an invention cannot be obtained in the United States if the invention was patented or described in a printed publication (or in public use or on sale in this country), *more than 1 year before the date of application for patent in the United States.* Thus, even if the applicant made the invention before the §102(b) event, she will be barred from being granted a patent if she does not file the patent application within 1 year of that 102(b) event. Note that §102(b) does not contain the phrase *by others* as in §102(a). Thus if an applicant publishes her own invention more than 1 year before filing the application for patent, she will be barred by §102(b) from being eligible to be granted a patent in the United States. Likewise, if a §102(b) event by another occurs, a patent applicant in the United States will not be able to antedate the §102(b) event.

In addition to the patent and publication bars, §102(b) also contains "public use" and "on sale" bars, if the activity occurred in this country but not if the activity occurred in a foreign country, but it does not precisely define what these two terms mean. Separating public from nonpublic use or sale of an invention is not always a trivial or easy endeavor. Often, scenarios for which this determination must be made relate to whether the inventor's own use of her invention constituted a public use and thus whether the 1-year clock started ticking. Unfortunately, there is no simple algorithm for determining whether a use is public or not, but there have been many cases that provide some general guidance. Ultimately, whether a use is public or not depends on the weighing of the many facts related to the alleged public use.

Some facts weighing against a finding of public use include the presence of a confidentiality agreement or obligation of secrecy.[25] Along these lines, a public use has been said to occur where an inventor has allowed another person to use the invention "without limitation, restriction or obligation of secrecy to the inventor."[26] The evidentiary purpose of the confidentiality agreement or order of secrecy is to demonstrate the level of control the inventor demonstrated over the invention as well as the purpose of the use. In this regard, the nature and location of the invention's use can also affect the outcome of the determination. Where the inventor's use is limited

---

[25]MPEP 2303.03(a) citing *Moleculon Research Corp. v. CBS, Inc.* 793 F.2d 1261, 1265 (CAFC 1986).
[26]MPEP 2133.03(a) quoting *In re Smith*, 714 F.2d 1127, 1134 (CAFC 1983).

to places where she had a "reasonable expectation of privacy" and the use was for the inventor's own enjoyment, that use would not be likely to constitute a public use.[27]

An important aspect of public use of an invention is that it does not require that the invention be in open display to be in public use; the only requirement is that the invention is being used in its "natural and intended way."[28] As a result, the sale or public use of a chemical mixture such as a paint, coating, or medicine would likely still be a public use even though the components in the mixture are not readily discernable to the public. In a similar vein, sale of a good made by a certain process is a public sale of the process and will toll the §102(b) time clock even though the process used to make the good could not be discerned from the good itself.[29]

In certain instances, it is necessary for an inventor to test an invention before filing the patent application, and that testing may require the use to be public. Some examples relevant to the chemical industry might include, for example, an additive to tar, asphalt, or concrete that is put in to improve the durability of the substance. In such a case, it might be necessary to test the substance under real-world conditions by actually using it to make a floor, road, or wall that is used according to its normal purpose to provide an indication of how well the substance will work in comparison with several different formulations over a long period of time to see which one works best. Likewise, to patent a compound as a drug, some indication of utility is required. Usually, this can be achieved by demonstrating the compound has activity in in vitro assays or animal models (i.e., the USPTO has very different criteria for demonstrating utility than does the Food and Drug Administration). However, in some instances, the establishment of a compound's utility might be possible only in a public human clinical trial. To assist in the determination of whether a use or sale is of a public nature for purposes of §102(b) or whether the use or sale is experimental, the courts have provided the following list of factors: (1) the necessity of the public testing, (2) the amount of control over the experiment retained by the inventor, (3) the nature of the invention, (4) the length of the test period, (5) whether payment was made, (6) whether there was a secrecy obligation, (7) whether records of the experiment were kept, (8) who conducted the experiment, (9), the degree of commercial exploitation during testing, (10) whether the invention reasonably requires evaluation under actual conditions of use, (11), whether testing was systematically performed, (12) whether the inventor continually monitored the invention during testing, and (13) the nature of contacts with potential customers.[30]

The three primary points of concern with the *on sale* aspect of §102(b) are whether a sale or offer for sale has actually occurred, whether the invention was ready for patenting, and whether the sale was primarily commercial or experimental.[31] Whether a sale or offer for sale has actually occurred can be determined by contract law.

---

[27]MPEP 2133.03(a) citing *Moleculon Research corp. v. CBS, Inc.*, 793 F.2d 1261, 1265 (CAFC 1986).

[28]MPEP 2133.03(a) citing *In re Blaisdell*, 242 F.2d 779, 783 (CCPA 1957).

[29]*D.L. Auld Co. v. Chroma Graphics Corp.*, 714 F.22d 1144, 1147–1148 (CAFC 1983).

[30]MPEP 2133.03(e)(4) citing *Allen Eng'g Corp. v. Bartell Industries, Inc.*, 299 F.3d 1336, 1353 (CAFC 2002).

[31]*Pfaff v. Wells Electronics, Inc.*, 525 US 55, 67 (1998).

Essentially, an offer can be deemed to have been made where an acceptance, whether actually made or not, would be legally binding on the offeror.[32] It is important to note that the sale referred to in §102(b) is a sale or offer for sale of the invention itself and not the rights in the invention. For example, a license granting marketing rights to a compound is not the same thing for purposes of §102(b) as a sale of the compound itself—the first activity would not toll the statute whereas the second will.[33] Unlike the public use portion of the §102(b) statute, the courts have found that a sale does not have to be public to constitute a sale.[34] In a similar fashion to what we saw for the public use portion of §102(b) where the invention does not need to be revealed to be in public use, the sale of the item does not require that all of the aspects of that item be understood or known by either of the parties to the sale; the sale of the item is sufficient.[35]

In reference to the second point, an item can be ready for patenting if it has already been actually reduced to practice. However, an actual reduction to practice is not necessary if the invention being sold is capable of being reduced to practice without extraordinary effort or skill—for example, if the inventor has prepared descriptions or drawings of such specificity to enable one of ordinary skill to practice the invention.[36]

Finally, as we saw previously for the public use bar, an experimental use can also negate an "on sale" §102(b) bar since there are instances when even a sale can be considered primarily experimental in nature. Obviously, the sale of the invention is not experimental per se, but the purpose of the transfer of the invention might be to allow use and observation of the invention to test its claimed features in various real-world settings. The question really turns on whether the purpose of the sale was primarily commercial or primarily experimental. Purpose implies intent on the part of the inventor, but one should appreciate that intent is inferred from the objective factual indices surrounding the sale or attempted sale.[37] The presence of sales lists, advertisements, demonstrations, etc. are all typical of commercial sales activities rather than experimental sales activity.[38]

In effect, §102(b) serves to balance the interest of the public with those of the inventor. Once a nonpatented invention is in the public domain, it is not unreasonable for the public to begin to take notice of that invention and begin to use that invention for its own purpose. If the invention is not patented and there is no record of a patent being filed on that invention, there would be no way for the public to be notified that the use of the invention could eventually become an infringement. An inventor might even try to take advantage of such a situation by publishing her own invention and waiting for a large segment of the public to begin to use and rely on the invention.

[32] MPEP 2133.03(b) citing *Group One, Ltd. V. Hallmark Cards, Inc.*, 254 F.3d 1041, 1048 (CAFC 2001).
[33] *In re Kollar*, 286 F.3d 1326, 1330 n.3 (CAFC 2002).
[34] *Hobbs v. United States*, 451 F.2d 849 (5th Cir. 1971).
[35] In *Abbott Laboratories v. Geneva Pharmaceuticals, Inc.* 182 F.3d 1315, 1319 (CAFC 1999), the Court found that the sale of a particular form of an anhydrous pharmaceutical barred a later patenting even though the parties to the purchase did not know about the form of the drug.
[36] *Pfaff v. Wells Electronics, Inc.*, 525 US 55, 68 (1998).
[37] Can you think of a good reason why intent might need to be inferred?
[38] See MPEP 2133.03(e)(2).

Then, like a hunter springing the trap, the inventor could file and obtain patent on the invention. The public would then have to deal with the effect of the patent on their continuing use of the now-patented technology. Section 102(b) limits this possibility by allowing the inventor only a 1-year lead time between the events listed in §102(b) and the ability to file for a patent. Thus a reasonable balance is achieved between allowing an inventor time to perfect his or her invention before filing an application for invention and the public interest in being able to rely on information available in the public domain.

3.3.b §102(c)

Section 102(c) is a little used provision but is still included in this section since a finding of abandonment under §102(c) acts as an absolute bar to a later filed patent application. Abandonment requires intent, either actual or inferred, on the part of the inventor to dedicate the invention to the public. Mere delay in the filing of a patent application, absent more, is not sufficient to impute the motive to abandon the invention onto the inventor.[39]

3.3.c §102(d)

Section 102(d) works as an absolute bar when a foreign patent application is filed more than 1 year before the U.S. filing *and* when that foreign application issues into a patent or inventor's certificate before the patent application claiming the same invention is filed in the United States.[40] Both conditions must be met. It is important, though, that it does not matter if the foreign patent *claims* the same subject matter as the subsequent U.S. patent application, rather the foreign patent needs only to contain description sufficient in its specification to fully support the material claimed in the later-filed U.S. patent application. For example, if an applicant or inventor filed a patent application in another country containing the description of a compound, a process for making the compound and a method of using that compound but claims to only the compound itself were issued in the foreign patent, the patent could still be good as a prior art reference under §102(d) against a later filed U.S. patent application that claimed only the process for making the compound since the earlier foreign patent application provided description that could have allowed a claim to the same subject material.[41] Section 102(d) is somewhat unique in that the prior art under §102(d) qualifies only if the invention was assigned to the same party as the later filed U.S. patent application.

---

[39] MPEP 2133.03(e)(7) citing *Moore v. United States*, 194 USPQ 423, 428 (Ct.Cl. 1977). Also see *Petersen v. Fee Int'l, Ltd.*, 381 F. Supp. 1071 (W.D. Okla. 1974).

[40] The term *patent* or *inventor's certificate* refers to an award of a right that entitles its holder the right to exclude another from using the claimed invention in that foreign jurisdiction. The name of the right is not as significant as the nature of the right itself.

[41] See *In re Kathawala* 9 F.3d 942 (CAFC 1993), where an earlier filed Spanish patent application that issued into a patent that claimed a process for making a compound but did not claim the compound. The later filed U.S. application claimed the compound but those claims were barred by the earlier filed Spanish application even though the Spanish application did not claim the compound. See also MPEP 2135.01.

The rationale for this section of the patent laws appears to be to ensure that inventors filing for a patent in foreign countries also promptly file for a patent in the United States. Rejections of a patent application under §102(d) are uncommon.

3.4 SECTION 102 REFERENCES IN SUPPORT OF OBVIOUSNESS REJECTIONS

As a general matter, the 102 sections listed in this chapter may be applied to show that a later filed patent application lacks novelty, is obvious, or both. A rejection of patentability under one or more of the 102 sections alleging that the claimed invention is not *novel* is usually referred to simply as a §102 rejection, referenced to whatever sections the prior art belongs to. For example, if a prior art nonpatent publication by another more than 1 year before the application in question was alleged to defeat the novelty of the applicant's claimed invention then a rejection under §102(b) could be brought; this would be called §102(b) rejection for lack of novelty.[42] For a rejection to be made solely upon a §102 reference, a *single* prior art reference must disclose every element of the patent application claim. If a combination of *multiple* prior art references is needed to disclose the entirety of the invention, this cannot be the basis for a novelty rejection. However, multiple references may be used to reject a patent application claim as being "obvious"—that is, the invention may be technically novel, but still not patentable.

Rejections alleging nonpatentability due to obviousness are referred to as §103 rejections and will be discussed extensively in Chapter 8. The general rule is that rejections made under §103 will rely on the same prior art references as do §102 novelty rejections, with a couple of narrow but important exceptions. Section 103 is composed of sections (a), (b), and (c) but most obviousness rejections pertaining to chemical cases are made under section (a), which states the statutory definition of obviousness.[43] In regard to our discussion of prior art, the narrow and important exceptions to this prior art rule fall under section (c)(1) and (c)(2):

U.S.C. 103 (c) (1):

> Subject matter developed by another person, which qualifies as prior art only under one or more of subsections (e), (f), and (g) of section 102 of this title, shall not preclude patentability under this section where the subject matter and the claimed invention

---

[42]Technically, this also qualifies as a §102(a) reference, but citing the reference under both sections would not add anything since the §102(b), rejection cannot be antedated even though the §102(a) reference could. However, citing a reference under §102(a) instead of §102(b), where either one qualifies, could make a substantive difference since the §102(a) reference might be antedated; whereas the §102(b) reference could not. Naturally the §102(b) section is applied in preference to section §102(a) when it is proper to do so.

[43]§103(a) provides: "A patent may not be obtained though the invention is not identically disclosed or described as set forth in section 102 of this title, if the differences between the subject matter sought to be patented and the prior art are such that the subject matter as a whole would have been obvious at the time the invention was made to a person having ordinary skill in the art to which such subject matter pertains. Patentability shall not be negatived by the matter in which the invention was made."

were, at the time the invention was made, owned by the same person or subject to an obligation of assignment to the same person;

U.S.C. 103 (c) (2):

For purposes of this subsection, subject matter developed by another person and a claimed invention shall be deemed to have been owned by the same person or subject to an obligation of assignment to the same person if—

(A) the claimed invention was made by or on behalf of parties to a joint research agreement that was in effect on or before the date the claimed invention was made;

(B) the claimed invention was made as a result of activities undertaken within the scope of the joint research agreement; and

(C) the application for patent for the claimed invention discloses or is amended to disclose the names of the parties to the joint research agreement.

Under the §103(c) subsections, we find that prior art defined by §102(e), (f), and (g) *will not* be applied against an organization or the members of a research collaboration regarding their own (102(e), 102(f), and 102(g)) prior art. Of particular importance regarding §103 (c)(1) and (c)(2) is that an organization (or collaboration under a joint research agreement) that files a patent application and then files another patent application before that first application being published will not have the earlier application used against them for obviousness when the patentability of the invention in the later filed application is determined. This is an important consideration since very often an organization, or one or more collaborators working together under a research agreement, will discover an invention and quickly file a patent application to cover what they have done while they continue to work in the same research area to further improve on their original filed invention. If they make such an improvement and file the patent application before the time the initial application publishes, even if the improvement invention is an obvious variation of what is disclosed in the earlier filed application, the earlier application can be ignored for the purpose of determining obviousness. This means that once a patent application is filed, the inventors (and their attorney) might wish to pay attention to the expected publication date to be certain to file, before that date, any additional applications that might be obvious from the disclosure of the first application and any other applicable art the earlier filed application might be combinable with. Once their patent application publishes, it is not only prior art under §102(e) traceable to the application's filing date but also might be available under section §102(a).[44]

---

[44]If the earlier patent application was to the same inventive entity then §102(a) and §102(e) would not apply since those sections apply to the work of another. However, exclusions under §103(c)/102(e) draw their usefulness from the fact that very often in an organization or within a group working together under a research collaboration, different or at least overlapping groups of inventors may work on related aspects of a project such that multiple patent applications may be filed over a period of time where each patent application lists a different inventive entity. In addition, the publication date of the first application is also important since it cannot be antedated in the patent world outside the United States.

3.5 DOUBLE PATENTING

In the section just covered, we learned that prior art rejections for *obviousness* under §103, citing categories of prior art under §§102(e), 102(f), or 102(g) can be removed when there was a common assignee for the prior art reference and the later filed patent application at the time the invention in question was made. Similarly, we discussed that certain other types of prior art can be removed under special circumstances such as when the inventive entity is the same (§102(e)). In cases when the prior art is a patent, there is an additional consideration that needs to be addressed briefly. So far, when we have reviewed patents or patent applications as prior art, we distinctly avoided differentiating the patent as a whole from the claims, pointing out that any section of the qualifying patent or application could be prior art. Likewise, we did not distinguish the specification from the claims when we discussed how those references might be removed, whether because they were by the same inventive entity (e.g., §102(e)) or because they were subject to a common assignee (§103(c)/102(e))—in either case the whole reference is removed all together. However, nothing explained so far is meant to suggest that the same invention can be patented more than once, it cannot. This means that the claims from a later filed patent application cannot issue if they contain the identical subject matter as the claims of another patent.

Let's use a hypothetical example to demonstrate the concept. A patent application filed on June 1, 2004, is published on January 3, 2006, by the PCT as WO06/0056791 and lists Thompson and Hickman as the two sole inventors. The patent application is published in English and designates the United States. The patent discloses and claims three molecules A, B, and C; a method of making the compounds; and the use of the three compounds to treat certain tropical diseases, including malaria. On December 7, approximately 11 months after the first patent application was published, Thompson and Hickman, again listed as the two sole inventors, file a second patent application that once again discloses and claims compound B together with additional compounds D and E. This patent also discloses methods for making and using the claimed compounds for treating tropical diseases, including malaria. Since the earlier, published patent application was to the exact same inventive entity as the later filed patent application, it does not qualify as prior art under §102(e) (nor §102(a) after it publishes) since the second application is not by another. Moreover, the first patent application was not published more than 1 year before the second application, so it is also not prior art under §102(b). For purposes of patentability then, it appears that Thompson and Hickman are home free since their earlier filed patent application is not prior art against their later filed patent application.

Does this mean they can patent compound B in their earlier case as well as their later case? The answer is no and the reason follows. Absent special circumstances, a patent is enforceable 20 years from the date of filing the application. So let's work a timeline for the two patents containing claims to compound B should they both be allowed to issue. The first patent application was filed on June 1, 2004, so a claim to compound B issuing from that patent would expire on June 1, 2024. The second patent application for compound B was filed on December 7, 2006, so a claim to compound B issuing from that patent would expire on December 7, 2026. The net effect is that Thompson and Hickman will have extended their patent rights beyond the normal 20 years to 22 years, 5 months. In terms of patents and public policy, this means that they will have increased the duration of their right without providing any additional benefit to the public, not a very sensible result.[45] As a practical matter, allowing such second patents to issue would also make for a cumbersome and inefficient process because each inventor would have plenty of motivation to file two applications for every invention to capture the extra time. If a social policy requiring more than 20 years for a patent was desired, then there would be more practical ways to accomplish that policy, such as legislatively making the time of exclusivity longer than 20 years.

So the law makes sense because it so happens that it is not legally permissible to patent the same invention twice. Section 35 U.S.C. 101, which sets the framework for United States patent law states:

> Whoever invents or discovers any new and useful process, machine, manufacture, or composition of matter, or any new and useful improvement thereof, may obtain *a* patent therefor, subject to the conditions and requirements of this title (emphasis added).

The use of the phrase *a patent* has been literally interpreted to mean "one patent" when linked together with the modifier *new*. Thus a new composition is afforded one patent. Recalling the earlier hypothetical, the issuance of more than one patent to the invention consisting of compound B would conflict directly with the language of §101; patenting compound B twice would not be allowed.

In the example set above, we explicitly defined the subject matter in the claims of the later filed patent application to be the same or substantially the same as the claims of the earlier filed patent application. If there are any substantive differences to the claims, then the statutory double patenting situation explained above is not relevant. A quick test of whether two claims are the same or substantially the same in this context is whether it is possible to literally infringe one of the claims without literally infringing the other. If yes, then the claims are not the same. For example, a claim to a chemical process that requires mixing A and B together and heating to a temperature of between 50 and 100 degrees for 3 hours is not the same as one that requires mixing A and B together and heating to a temperature of between 60 and 100 degrees for

---

[45] If double patenting were allowed, the extra time granted could be much greater under a less common but nevertheless very plausible scenario. If Thompson and Hickman filed their first case in the United States with a request to not publish the patent application (this can be done by agreeing to not file a patent application in another country that requires publication of applications), then the patent application would not publish until it issued. As a result, the first patent application would not be prior art under §102(b) against Thompson and Hickman until 1 year after it issued. Normal patent prosecution in the United States can take many years until an application issues into a patent and if the applicants were not especially motivated to move the process along, maybe even 5 years or longer. So under this alternate scenario, Thompson and Hickman would wait until their earlier application issues and then wait another 11 months, after which they would file their second patent application—this would give them an additional 5 years and 11 months of patent life for the same compound, extending their patent term to a total of almost 26 years from the filing of the first patent application.

3 hours; it is possible to practice the first claim without infringing the second simply mixing A and B together by heating between 50 and 59 degrees for 3 hours.

3.6   OBVIOUSNESS-TYPE DOUBLE PATENTING

It makes sense why the *same* invention cannot be patented twice, but what about obvious variants of the invention? As we just learned, the same invention cannot be patented twice because it leads to an extension of a patent monopoly that does not provide the public with any additional benefit. More important, we saw how the prohibition against double-patenting has statutory legitimacy since it traces its lineage to 35 U.S.C. 101. However, §101 does not provide any prohibition against multiple patents to inventions that are obvious variants of each other. Nevertheless, a judicially created doctrine of "obviousness-type" double patenting has evolved that governs these situations. The doctrine originates from the recognition that allowing obvious variants of the same patentable subject matter would effectively extend patent life for inventions that are different but still not patentably distinct. As a matter of public policy, allowing such extensions for obvious variations of previous inventions would be counterproductive to the purposes of patent law, which is to reward true innovation but not obvious changes to previous inventions. In statutory double patenting, the solution is that only one patent can issue to the same invention, the second is not allowed. For obviousness-type double patenting, the solution is not quite as severe to the patentee. In obviousness-type double patenting, an obvious variant of an earlier-issued patent can in some cases still be patented but with the catch that the assignee of the two patents must agree that the term of the second patent is limited in terms of its length of enforceability to the same term as the earlier-term patent. This agreement is known as a *terminal disclaimer* and works as a compromise between an unreasonable extension of patent life for an obvious variant while still allowing the party holding the patents the ability to capture the full scope of the claimed inventions. The terminal disclaimer also prohibits separate assignments of the two or more patents that are terminally disclaimed. The prohibition against the separate assignment serves to prevent an assignee from splitting a group of terminally disclaimed patents and assigning them to different parties. If such a splitting of assignments was allowed, then one infringer could have multiple parties pursuing them for infringing what in effect is one invention. While this would not extend the term of the patent, it could multiply the value of the patent by separate assignments despite the fact that there are not multiple distinct inventions involved.

To better envision an obviousness-type double patenting scenario, let's briefly rework the facts from the hypothetical of our statutory double patenting example just stated. In that example, we explained that attempting to claim the exact same compound twice (compound B) was not permissible, even though the earlier patent application was not §102 prior art. In this earlier scenario, we stated that the later patent application contained additional compounds that were not disclosed in the earlier patent application—compounds D and E. Since we have not yet discussed the analytical framework for obviousness, let's assume that compound D is an obvious variant of the earlier-claimed compound B. Nevertheless, the inventors wish to claim the compound in the later patent application, but if they do so, they will receive a provisional obviousness-type rejection. If and when the earlier patent application issues as a patent, the provisional rejection will be converted to an actual rejection in the later patent application. If there are no other impediments to the later patent application issuing, the applicant may acquiesce to a terminal disclaimer and have the later patent to compound D issued. The terminally disclaimed patent will list on the cover page that it is subject to a terminal disclaimer and list the earlier patent to which its term is tied. Despite the term limitation on compound B, the patentees will still have an enforceable right specifically with respect to that compound that they otherwise would not have had.

3.7   PRIOR ART HYPOTHETICAL EXAMPLE 1

To clarify our understanding of prior art, let's consider a couple of hypothetical examples. The first hypothetical tells us the story of a chemist working for the Acme Major Chemical Corporation who conceived of a compound on May 6, 2005. In addition to the conception of the compound, he drew up a synthetic scheme that was likely to be useful for producing that compound; he recorded his conception and proposed synthesis in his lab notebook and had it witnessed and signed, all on May 6, 2005. He believed the compound would have good activity as a serotonin-reuptake inhibitor and therefore be useful as an antidepressant. He immediately ordered the starting materials, and when they arrived, began to prepare the compound. He completed the compound synthesis on June 3, 2005, and submitted it for internal testing to see if it was active. Fortunately, the compound demonstrated good activity in both binding and functional cell assays believed to be predictive of in vivo antidepressant activity; those tests were completed on June 17, 2005. Encouraged by these early results, a number of in vivo studies were planned to take place over the coming year or so; the company had an orderly progression of in vivo assays, starting with certain rat models and proceeding to nonhuman primates. If the compounds were deemed desirable after all of the in vivo tests, a broad counterscreen panel was employed, and if everything looked acceptable, a patent application was drafted and filed. In June 2006, all of the in vivo results were analyzed, and it was decided to advance the compound. On July 15, 2006, the counterscreen results were completed and, interestingly, the compound demonstrated surprisingly potent anticholinesterase activity. When the chemist saw the results, he realized that the compound could be useful for cognition enhancement in the early stages of Alzheimer disease. He thought the serotonin-reuptake inhibition might be supportive of that indication as well since many Alzheimer patients suffer from depression. On September 29, 2006, his company filed a patent application in the U.S. Patent office containing the three claims shown in Figure 3.1.

There are two questions we will explore for this scenario. First, on what date (or dates) was the chemist's reduction to practice and were they actual or constructive reductions to practice? Second, what date(s) of invention can the chemist establish for each of the claims?

We claim:

1. A compound having the structural formula A:

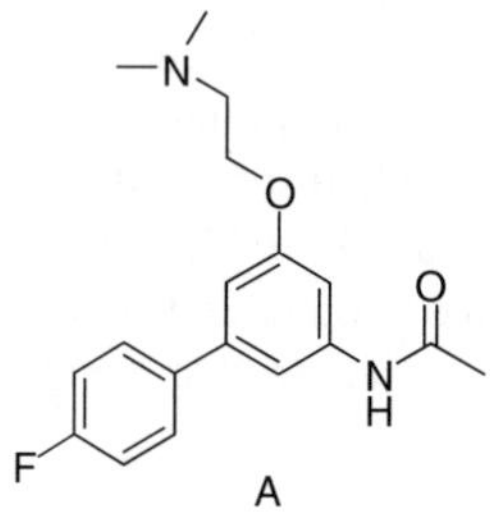

2. A method of treating depression comprising the administration of the compound according to claim 1 to a person in need thereof.

3. A method of enhancing cognition comprising the administration of the compound according to claim 1 to a person in need thereof.

FIGURE 3.1 Claims to compound A and methods of use thereof.

To determine on what date the reduction to practice took place, it first helps to understand that there are actually three inventions and not one. The first invention is the compound itself, the second is the method of using the compound for treating depression, and the third is a method of using the compound as a cognition enhancer. Let's take them each in turn. We know that the compound was conceived on May 6, 2005, and its synthesis was complete on June 3, 2005. Earlier we learned that an actual reduction to practice occurs when an invention is shown to work for its intended purpose. In general, the reduction to practice for a compound is complete when the compound is identified, a method of making it is available (or the compound is made) and a credible utility established. As we have learned, the utility of a pharmacological compound can be established through testing of the compound in in vitro assays that are at least reasonably predictive of in vivo activity. In the present case, the reduction to practice most likely took place upon the in vitro testing of the compound that established activity predictive of anti-depressant activity.[46] At this point, the

---

[46]One could make the argument that the method of treating depression as claimed in the patent is not directed at cells but rather humans; therefore, the reduction to practice would take place when the actual treatment of humans began. However, a reduction to practice requires testing only sufficient to show that the invention will likely work for its intended purpose. In this instant, the in vitro testing was believed to be reasonably related to the desired human pharmacology, and so it would be sufficient to establish an actual

compound's utility was established and recognized and so a reduction to practice for the compound took place on June 17, 2005. With respect to the method of treating depression, it should be fairly evident that the reduction to practice also took place on June 17, 2005. On that date, the method was credibly established.

The second method claim was evidenced by an actual reduction to practice when the testing to establish the anticholinesterase utility took place: July 15, 2006.

We now turn to the second question, What was the date of invention for each claim? The date of invention is the date of the reduction to practice, unless there was a prior conception followed by a period of uninterrupted due diligence from the period of conception to the time of the reduction to practice. If there was a prior conception but the due diligence linking the conception to the reduction to practice was not continuous, the invention date would be the period when the due diligence began that was continuous up to the time of the reduction to practice. In order for the conception to be complete, the entire claimed invention must be envisioned such that no additional inventive faculty needs to be invested. In this hypothetical, we are told that there were three claimed inventions: the compound, the method of using the compound for the treatment of depression, and the method of using the compound for the enhancement of cognition. It appears from the facts given that the compound was conceived on May 6, 2005, and that the conception was complete since the inventor knew the structure he wanted to make, the process for making it, and a practical use for the compound. Furthermore, he diligently pursued the synthesis since he immediately ordered the starting materials for the synthesis and when the starting materials arrived he started to work making the compound right away and actually completed the synthesis fairly quickly. Thus the date of invention of the compound can be traced back to the conception date, May 6, 2005. In the patent application more than just a compound was claimed, two methods of using the compound were claimed as well. Those two claims require both the specific compound and the use of that compound in the claimed method. The method of treating depression was part of the chemist's original idea of how to use the compound, and the compound did show activity in assays that were believed to be predictive of the intended utility. To show diligence from the time of conception, the compound synthesis will have to been diligently pursued as well as its testing in assays to predict whether it will work for the intended method of treatment. Returning to the facts of the hypothetical, we see that the compound was submitted for testing right after the synthesis was completed, and the testing in the two in vitro assays believed to be predictive of the desired activity was completed in 2 weeks from the time the compound was submitted for testing, not an unreasonable period of time, so the invention date for the treatment of depression is also the conception date of May 6, 2005.

This leaves the method of using the compound for the enhancement of cognition in Alzheimer disease. As we discussed before, a conception date can be the date of

---

reduction to practice for the method of treating depression. In multiple cases, the Court has founding in vitro testing to be sufficient to establish a utility for a compound. See in particular *Cross v. Izuka*, 753 F.2d (Fed. Cir. 1985) and *In re Brana*, 51 F.3d (Fed. Circ. 1995) along with further discussion in the MPEP §2107.01.

invention where the conception is complete; all of the elements in the claim must be appreciated or recognized for the conception to be complete. In the earlier portion of this problem, we already learned that the compound was conceived on May 6, 2005. However, not all of the elements of claim 3 were recognized or appreciated at that time because there was no expectation at the time that the compound was conceived that the compound would be useful in a method for enhancing cognition. The conception of claim 3 was not complete at the same time as it was for claims 1 and 2. Rather, it was when the compound was tested for antiholinesterase activity that the recognition of its special utility was appreciated and that occurred on July 15, 2006. Unlike the method of claim 2, the method to treat cognition was not appreciated or recognized until the compound actually demonstrated that unexpected activity. As a result, we cannot trace an earlier conception date to that method, but rather, the invention date was the date that the reduction to practice and recognition of the unexpected utility occurred, July 15, 2006.

3.8 HYPOTHETICAL EXAMPLE 2

Since we now know the invention dates for the three claims, let's extend the preceding hypothetical. In particular, we are going to use the various dates established for the inventions and the patent application filing to figure out how certain types of prior art might affect the ability of our chemist to patent his three inventions. To fully cover this topic, we will consider seven different scenarios:

1. On October 1, 2005, a paper written in a Japanese chemical journal (written in the Japanese language) was published, describing the synthesis and structural identification of the exact compound A described in claim 1. The publication did not detail or suggest any potential biological activity.

2. The same facts as part 1) except the journal article's publication date was September 24, 2005.

3. On July 4, 2006, a paper written in an Australian chemical journal was published, describing the synthesis and structural identification of the exact compound described in claim 1. The publication also mentioned that the compound demonstrated anticholinesterase activity, and its potential use for cognition enhancement was described in excruciating detail. No other biological activity was disclosed.

4. On May 11, 2005, Dr. Klaus, a German inventor working in Germany, conceived the same compound described in claim 1. He diligently worked on making the compound until June 3, 2005, when he completed the synthesis. On June 11, 2005, he had submitted the compound for testing as a serotonin-reuptake inhibitor, and on June 15, 2005, he received notice that the compound was active. On July 17, 2005, Klaus filed a patent application in the United States with a description of the compound and its preparation as well as a description of its use as an antidepressant, including the supporting data for that use. Claims

were included that were directed to the compound as well as a method of using the compound as an antidepressant.

5. On January 4, 2005, a chemist in Germany, Dr. Warner, conceives the exact same compound as the compound of claim 1 of our hypothetical patent application. The conception was duly recorded and witnessed, and it was also noted that the compound should be good for treating depression as well as cognition enhancement. The compound was soon thereafter synthesized and tested for activity in both in vitro and in vivo models believed to be predictive of treating depression and cognition enhancement. Analysis indicating that the compound was very likely to work for the intended indications was completed by February 17, 2005. On March 7, 2005, a patent application was filed in Germany as a German national patent application. A PCT patent application was filed on March 1, 2006, which claimed priority to the earlier national filing. The PCT patent application was published (in the English language) on September 15, 2006, and designated a number of countries, including the United States. The PCT application was subsequently abandoned and a national application was never entered in the United States. A German patent to the claimed subject matter eventually issued on April 1, 2007.

6. Soon after conceiving the invention, our ever-optimistic Acme chemist sends an abstract to the American Chemical Society for an upcoming meeting. The abstract is accepted online immediately and posted on the Internet the same day. The abstract describes the proposed preparation of the compound together with a description of its possible utility as an antidepressant. The chemist lists his name and two of his lab associates who will help him in making and testing the compound. The date the abstract is posted on the Internet is June 1, 2005.

7. Apparently our chemist was unaware of this but on April 1, 2006, a different chemist in his same organization but at a different research site had submitted a patent application in the United States describing and claiming an obvious variation of his compound together with methods of using the compound to treat bacterial infections.

To discuss the impact of these potential prior art references, it will be helpful to have the most relevant dates for our chemist's invention conveniently listed as in Table 3.1.

TABLE 3.1   The Relevant Dates Needed to Determine the Impact of Prior Art References on the Patent Application Filing

Item	Relevant Date
Invention date of claim 1	May 6, 2005
Invention date of claim 2	May 6, 2005
Invention date of claim 3	July 15, 2006
Filing date of patent application	September 29, 2006

The facts for part (1) of this hypothetical indicate that the Japanese publication occurred on October 1, 2005, 2 days day short of 1 year before our chemist filed his patent application (what a coincidence!). Does it matter that the prior art publication is in Japanese? The answer is no, the language of publication is of potential consequence only for §102(e), but this is not a patent or patent application so §102(e) does not apply. In this instant, we simply have a published paper appearing less than 1 year before our chemist's patent filing. Because it is less than 1 year, §102(b) will not apply. Glancing through the sections, we see that only §102(a) could possibly apply. Notice that §102(a) applies to a publication anywhere and does not include a language limitation. Fortunately for the chemist, §102(a) is not an absolute bar to patentability, rather he still has a chance to establish that his invention occurred before the publication date of the Japanese paper. To do this successfully, he will need to submit an affidavit (preferably with the assistance of counsel) establishing that he invented the compound before the publication date of the paper, which he should be able to do given the documentation of his conception date and the fact that he was diligent from conception through the reduction to practice.[47] In regard to the claimed methods of use, the inventor will not need to establish his invention date because those method of use claims are probably not affected because the prior art makes no reference to any method of using the compound.

---

[47] Affidavits that testify to an invention date can be filed according to the instructions described in MPEP Chapter R Patent Rules §1.131 and is reproduced in its entirety below:

§ 1.131 Affidavit or declaration of prior invention.

(a) When any claim of an application or a patent under reexamination is rejected, the inventor of the subject matter of the rejected claim, the owner of the patent under reexamination, or the party qualified under §§ 1.42, 1.43, or 1.47, may submit an appropriate oath or declaration to establish invention of the subject matter of the rejected claim prior to the effective date of the reference or activity on which the rejection is based. The effective date of a U.S. patent, U.S. patent application publication, or international application publication under PCT Article 21(2) is the earlier of its publication date or date that it is effective as a reference under 35 U.S.C. 102(e) Prior invention may not be established under this section in any country other than the United States, a NAFTA country, or a WTO member country. Prior invention may not be established under this section before December 8, 1993, in a NAFTA country other than the United States, or before January 1, 1996, in a WTO member country other than a NAFTA country. Prior invention may not be established under this section if either:

(1) The rejection is based upon a U.S. patent or U.S. patent application publication of a pending or patented application to another or others which claims the same patentable invention as defined in § 41.203(a) of this title, in which case an applicant may suggest an interference pursuant to § 41.202(a) of this title; or

(2) The rejection is based upon a statutory bar.

(b) The showing of facts shall be such, in character and weight, as to establish reduction to practice prior to the effective date of the reference, or conception of the invention prior to the effective date of the reference coupled with due diligence from prior to said date to a subsequent reduction to practice or to the filing of the application. Original exhibits of drawings or records, or photocopies thereof, must accompany and form part of the affidavit or declaration or their absence must be satisfactorily explained.

In part (2) we have the same facts as part (1) except that the publication date has been moved one week earlier to September 24, 2005. What a difference a week makes! The reference has a publication date more than 1 year before the patent application filing date and is now a §102(b) reference instead of a §102(a) reference. This means there is no possibility of proving prior invention, and the reference is admissible for all that it discloses. Since the reference provides an enabling disclosure for the preparation of the compound, claim 1 of the chemist's patent application is no longer novel in view of this §102(b) reference and thus cannot be patented.[48] The novelty of claims 2 and 3 are not affected by the reference because it does not contain any description or teaching regarding the use of the compounds for treating depression

---

[48] As an aside, it is common for a patent applicant who is describing a compound with utility to claim not only the compound itself but also the method of using the compound for the desired utility and anything that might be useful together with the compound in a combination claim. For example, our chemist discovered what he thought was a new compound and a new method of use. His ultimate goal for the compound will be to put it into a pill, capsule, liquid vehicle, etc. As a practical matter, almost any medicine will be administered not as the compound itself but in combination with one or more pharmaceutical excipients (additives) to improve the stability, formulatability, bioavailability, appearance, taste, etc. The drug together with all the excipients makes up the pharmaceutical composition. One of the goals in patent protection is to get the broadest (and narrowest) claim coverage possible. The way that this is achieved depends on the particular area or subject matter. In the case of novel compounds being used as medicines, one of the broadest possible protections is obtained by a claim comprising the compound (or group of compounds) itself. This means no additional limitations are included in the claim. Thus if somebody makes, uses, or sells any amount of the compound, they literally infringe the patent. However, broader claims have a downside; not only is a broader claim more easily infringed (good thing for the patent holder) but is also more easily rendered not patentable due to a lack of novelty or a finding of chemical obviousness (bad thing for a patent holder or patent applicant). As a result, it is common to claim an invention in a continuum from the broadest protection to the narrowest. In our example, the claims leaped from the broad (compound itself) to the much narrower (claims describing the specific uses of the compounds for treating depression or enhancing cognition). Returning to our hypothetical piece of prior art, we saw that the Japanese chemical journal disclosed the compound and a method of making it but nothing more. In part (2) of this hypothetical, the journal article was published more than 1 year before our chemist's patent application, making his claim to the compound itself not patentable due to a lack of novelty over that prior art reference under §102(b). While the claims to the methods of use are still novel, the chemist now does not have any claims to compositions containing the compound. If somebody decided to make this compound and sell it as a drug for a method of treatment other than the chemist's narrowly claimed methods, the chemist probably would not be able to stop her. For example, imagine if another drug manufacturer were to start selling the same dosage amount in a pill similar to his, but that manufacturer marketed it for putatively a different purpose, say preventing colds. Now further imagine that manufacturer were to sell her tablet for half the price. Assuming many of the targeted consumer base would understand it was the same active component, they may buy the cheaper pill and still use it for treating depression or enhancing cognition (an off-label use). Though technically those consumers would be infringers of our chemist's use claims, practically there would be no viable enforcement mechanism. To avoid such a scenario, our chemist might have claimed the compound in ways that would have avoided prior art references disclosing the compound but still allow him additional scope of protection. He might have claimed the compound together with at least one or more common pharmaceutical excipients, or he might have claimed the compound in a generic pharmaceutical dosage form (pill, tablet, syrup, etc). Since the prior art did not teach the use of the compound for any medicinal uses, such broad claims to the compound as a pharmaceutical composition should not be obvious (in the legal sense). Then even if a manufacturer were marketing the drug for a different use from what the chemist had claimed, he would still have an enforceable claim against them.

or in cognition enhancement. So he can still patent the *use* of the compound in these diseases, but he can no longer patent the compound itself.

In part (3) we learned that not only was the preparation and identification of the compound disclosed but its activity as an anticholinesterase useful for cognition enhancement was disclosed as well. Since the effective publication date of the journal article was July 4, 2006, less than 1 year before the filing date of the patent application, the reference is citable only under §102(a). This paper does not affect the chemist's ability to patent the compound itself. He still invented the compound before the publication date of the reference and a §102(a) reference can be antedated. Likewise, the method of using the compound as an antidepressant will probably not be affected by the paper's disclosure of anticholinesterase activity for at least two reasons. First, the chemist can show prior invention of this method (May 6, 2005) and second, there is nothing in our hypothetical to suggest that a prior art disclosure of the anticholinesterase activity of the compound should affect the patentability of a method claim for treating depression. The only issue remaining is the patentability of the third claim in view of this reference. The prior art reference discloses the same method described in the third claim—a method of enhancing cognition using the claimed compound. We have already determined that this method could be attributed an invention date of July 15, 2006. Since the prior art publication was already published before July 15, 2006, it is not possible for the chemist to swear behind the reference, and as a result, the §102(a) rejection will stand for cognition enhancement.

Does this mean that our Acme chemist should give up on any notion of developing the compound as a cholinesterase inhibitor? The answer is probably not. This is not a situation in which the specific commercial use of the compound is blocked by what is in the prior art because the prior art was a reference in a trade journal. It is not somebody else's valid patent that is blocking them. Rather, the prior art reference works only to negate their own patentability for the specific subject matter but not their own freedom-to-operate. Assuming they get claim 1 issued, they still will have the ability to exclude others from making, using, or selling the specific compound for any purpose, including the treatment of cognition. So what is the point of trying to get claim 3 if they can already block anybody by having the claim to the compound? The first reason is defensive in nature. If Acme had filed their patent application but did not disclose or claim the methods of enhancing cognition, they could leave themselves open to another competitor seeing their patent application when it publishes and recognizing that the compound might be good for cognition enhancement. If the competitor then was able to obtain a patent to that specific method with that specific compound, they could block Acme from later trying to commercialize that method. By Acme filing the methods and the compound, they have a better chance of preempting their erstwhile competitors from staking out claims in their method of use space. Second, it is not always possible to know what prior art might come up during the course of patent prosecution or patent litigation. While you may think you have a novel compound and strong claim to its structure, it is possible that for some reason the claim to the structure might be found to be nonpatentable or invalid but the claim to the method might still be patentable or valid. In such a situation, the method claim might serve as an enforceable backstop in situations where the compound claim falters.

In part (4) we are not dealing with a publication per se but rather a competition between two different inventive entities (a §102(g) scenario) but the patent office has no way of knowing who invented the subject matter first absent conduction of an interference proceeding. As a practical matter, what is likely to happen in this scenario is the USPTO will become aware through its internal processes that two patent applications claiming the same subject matter are co-pending. Assuming the facts described in our hypothetical were provable in the interference proceeding, the Acme inventor will win the priority contest since he conceived the compound and method of using it to treat depression first and was diligent through its reduction to practice. He can rightfully claim his date of conception (May 6, 2005) as his date of invention, whereas Klaus can claim May 11, 2005. Klaus will thus receive a rejection based on §102(g), and Acme will receive a patent with all three claims.

In part (5), we learn that a German chemist apparently invented the subject matter first, having conceived of the invention on January 4, 2005, and reduced it to practice by no later than February 17, 2005. Given the short time between the actual conception and the compound synthesis and testing, one can assume adequate due diligence was exercised such that the German chemist's invention date would be the date of conception. We also learned that he filed a patent application to the identical subject matter as a German patent application in the German patent office on March 7, 2005. Subsequently, a PCT international application containing the same subject matter was filed on March 1, 2006, and that application properly claimed priority to the earlier filed German patent application. The international PCT application designated the United States among many other countries and was published in the English language on September 15, 2006. The patent application was then abandoned in the United States though it eventually issued as a patent in Germany on April 1, 2007.

This hypothetical contains a number of relevant facts and dates and requires us to analyze the different activities under a number of different 102 sections. The most logical way to proceed would be to begin chronologically from the beginning. To help organize our analysis, we will list all of the potentially relevant dates of the prior art in chronological order.

1. Conception date: January 4, 2005
2. Reduction to practice (actual): February 17, 2005
3. Invention date: January 4, 2005
4. German patent application filed: March 7, 2005
5. PCT international application filed: March 1, 2006
6. PCT international application published (in English): September 15, 2006
7. German national application issues: April 1, 2007

As we have already learned, prior art does not mean just publications and patents but can be knowledge, use, and sales as well; it can also be proof of a prior invention by another party under the appropriate circumstances. Accordingly, let's begin by examining the prior art effect of the German invention itself. As we can see from the facts given, the German chemist invented the same subject matter before the date of the invention of the subject matter by the Acme chemist (January 4, 2005, for all the subject matter of the German invention vs. May 6, 2005, for the compound and method of treating depression claim and July 15, 2006, for the method of enhancing cognition claim for the Acme chemist's invention). The prior invention by the German chemist must first be considered for its prior art effects, if any, under 102 sections. Under §102(a), we see that the "known and use" provision applies only to knowledge and use in this country. From the facts given, the German chemist conceived and reduced to practice his invention entirely in Germany so it appears that §102(a) will not apply to his inventive acts.[49]

The only other 102 section that might apply to the prior inventive acts of the German chemist is §102(g), but it will not. Unlike in part (4), we do not have an interference situation. In the previous hypothetical, the German inventor was pursuing a patent in the United States and found himself in a contest to prove who invented the subject matter first. We learned that the German inventor abandoned his patent application in the United States and thus could not be involved in an interference. As a result of this analysis, the date of the act of the invention itself will not be relevant because none of the 102 sections apply to our fact scenario.

Beyond the act of invention itself, there were additional acts/information that took place regarding the German invention, which must be considered for prior art purposes as well. First, he filed a patent application in the German patent office on March 7, 2005, which issued on April 1, 2007. Although this entire portion of the record relates to the German application only, a quick review of the 102 sections indicates that a foreign patent can be prior art under certain circumstances. Let's review those in view of the relevant facts. First, section §102(a) provides that prior art includes inventions patented in a foreign country before the *invention date* by the applicant. For the purpose of §102(a), the term *patented* refers to the date that rights of exclusivity are awarded to the applicant, generally when the patent is granted. In this case, the invention was patented in Germany on April 1, 2007. The Acme patent application describes three inventions, the first two with an invention date of May 6, 2005, and the third with an invention date of July 15, 2006. So the subject matter of all three claims was invented before the patenting by the German inventor.

Section §102(d) applies to the foreign filing and issuance of a patent application by the inventor himself or his assignees. This means that in the instant hypothetical, the filing and issuance of the German patent is not an issue because the foreign application and issued patent were to a different party.[50]

So the German act of invention is not a problem, the German patent is not a problem, but what about that PCT patent application filed on March 1, 2006, which published in the English language on September 15, 2006? A published patent

---

[49]In any event, the use or knowledge requires a public use or knowledge, and the prior invention by the German chemist, even if performed in the United States, would not meet the public knowledge or use bar; there are no facts indicating that the invention was *publicly* known or used.

[50]The timeframe of the German patent application would disqualify it even if it were to the same party. Even though the German patent application was filed more than one year prior to the filing of the patent application by the Acme chemist, it did not issue until after the Acme chemist's patent application was already filed. §102(d) requires the foreign application to be filed more than one year prior to the US filing and that the foreign application issue prior to the filing in the United States.

application is a prior art publication under §102(a), good for all that it discloses (claimed and unclaimed subject matter) as of the date it publishes. However, since it is §102(a) prior art, it can be antedated by a showing of prior invention. The Acme inventor can establish prior invention for all three claims (May 6, 2005 and July 15, 2006 invention dates), so the reference can be removed as §102(a) prior art by an appropriately filed affidavit to the USPTO. Since the patent application was not published more than 1 year before the Acme chemist's patent filing, §102(b) does not apply.

Remember though that patent applications can also be cited as §102(e) prior art; this is where things really start to get interesting. Let's recall that section §102(e) allows a patent application to be cited as prior art as of the date the application is *filed*, not just the date the application is *published*. Under our present facts, we know the first application in the chain was filed as a German patent application on March 7, 2005, well before our inventor's date of invention. The PCT application was filed on March 1, 2006, and it properly claimed priority to the earlier German application. Section 102(e) provides in part that an "international patent application filed . . . shall have the effects for the purpose of this subsection of an application filed in the United States . . . if the international application designated the United States and was published . . . in the English language." We know from the given facts that the application was an international application that designated the United States and it was published in the English language. This means that the patent application itself was §102(e) prior art as of the date it was filed, March 1, 2006. Recall that §102(e) prior art can be antedated, just like §102(a) prior art. Since the Acme chemist's first two claims have an invention data of May 6, 2005, the PCT application with a filing date of March 1, 2006 can be antedated for these claims. However, the claim to cognition enhancement has an invention date of July 15, 2006. As a result, the international patent application to the German chemist is prior art under §102(e) for claim 3 and cannot be antedated.

What about the earlier patent application filed in Germany on which the international patent application properly claimed priority to? If this earlier foreign patent application was effective as a §102(e) prior art reference, the first two claims of the Acme invention would also be rendered notpatentable since the German national patent application was filed on March 7, 2005, almost 2 months before the invention date assigned to the first two claims of the Acme chemist's patent application. This is not the case because the prior foreign application is not given any prior art effect under §102(e). The earliest prior art date can only be afforded to an international patent application designating the United States or to an earlier filed U.S. patent application from which priority is properly claimed. If the facts of our scenario had been changed slightly so that the initial patent application was a provisional patent application filed in the U.S. office, then it could have served as prior art under §102(e) as of the date it was filed. Finally, an important point to note is that the patent application need never issue as a patent in order for it to be used as §102(e) prior art. In this situation, the Acme chemist will not obtain a cognition use patent in the United States because of the §102(e) rejection, and the German chemist will not obtain a cognition use patent because the application was abandoned. Further, after March 1, 2007, no one else may obtain a cognition use patent for that compound because the PCT application will then pose an absolute statutory §102(b) bar, since it will have been 1 year since it was effectively published (i.e. filed).

Unfortunately, scenario (6) is very common. The inventor publishes the invention before filing the patent application. In general, §102(a) or §102(b) will be cited as a basis for rejection, depending on the authorship of the prior publication and how much time elapsed between the publication and the patent application filing. By the present facts, our erstwhile inventor's patent application was filed more than 1 year after the abstract was published. As a result, the published abstract is prior art under §102(b) and is good for all that it discloses. Our Acme chemist cannot antedate 102(b) prior art. Notably, the published abstract may not affect the claim to the use of the compound for cognition enhancement since the abstract did not disclose, teach or even suggest such a use.

Finally, part (7) describes another common situation in which an inventor in the same company has filed a patent application describing and claiming an obvious variant of a later claimed compound from the same organization. The method of treatment in the earlier filed application should not affect the later claimed methods. Under the limited facts presented, it appears that the prior filed patent application is potentially prior art under §102(e) only. By the time the earlier April 1, 2006, application describing and claiming the obvious variant of his compound publishes, his application will have already been filed (on September 29, 2006) so the April 1, 2006, application will not be prior art under §102(a) or §102(b). The section that would be cited is §102(e), which relates to prior filed patent applications. Since this piece of prior art is citable for obviousness but not novelty—that is, it would be a §102(e)/§103 rejection—the Acme chemist has an escape key. Section 103(c) provides that obviousness rejections citing patent applications under §102(e) (or f or g) can be removed when the patent application in question is subject to an assignment to the same entity as the patent application being cited as prior art. So no worries, right? Well, not exactly. As it was explained in the hypothetical, the earlier April 1, 2006, patent application not only describes but also claims the compound that makes the later filed compound obvious. This situation should result in an obviousness-type double-patenting rejection where by if the earlier compound is allowed to issue as a claimed compound, our Acme chemist's claim to the obvious variant will have to include a terminal disclaimer pegged to the date of the earlier filed patent application. So what this means is that while the patent application filed on September 29, 2006, would normally provide coverage until September 29, 2026, in this case it may carry a terminal disclaimer and only run until April 1, 2026, the date of the earlier filed application.

We hope these scenarios have demonstrated some of the principals regarding what material can constitute prior art. Unfortunately, it is not always sufficient to compare the publication dates of prior art to the filing date of your own patent application. At a minimum, however, you can now at least recognize some of the more salient issues and know when a more detailed analysis might be required.

CHAPTER 4

Inventorship

Possibly one of the most contentious work experiences for many chemists is the naming of inventors on patent applications. Not surprising, those whose names are left off patent applications can feel slighted, believing their contribution to the application has been ignored or otherwise marginalized. Likewise, some inventors might often have less than charitable feelings toward other listed inventors who they feel did not make significant contributions to the invention. These feelings should come as no surprise when one considers the personal stakes involved. At a minimum, one's professional résumé of chemical experience is to some extent based on the presumed productivity that inventorship represents. Furthermore, as we soon will see, the co-inventors of a patent are initially the co-owners of this intellectual property and thus a significant financial incentive can be in play as well. Even in cases in which inventors are obligated to assign their inventions to another according to a contractual agreement (e.g., employment contract), there still may be financial inducements to a co-inventor in the form of a royalty sharing arrangement or other internal or even external recognition awards for co-inventorship. Even from a standpoint of basic fairness, we all have notions of what kind of contributions should qualify one for inventorship, and not surprising, those types of contributions usually resemble our own! However, unlike the listing of authorship on research publications that is usually left up to the discretion of the corresponding or lead author, the proper listing of inventors is a legal requirement of U.S. patent law. So into the maelstrom of the often-competing interests and opinions of the many inventors, wannabe inventors, financially vested parties, and so forth, warily steps the patent agent or attorney, providing her legal interpretation of who are the actual inventors. Before getting to the essentials of what it takes to be an inventor on a chemical patent, let's first take a quick look, from the legal standpoint, of why it is important to properly list inventors.

4.1 INVENTORSHIP AND OWNERSHIP OF U.S. PATENTS

Inventorship determines initial ownership of the claimed invention and thus properly determining and listing inventorship is in some cases every bit as important as properly listing the names of the owners on a deed to a very valuable property. The ownership

The Chemist's Companion Guide to Patent Law, by Chris P. Miller and Mark J. Evans
Copyright © 2010 John Wiley & Sons, Inc.

of a patent vests, at least initially, in the inventor. If there is more than one listed inventor, than each inventor becomes an equal, joint owner of the entire invention. This means that neither the relative weight of each individual inventor's contribution nor the order in which the inventors are listed on the patent application has any legal significance with respect to the initial ownership of the invention.[1] Furthermore, under the law, joint ownership of an invention means joint ownership in the entirety. Any inventor of the patent can grant, assign, license, or otherwise dispose of his ownership right as he sees fit. Even if an individual co-invented the subject matter of just one claim among many in a patent, that individual's right applies to the entire patent.[2] Since the holder of a patent has the right to exclude others from practicing the patented invention, and each inventor has a right to the entire patent, a single listed inventor could assign the right to the entire patent to others as he might see fit, without consulting with or remunerating his co-inventors.[3]

Does this sound like a recipe for disaster? It might be if one is not aware of the ramifications of co-inventorship and does not prepare in advance. In practice, most astute researchers will have an agreement in place before setting out on a joint venture. The most typical situation occurs where inventors work for a company or corporation. In exchange for their employment (e.g., salary), employees typically agree in advance to assign the rights of any of their inventions to the company. This is standard in the employment contracts of most companies in which intellectual property or patents make up or are likely to make up any of the value of the company's business. A second example is the establishment of a collaborative research agreement between entities, in which establishment of patent ownership rights at the outset can be critical.

An assignment of less than the entire rights of the invention by the patent applicants is more commonly referred to as a license.[4] It is possible that according to some

---

[1] See, for example, *Fina Technology, Inc. v. Ewen*, 265 F.3d 1325, 60 U.S.P.Q.2d 1314 (Fed. Cir. 2001), "the particular order in which the names appear is of no consequence insofar as the legal rights of the joint applicants are concerned."

[2] 35 U.S.C. §116 paragraph 1 provides the following: "When an invention is made by two or more persons jointly, they shall apply for patent jointly and each make the required oath, except as otherwise provided in this title. Inventors may apply for a patent jointly even though (1) they did not physically work together or at the same time, (2) they did not make the same type or amount of contribution, or (3) each did not make a contribution to the subject matter of every claim of the patent."

[3] 35 U.S.C. §262: "In the absence of any agreement to the contrary, each of the joint owners of a patent may make, use, offer to sell, or sell the patented invention within the United States, or import the patented invention into the United States, *without the consent of and without accounting to the other owners*" (emphasis added).

[4] A license is a contractual obligation between parties and as mentioned in the text, typically refers to an assignment of rights of less than the entire invention. For example, one might license the right to a compound but only for a certain use. Or one might license another's coating for a limited amount of time or for use in a limited geographical area. Licenses are typically customized to the particular desires of the parties in the same way any other contract is. The party granting the license is referred to as the *licensor* and the party acquiring the rights is the *licensee*. Licenses can be exclusive or nonexclusive. Exclusive rights mean that one can use the licensed rights without fear of competition from others granted the same or overlapping rights. Of course, the licensor might still grant additional licenses, but those licenses would be directed to different rights. The licensor could grant an exclusive right to the invention to one party but limited to a particular geography while granting the same set of rights to another party but limited to a

employment agreements, the inventor is allowed a share of any of the profit that the invention makes. For example, universities in the United States typically provide a sharing arrangement with employees (e.g., professors) for profitable inventions discovered by the employee and assigned to the university.[5] While one theoretically has the right to bargain with his future employer regarding what rights he will retain to any of their future inventions, as a practical matter this is not done except in very special situations.[6]

Where invention ownership issues arise outside of the employer/employee venue, contracts between the parties involved will govern the arrangement much as the employment contract does in the employer/employee venue. In some instances, one of the parties will agree by contract to assign any future patent rights to the other party, often in exchange for a predetermined royalty rate on any products, possibly in

different geography. Licensees would then have an exclusive right to their particular geographic location. In addition to the exclusive/nonexclusive classification, licenses are also typically classified by whether they can be further transferred by the licensee to others (and in the process making the licensee a licensor too!). A licensee can still have standing to challenge the validity of a patent that they have licensed (*MedImmune, Inc. v. Genetech, Inc.* 549 U.S. 118 (2007)). This means that taking a license to somebody's patented technology does not rule out the possibility of still challenging the patent in a federal court. Until recently, it had been held that the fact that a licensee had a license to the patent removed them from having the standing to sue because they were not in imminent risk of a patent infringement suit. To challenge the patent, the licensee would have had to break the license off and risk an actual infringement suit (and damages) to have their case heard in a federal court; *MedImmune, Id.* changed that.

[5] It is interesting that in certain countries (e.g., Germany, Japan) laws provide that if an employee invents and assigns his invention rights to his employer, then the private employer must include compensation in addition to the employee's regular compensation. For example, Shuji Nakamura (who invented the blue LED technology used for lights on many different electronic devices) was recently awarded the equivalent of almost $190 million in a landmark Japanese court case. Although the amount was reduced to about the equivalent of $8 million on appeal, the case still illustrates the difference between a private party's freedom to contract in the United States versus some other countries, where government regulation can play a greater role.

[6] The ability of an employee to bargain in such a situation probably would connect most directly to the amount of leverage that employee brings to the table. In the typical situation, the employer simply presents the employment contract form to the employee and the employee signs it. In practice, even in the absence of an explicit employment contract assigning an employee's rights to the employer, such an obligation may still be found, depending on the particular situation. For example, a court called in to resolve a dispute between an employee and employer might still find that an obligation to assign exists when an employee handbook describes such an obligation. Even less explicitly, the nature of the relationship of the employee to the employer itself might create an implicit obligation to assign by the employee. For example, if the employee is an officer or high-ranking person in the company, a fiduciary duty may be assumed with attendant loyalty to the employer dictating an assignment of inventions to the employer. An implied agreement may also be found where the employee was hired to solve a particular problem whose solution becomes the ultimate subject matter of the patent in question. Even if an invention is outside of the scope of the employee's job description, the employer may in some cases, depending on state law, acquire shop rights in the invention if the invention was made on company time or with company resources. Under shop rights, the employer obtains a royalty-free, nonexclusive right to use the invention. However, the employer cannot transfer the shop rights and cannot prevent other parties from using the invention. Since the employer normally has significant leverage at the time of the offer of employment, they are best advised to request the employee sign an agreement at the outset and avoid any uncertainties down the road. It is important that this same advice also extends to consultancy arrangements, collaborations, and the like.

addition to various milestone and support payments. Alternatively, the parties might contract in advance not to license any co-invented subject matter to any other parties absent agreement from all of the patent holders. No matter what the arrangement, it is important that all the parties involved appreciate the fact that each inventor is a joint owner of the entire invention and thus plan their agreement accordingly.

When the rights to an invention are assigned from one party to another, that assignment needs to be recorded at the USPTO in order to function as notice of ownership. This is important because if the assignment is not recorded, then it is possible that the original assignor could assign the patent rights to a different party, in effect selling the same invention more than once.[7] If the second party is not aware of the first assignment, and the first party does not record his assignment within 3 months with the USPTO, it is possible the first purchaser could lose his rights to the second purchaser.[8] Many inventors may have had the experience of marching down to one of their company's patent administrators or paralegals and signing forms of assignment, which effectively pass possession of their invention from them to their employer.[9] The signed assignment grants the employee's rights in the invention to their employer, and the employer will subsequently record this at the USPTO. Should the employer then assign the invention to yet another party, that assignment will also need to be recorded at the USPTO as will any further assignments. Licenses are not required to be recorded at the USPTO although they may be if desired by either or both of the parties to the license.

While we have seen that many of the ownership issues that arise from inventorship may be obviated by carefully assigning the rights to the invention in advance, this does not completely eliminate the need for proper determination of inventorship. For example, contracts assigning invention may still allow for the inventors to receive a certain percent of royalties should the invention become profitable (e.g., university inventors). However, as we will see in the next section, proper determination of inventorship goes beyond the issue of ownership; proper inventorship goes right to the heart of the validity of the patent itself.

---

[7]While such a second sale could very well be fraudulent, creating a cause of action against the duplicitous seller and recovering damages or restitution can be difficult.

[8]The primary reason for granting the second assignee the right is that he would not have any way to know that the invention had already been assigned to somebody else unless it was officially recorded. The first party to be assigned the invention has the power to record the assignment and thus the onus is on him to do so to let others know that he owns the invention. Assuming the second buyer has bought in good faith (did not know of the first sale) it would be unfair to punish him when he had no way of knowing that the invention had already been assigned to another party.

[9]In some cases, the inventor may receive a nominal amount of money (e.g., a silver dollar) upon assignment of his invention to the company. This gesture, largely symbolic in a present-day employment context, represents what is referred to as *consideration* in common law contract parlance. Under English common law from which American law owes its roots, a contract requires valuable consideration to pass hands (i.e., quid pro quo, which translates from Latin as "something for something") in order for the contract to be valid. A simple promise to do something without an exchange of value might fall within the realm of a "gratuitous promise," which is largely unenforceable absent special circumstances (and even then only enforceable to a point). In the usual circumstance, the employee is probably already bound by the employment contract to assign the invention to the employer so for that reason alone, separate consideration for signing the assignment is not necessary. Can you guess what the consideration is for the employee's signature on the employment contract?

4.2 PATENT VALIDITY AND CORRECT LISTING OF INVENTORSHIP

As we saw in the previous section, properly determining inventorship is important because inventorship determines initial ownership and all of the attendant financial considerations that go with it. Listing proper inventorship is also critically important because *improper* inventorship means that the inventors listed are not *the* actual inventors of the subject matter. If you are thinking to yourself, "so what?" then you need to recall that §102(f) of the U.S. patent code provides that: "A person shall be entitled to a patent unless ... *he* did not himself invent the subject matter sought to be patented" (emphasis added). In §102(f), the word *he* refers to the inventive entity (all of the inventors) named on the patent.[10] Thus if an inventor has been left off the patent (referred to as "nonjoinder" of a proper inventor), then the listed inventors did not invent the subject matter of the claimed invention entirely "themselves" since they did not invent all of the subject matter of the claimed invention.[11] *Such a patent is invalid.* Likewise, if a patent is granted that lists one or more persons who were *not* actually inventors, then "he" would not have invented all of the subject matter sought to be patented since some of the listed inventors would not actually be inventors (this is referred to as "misjoinder"). *Such a patent is also invalid.* If this seems a harsh remedy for what can be a difficult determination to make (as we soon will see), then you will be relieved to know that mistakes in inventorship can be corrected both before and after issuance of the patent.

Inventorship correction before the issuance of the patent is preferred and is governed by 35 U.S.C. §116 of the U.S. patent law which provides in part:

> Whenever through error a person is named in an application for patent as the inventor, or through an error an inventor is not named in an application, and such error arose without any deceptive intention on his part, the Director may permit the application to be amended accordingly, under such terms as he prescribes.

In practice, inventorship correction in a patent application depends on the circumstances for the correction, but in most cases requires a statement from each person being added and/or deleted that the error occurred without deceptive intent.[12] In cases where an inventor or group of inventors refuse to sign such a statement or are otherwise not able to sign, it is possible for the assignee (assuming the invention has been assigned to another party such as an employer) to file a petition to the

---

[10]See, for example, *A.F. Stoddard & Co. v. Dann*, 564 F. 2d 556, 566–567 (D.C. Cir. 1977).

[11]See *Pannu v. Iolab Corp.*, 155 F.3d 1344 (Fed. Cir. 1998) citing *Trans-World Mfg. Corp. v. Al Nyman & Sons, Inc.*, 750 F.2d 1552, 1562(CAFC 1994) ("a jury's conclusion of invalidity, based on its specific findings that the sole named inventor had in fact co-invented the claimed design and 'deceptively intended not to disclose that co-inventorship in its application to the Patent Office ... reflected the bar in 35 U.S.C. §102(f) against the issue of a patent to an applicant who did not himself invent the subject matter sought to be patented' ").

[12]The specific circumstances under which corrections may be made are provided for in 37 CFR 1.48 and in brief depend on whether the correction is after the oath/declaration of inventorship have been filed (1.48 a), the correction is due to a change in claim scope during prosecution of a nonprovisional patent application (1.48 b and c), or the correction is adding or deleting inventors from a provisional application (1.48 d and e).

USPTO requesting a suspension of the rules.[13] As an alternative to the correction of inventorship provisions provided for under 35 U.S.C. §116, applicants may also file a continuation application wherein they simply execute a new oath/declaration of inventorship listing the correct inventors.[14]

Corrections to an issued patent maybe made according to provision 35 U.S.C. §256 of the patent laws, which states

> Whenever through error a person is named in an issued patent as the inventor, or through error an inventor is not named in an issued patent and such error arose without any deceptive intention on his part, the Commissioner may, on application of all the parties and assignees, with proof of the facts and such other requirements as may be imposed, issue a certificate correcting such error.
>
> The error of omitting inventors or naming persons who are not inventors shall not invalidate the patent in which such error occurred if it can be corrected as provided in this section. The Court before which such matter is called in question may order correction of the patent on notice and hearing of all parties concerned and the Commissioner shall issue a certificate accordingly.

Due to the peculiar wording of the statute, §256 has been interpreted as applying a different standard of correction depending on whether the situation is one of misjoinder or nonjoinder. In the case of misjoinder ("through error a person is named in an issued patent"), the correction may be made even when there was deceptive intent on the part of the misjoined party (the improperly listed inventor).[15] In its statutory construction of §256, the Court in *Stark v. Advanced Magnetics, Inc.* held:

> In sum, to avoid redundancy and give meaning to the entire section, this court construes the term "error" to extend to mistakes, whether deceptive and dishonest or merely uninformed and honest. Therefore, section 256 allows deletion of a misjoined inventor whether the error occurred by deception or by innocent mistake. As well, the section allows addition of an unnamed actual inventor, but this error of nonjoinder cannot betray

---

[13]35 U.S.C. §118 provides that: "Whenever an inventor refuses to execute an application for patent, or cannot be found or reached after diligent effort, a person to whom the inventor has assigned or agreed in writing to assign the invention or who otherwise shows sufficient proprietary interest in the matter justifying such action, may make application for patent on behalf of and as agent for the inventor on proof of the pertinent facts and a showing that such action is necessary to preserve the rights of the parties or to prevent irreparable damage; and the Director may grant a patent to such inventor upon such notice to him as the Director deems sufficient, and on compliance with such regulations as he prescribes." 37 C.F.R. 1.183 provides in part: "In an extraordinary situation, when justice requires, any requirement of the regulations in this part which is not a requirement of the statutes may be suspended or waived by the Director or the Director's designee, *sua sponte*, or on petition of the interested party, subject to such other requirements as may be imposed." (*Sua sponte* means "on its own will," generally referring to an administrator or judge performing an action without being requested by any of the involved parties.)

[14]The only requirement is that at least one inventor from the first application be retained in the oath/declaration in the continuing application.

[15]As we shall soon see in the case of *Frank's, supra*, even though a correction to such a patent can be made under §256, a finding of inequitable conduct can still invalidate a patent.

any deceptive intent by that inventor. In other words, the statute allows correction in all misjoinder cases featuring an error and in those cases where the unnamed inventor is free of deceptive intent.[16]

Using this interpretation of §256, we might come to the conclusion that most cases of improper inventorship can be corrected without invalidating a patent. For example, deceptive intent on the part of an improperly listed inventor could be corrected according to §256. This means that one who lied about his or her contributions to be listed as a co-inventor could still have the patent corrected to reflect the true inventorship. Likewise, a nonjoinder error resulting from an inventor or group of inventors who deceived the patent office to keep another inventor or group of inventors from being listed on the patent could still be correctable under §256 since the deceptive intent was not by the nonjoined inventor (the clause reads "*an inventor* is not named in an issued patent and such error arose without any deceptive intention on *his* part"). Now that you appreciate the financial incentives that flow from inventorship, you also appreciate the possible motivation for someone to practice deception in order to be listed on a patent since that makes him a co-owner. Likewise, you can appreciate the motivation for an inventor or group of inventors to deceive the patent office to keep additional inventors from being listed on the patent. It's interesting that §256 focuses only on deception by an individual to keep his own name off a patent, something true co-inventors do not often like to do.

The federal Courts have sometimes dealt with these apparent gaps in the inventorship correction statutes through findings of inequitable conduct[17] as seen in the case of *Frank's Casing Crew & Rental Tools, Inc. et al. v. PMR et al. Technologies.*[18] This case relates to reissue patent 34,063 (the '063 patent) originating from US 4,738,145 (the '145 patent). The invention disclosed and claimed in the '063 patent has particular utility in the oil and gas drilling business and relates to a method of determining when two sections of pipe have linked up. The claim at issue in the appeal is the method claim 30:

30. A method of monitoring torque conditions during the make-up of a threaded tubular connection, wherein shoulder contact is formed in the threaded connection, comprising the steps of:

(a) sensing the torque imposed on the connection during makeup;

(b) monitoring the torque conditions during said step of sensing the torque to detect if a satisfactory threaded connection is obtained;

(c) establishing a supplemental torque level for an acceptable minimum torque differential after shoulder contact; and

(d) displaying the sensed torque during make-up.

The named inventors on the patent were the Vincent brothers (Larry and Darrell) and Shaunfield. During the course of the development of the technology in question,

---

[16]*Stark v. Advanced Magnetics, Inc.*, 119 F.3d 1551 (CAFC 1997).

[17]Inequitable conduct was discussed in Section 2.3.

[18]*Frank's Casing Crew & Rental Tools, Inc. et al. v. PMR Technologies, et al.*, 292 F.3d 1363 (CAFC 2002).

the Vincent brothers, through their company Tubular Make-Up Specialists, Inc. (TMS), hired Peter Weiner as a consultant to help them develop torque monitoring instruments. Weiner was very experienced in torque monitoring and was a named inventor on several patents in the field. Before starting the TMS corporation the Vincent brothers had no experience in the torque monitoring business. The Vincent brothers, working through a law firm, filed the patent application that eventually became the '145 patent, which was later reissued as the '063 reissue patent; the patent did not list Weiner as an inventor. During the time that the Vincent brothers (acting through TMS) were preparing the patent application, the relationship between Weiner and TMS had deteriorated. Several years of Weiner invoices to TMS had gone unpaid and furthermore, TMS failed to redeem stock options exercised by Weiner that would have given Weiner a 25% share in TMS. Soon after the patent application was filed, the Vincent brothers settled their dispute with Weiner without telling him about the filing of the patent application. Shortly after issuance, the '063 patent was licensed to PMR Services, Inc. who subsequently sought to enforce and/or license the patent to a number of other companies, including Frank's Casing Crew and Rental Tools, Inc. (Frank's). Frank's, apparently not wanting to pay money to license the patent, then brought an action in the federal district court, arguing, among other things, that the '063 patent was invalid due to improper inventorship. Frank's also argued that Weiner was the only true inventor and that the inventorship should be corrected to reflect that fact so that the patent, as corrected, would be valid. As it turns out, Frank's had an interest in the patent being held valid and enforceable because Weiner had assigned any rights in the patent *that he might have* to Frank's. Thus the best result for Frank's would have been for the Vincent brothers to be removed as co-inventors and Weiner to be listed as sole inventor, and the patent be held valid and enforceable. Under such a scenario, Frank's conceivably could have used the invention as well as enforced it against other parties.[19]

During the course of the trial, the District Court found that Weiner was an "innocent co-inventor" since he was an inventor or co-inventor of a torque-monitoring device, the TMS-1000, which was the subject of the '063 patent. According to the District Court, "Darrel [sic] and Larry Vincent deliberately concealed Dr. Weiner's involvement in the conception of the invention . . . [and] engaged in a pattern of intentional conduct designed to deceive the attorneys and the patent office as to who the true inventors were."[20] The District Court inferred intent from the fact that TMS, in its settlement with Weiner, paid him the consultancy fees but did not recognize his stock options, which would have given him a 25% ownership share of TMS. Presumably, had Weiner known about the filing of the patent application, he may have had additional leverage in the negotiations and further, he may have been more interested in not bargaining away his equity position in the company due to the value of the

---

[19]It was not clear from the discussion in the Court of Appeals of the Federal Circuit's opinion what argument Frank's was making for removing Schaunfield as a co-inventor. Schaunfield was not listed as an inventor on the original '145 patent although he was an inventor on the reissue '063 patent.
[20]CAFC quoting the District Court, 292 F.3d 1363 (CAFC 2002).

invention claimed in the patent application. A quick review of §256 would indicate that it should have been at least possible to add Weiner's name to the patent since he was a non-joined inventor and there was no apparent deception on *his part* in leaving his name off the patent. However, despite the possibility of correction under §256, a patent may still be held to be unenforceable due to inequitable conduct during the prosecution of the patent application. As it turns out, that is what happened in this case; the '063 patent was held unenforceable by the District Court due to inequitable conduct by the named inventors.[21] Upon appeal, Frank's argued that Weiner was the sole inventor and as the sole inventor, the inequitable conduct of the Vincent brothers should not spoil the patent for the one true inventor. In other words, Frank's argued that if the Vincent brothers were not true inventors then their inequitable conduct should not go against the enforceability of the patent. The CAFC was not persuaded by their argument,

> [I]t was the Vincents who sought a patent on the invention, regardless of whose invention it was. Thus, their inequitable conduct during prosecution of the application leading to the patent renders the patent unenforceable, just as the conduct of an attorney who participates in the prosecution of a patent application may render a patent unenforceable. We have explained that "if unenforceable due to inequitable conduct, a patent may not be enforced even by 'innocent' co-inventors. *One bad apple spoils the entire barrel* [emphasis added]. Misdeeds of co-inventors, or even a patent attorney, can affect the property rights of an otherwise innocent individual."[22]

From the decision in *Frank's*, its clear that even when inventorship might be correctable under §256, the issue of inequitable conduct may be dispositive where the court finds deceptive intent.

Even when §256 can be meaningfully applied (e.g., no inequitable conduct rendering the patent unenforceable), it does not operate automatically, and thus a patent does not avoid invalidation simply because it might be corrected; it must be corrected.[23]

---

[21] Leaving an inventor off is material because it results in an invalid patent. The intent was shown, at least in part, by the clear financial incentive the Vincent brothers had for leaving Weiner off. As you will recall, materiality and intent together make up inequitable conduct.

[22] Franks, *id*, citing *Starks v. Advanced Magnetics, Inc.*, 119 F.3d 1551, 1556 (CAFC 1997).

[23] As we saw in *Frank's*, a litigating party who is opposing the patent holder will often attempt to invalidate the patent. A sometimes asserted allegation is that the patent is invalid because it improperly lists inventors. Once issued, a patent is presumed valid unless proven not valid by "clear and convincing evidence," which is a higher evidentiary burden than the typical civil bar of a "preponderance of the evidence." Since a patent with improper inventorship is invalid, the party asserting that the patent improperly lists inventors must prove the improper listing by clear and convincing evidence. Properly listing inventors in the first instant is the best defense to this allegation and a legitimate, bona fide effort at determining inventorship in the first instant can go a long way toward meeting that goal. To correct inventorship under §256, the burden is on the patentee. See, for example, *Pannu v. Iolab Corp.*, *id*: "the patentee must claim entitlement to relief under the statute and the Court must give the patentee an opportunity to correct the inventorship. If the inventorship is successfully corrected, section 102(f) will not render the patent invalid. On the other hand, if the patentee does not claim relief under the statute and a party asserting invalidity proves incorrect inventorship, the Court should hold the patent invalid for failure to comply with section 102(f)."

4.3 DETERMINING INVENTORSHIP

Now that we have examined the important relationship between patent inventorship and initial ownership as well as the relationship between correct inventorship and patent validity, we can better appreciate the importance of determining inventorship correctly in the first place, so as to avoid any uncertainties that accompany a patent with improperly listed inventors. Unfortunately, inventorship determination is one of the "muddiest concepts in the muddy metaphysics of patent law"; it deigns to draw clear boundaries around subject matter that is often amorphous and subjective.[24] Inventorship issues typically arise when joint inventorship is being considered and the question revolves around whether one or more individuals' contributions qualify them as joint inventors.[25]

In this vein, it is vital to realize that "getting inventorship right" begins with understanding that the *issued claims* determine the scope of the inventorship inquiry. If an individual contributed material that was part of the patent application but ultimately was not important to the issued claims, she is not an inventor.[26] Once the subject matter of each claim is determined, that subject matter is then compared to each prospective inventor's contribution to see if that contribution is inventive. The inventive contribution to the claims thus holds the key to unlocking this mystery of inventorship, and as we will soon see, not all contributions to the invention are inventive contribution; not all contributions are created equal.

As we learned earlier, inventions can be broken down into two parts: *conception* and *reduction to practice*. *Conception* has been defined as "the complete performance of the mental part of the inventive act."[27] One might think of the conception portion of the invention as the idea to do the thing that eventually becomes the invention. Sometimes the bulb burns very brightly in an individual's head, and the conception for an invention comes about in a single, well-illuminated moment. Other times, the bulb glows only intermittently, perhaps borrowing its light from several individuals over a course of time. As a result, it is often the case that a claim to an invention

---

[24]"The exact parameters of what constitutes joint inventorship are quite difficult to define. It is one of the muddiest concepts in the muddy metaphysics of the patent law." *Mueller Brass Co. v. Reading Industries, Inc.*, 352 F.Supp. 1357, 1372 (E.D.Pa. 1972).

[25]Note that 35 U.S.C. §116 provides that "Inventors may apply for a patent jointly even though (1) they did not physically work together or at the same time, (2) each did make the same type or amount of contribution, or (3) each did not contribute to the subject matter of every claim."

[26]During the course of prosecution of the application, the claimed invention may change considerably. In some instances, the USPTO imposes a restriction requirement meaning that the initial patent application might be divided up into several different applications, where each application putatively claims a different invention (see Section 2.1). For example, a patent application might be filed initially claiming eight different, patentably distinct classes of compounds. In response, the USPTO might issue a restriction requirement requiring the applicant to elect which one of the classes of compounds he wishes to pursue in that application. The claims to the other classes of compounds must then be canceled, but can be pursued in separate applications, referred to as divisional applications. The inventorship of each of the classes of compounds in each of the applications must be listed properly or any patent issued from that application will be invalid (although correctable if meeting the provisions of §256).

[27]*Townsend v. Smith*, 36 F.2d 292, 295 (CCPA 1930).

requires the conceptual input of several inventors over a period of time. In either event, conception is completed when each and every element of the claimed invention has been mentally contemplated.

As we already learned, reduction to practice can be actual or constructive. Actual reduction to practice refers to the production of the invention in a physical, tangible form that contains every element of the claim corresponding to that invention.[28] Furthermore, an actual reduction to practice requires that the invention has been sufficiently tested to demonstrate that it will work for its intended purpose. In contrast, a constructive reduction to practice means that the invention is described in such a way that one of ordinary skill in the art can make and use the invention without undue experimentation, even though an actual working example has not been prepared— perhaps a well-detailed scheme or drawing is sufficient. In this case, one does not actually need to physically make the invention. A constructive reduction to practice normally occurs on the filing of a patent application since the patent application must, to be valid, provide sufficient instruction to allow one of ordinary skill ordinary skill in the art to make and use the invention without undue experimentation in the same way as a constructive reduction to practice requires.

Although it takes both conception and reduction to practice to make an invention, the two parts are not treated equally in terms of determining inventorship. In fact, a person cannot be an inventor unless they contributed to the conception of the claimed invention because "conception is the touchstone of invention."[29] However, the mere wish for the achievement of a particular result absent anything else is unlikely to qualify a person for inventorship. As was stated in a federal court decision where this issue was considered:

> The party claiming conception of an invention must show that it was complete and operative and such as would enable a person skilled in the art to reduce the conception to practice without any further research or exercise of the inventive skill. It is not sufficient, therefore, to show that a party claiming an invention has conceived a result to be obtained; the patentable thing is the means provided and disclosed by him to accomplish that result.[30]

Likewise, merely providing technical assistance to facilitate the reduction to practice of an invention does not qualify one as an inventor unless the reduction to practice itself required inventive contribution that became part of the claimed invention. In a common scenario that often occurs in the chemical industry, one person may propose a compound or set of compounds or a particular composition that she wishes to see made and tested for a given property or activity. The actual production of the compound or set of compounds may ultimately be carried out by a different person. At first blush, one might guess that the person who proposed the compound(s) for the particular activities is the one who conceived the invention and the chemists synthesizing the desired compounds were technical support providing the actual reduction

[28] *Wetmore v. Quick*, 536 F.2d 937, 942 (CCPA 1976).
[29] *Burroughs Wellcome Co. v. Barr Labs, Inc.*, 40 F.3d 1223, 1227–1228 (CAFC 1994).
[30] *Land v. Dreyer*, 155 F.2d 383, 387, 69 U.S.P.Q. 602, 605 (CCPA 1946).

to practice. Under this assumption, the person proposing the target compound(s) is likely the sole inventor. However, let's recall that it is the patent claims that define the inventorship. Are the claims directed to the compounds only, or are there claims directed to processes for making the compounds as well? If the latter is true, it is possible that the persons making the compounds are co-inventors, provided they contributed to the *conception* of the process claims.[31] Likewise, should the preparation of the compounds require skill beyond what is routine in the art or should the production of those compounds require additional research such that the means to achieve the intended result (the compound itself) is not present in the original conception, then the conception is incomplete without the additional experimental input. As a result, it is possible that the provider of that additional experimental input is also a provider of the conception of the invention and therefore a co-inventor of the invention as well.[32] Let's consider an inventorship hypothetical which will hopefully provide some color to the previous discussion.

Several chemists are working together on a top-secret project being conducted at the Acme "Miracles Happen Here" Drug Factory. The chemistry team consists of one

---

[31]As we will see later, the patentability of processes for the preparation of compounds can sometimes mirror the patentability of the compounds used and/or made in the process itself. Thus if a straightforward, routine conversion is used to produce a patentable compound or composition, then a claim to that process might also be patentable though no per se rule regarding obviousness is in effect (*In re Ochiai*, 71 F.3d 1565 (CAFC 1995)). However, the question remains as to whether the person designing and/or implementing that process would be an inventor or co-inventor to a claim of that patentable process. One might reasonably argue that absent any evidence of further inventive contribution by the person carrying out the process, she should not be a co-inventor even as to a patentable claim directed to the process for making the compound or composition. This position could be defended on the basis that the patentability of the process was due solely to the patentability of the compound made by the process and where the compound was conceived of by a different person, that different person is the inventor of the claimed process as well. However, should the process require additional inventive contribution such as use of a nonroutine route, reagent, starting material, or intermediate that was not part of the original conception and was not routine in the art, then the case of co-inventorship for the chemist who reduced the process to practice becomes much stronger.

[32]MPEP 2138.04 2100–113 citing *Oka v. Youssefyah*, 849 F.2d 581 (CAFC 1988) "conception of a chemical requires both the idea of the structure of the chemical and the operative method of making it." Inventorship determinations are not subject to easy generalizations or bright line rules. For example, a chemist's contribution to the subject matter covered by a dependent claim does not per se rise to the level of co-inventorship even where she conceived the subject matter of the additional limitation added by the dependent claim. In the recently decided *Nartron Corp. v. Schukra U.S.A., Inc.*, 558 F.3d 1352, 1358–59 (CAFC 2009), the Court stated:

> Borg Indak asserts that Benson was the inventor of the sole feature added by claim 11. *However, a dependent claim adding one claim limitation to a parent claim is still a claim to the invention of the parent claim, albeit with the added feature; it is not a claim to the added feature alone.* Even if Benson did suggest the addition of the prior art extender to what Nartron had invented, the invention of claim 11 was not the extender, but included all of the features of claims 1, 5, and 6, from which it depends. It has not yet been determined whether Benson contributed to the invention of claim 1 (although he does not claim to be a co-inventor with respect to claims 5 and 6). *If Benson did not make those inventions, he does not necessarily attain the status of co-inventor by providing the sole feature of a dependent claim.* See [*Hess v. Advanced Cardiovascular Sys., Inc.*, 106 F.3d 976, 980–81 (CAFC 1997)] (emphasis added).

project team leader (Gus), one lab leader (Curt) and two synthetic lab associates (Mike and Charles). Collectively, the team is trying to design compounds that will bind to and selectively modulate the "go-fast receptor." One day at a meeting of the chemical team, Curt proposes a series of compounds for preparation and testing against the go-fast receptor and if successful, further animal testing in various models as needed. Gus disagrees with Curt but after a heated discussion, acquiesces and finally even states: "Actually that is a really good idea, I strongly recommend you proceed with the compounds and make sure that you let me know if you need anything to make it happen."

Within weeks of that meeting, Curt, Mike, and Charles begin to investigate routes for the production of the desired compounds. Curt proposes one route to the target compounds that uses known starting materials and proceeds by straightforward routes to the desired novel targets. A second route, proposed by Mike, uses novel starting materials that need to be prepared by chemical routes not directly known in the art although plausible via the scheme he has proposed. Curt approves the route because it might allow access to different members of the chemical series and it also increases the chances of getting at the target compounds if the more conventional route fails. After several weeks of diligent work, Charles, working on the route proposed by Curt, produces some of the target compounds without any problems, and Mike's proposed route produces some additional synthetic targets. In routine binding experiments, most of the compounds are found to have affinity for the go-fast receptor with some demonstrating fairly exciting potency. The lead biologist (Susan) directs her group to test the compounds in standard in vivo models associated with the target activity where many of the compounds again prove to have significant activity (it seems the compounds make the rats run much faster than normal rats). Curt explains the results to Gus who exclaims: "Great, let's put together a record of invention, I think we've really got something here."[33] A record of invention is drafted that documents the technical details of the work together with the biological results. From the record of invention and additional background provided from the Curt, the project leader, the patent attorney drafts a provisional patent application that describes the group's work. A list of potential inventors is put together that includes the entire chemistry team as well as the lead biologist. The patent attorney assigned to that therapeutic area reviews the record of invention and decides to conduct interviews with each of the putative inventors, starting with the three chemists who have actually synthesized the target compounds. The three chemists are each interviewed separately:

Patent attorney: "Charles, I see you take credit for the synthesis of 30 out of the 38 compounds that have been specifically claimed in this patent application, can you explain to me exactly how you contributed to the claimed material in this patent application?"

---

[33] Different organizations have different formats for the record of invention (i.e., invention disclosure) but some common threads include a brief description of the invention, key dates including conception and reduction to practice, references to key notebook pages, and a list of proposed inventors and their contact information. This document serves as an important record as well as a helpful aid to get the patent agent or attorney started on drafting the patent application.

Charles: "Well, it goes like this . . . you see, I work harder and I am more effective than anybody else in the lab. I am the only guy that knows how to walk and chew gum (though I don't chew it in the lab) at the same time. I keep the HPLC running, and everybody comes to me when they have a problem isolating their compound. I make a lot of compounds because I set up a lot of reactions and purify them in parallel. That's why I'm one of the most productive guys in the department."

Patent attorney: "Well that's pretty impressive but can you tell me whose idea these compounds were?"

Charles: "Well, I got my target list and synthetic scheme from Curt, our lab leader. But don't think that means the compounds just make themselves. I had to play around with the reaction conditions to get the reactions to work, some of them go fine at room temperature but others I had to heat the heck out of. You know, some of these compounds aren't easy to isolate either. Without my hands on this HPLC you could bet that it would take a lot longer to get those same compounds made and tested."

Patent attorney: "Anything else you might want to add?"

Charles: "Yeah, don't rip me off me like last time. I need to list some patents on my résumé because there have been a lot of lay-off rumors going around and I want to be ready in case my number gets called."

The patent attorney next talks to Mike about his contributions to the work.

Patent attorney: "Mike, I see you take credit for 8 out of the 38 compounds that are claimed in this patent application, can you explain to me exactly how you contributed to the project?"

Mike: "Well, I made all of the indoles on the project. Curt's scheme wouldn't have worked to get them because they would have disintegrated in the deprotection step, so we needed a different scheme. I proposed the scheme because I thought it would be a cool way to make the compounds. Not much is known about that type of coupling reaction but I thought it was worth a shot. I love coming up with new reactions."

Patent attorney: "Anything else"?

Mike: "Well, I think it is important to appreciate that without my scheme, we could not have synthesized the indoles, which are some of the better compounds from this work. Using my chemistry, we were able to explore areas in the receptor we could not have otherwise. Once those areas were opened up, Curt gave me free rein to analogue the series based on whatever starting materials I could buy. For example, the 3-methyl compound came from a commercially available starting material that I picked out since I knew I could get the chemistry to work. As it turned out, although the compound was not the very best of the group, it did have some good properties."

Patent attorney: "Thanks!"

The patent attorney next calls in the lead biologist, Susan, on the project:

Patent attorney: "Susan, I see you're the lead biologist for this project. You've had a chance to review the record of invention prepared by the chemists as well as the draft claims, can you tell me how you contributed to this work?"

Susan: "Well, let me give you a little background before we proceed. Many other companies have worked on this receptor, but our management thought we also needed to have a presence in this area and so I proposed the screening paradigm that we are currently following. First we test the binding, then we have our cellular assay to determine the functional activity, and then we have the standard in vivo models of activity. If a compound meets or exceeds our predetermined cut-offs, we select it for a preclinical advancement candidate and proceed to look at other properties such as pharmacokinetics, safety, etc."

Patent attorney: "So the basic utilities we are screening for are already out there in terms of what is known?"

Susan: "Yes, I would say that. Although we do have some newer cellular constructs we are looking at that we think might be a better way to characterize the compounds."

Patent attorney: "Make sure you write that work up when you've finalized it and maybe we can file something on it down the road".

Susan: "Great, more work. Just what I need. [Laughs.] I'll be in touch."

The patent attorney next calls in the lab leader (Curt).

Patent attorney: "Curt, I see you're the lab leader that was involved with this project. You've had a chance to review the record of invention prepared by the chemists as well as the draft patent application and claims, can you tell me how you contributed to this work?"

Curt: "Well, basically the idea for this template was mine. I also came up with the main chemistry plan, though Mike did come up with the other scheme that provided the indoles. Are you trying to determine the inventorship?"

Patent attorney: "Yes, I think I have gotten enough—"

Curt: "Well, I didn't actually make any of the compounds. I've been busy with so much lab management responsibilities and meetings. It's gotten to be nothing but meetings around here. I used to actually make compounds but—"

Patent attorney: "Oh that's okay, I think I've gotten what I needed from you, thanks."

The last interviewee is the chemistry project team leader, Gus.

Gus: "I guess I'm here to sign some invention forms?"

Patent attorney: "You mean the assignments?"

Gus: "Yeah, whatever they are. I want to get the silver dollar, I promised my youngest son I would give him the next one I get."[34]

Patent attorney: "Well not so fast, we haven't filed the case yet. We are just doing the inventorship determination."

Gus: "Well, I'm the project team leader from which this invention originated. What else do you need to know?"

Patent attorney: "Well, I need to know how you contributed to this invention. You've had a chance to review the record of invention as well as the draft claims, can you tell me how you contributed to those claims?"

Gus: "I gave them the go ahead. I even told Curt that it was a good idea and that he should proceed with it. Not only that, I defended the idea to my upper line management. I took a lot of heat for it."

Patent attorney: "Yes, but as to the actual substance of the claims, was any of it your idea?"

Gus: "Well no, but without my buy in and advocacy, this project would have been shut down before it even got off the ground!"

Patent attorney: "Sure, I believe that. Is there anything else?"

Gus: "When do I get my silver dollar?"

Patent attorney: "I'll get back to you on that".

Gus: "Hey, it's been nice chatting but I've got a meeting to attend."

Do any of these scientist's situations sound familiar? If so, you might be curious to see how the inventorship analysis turns out.

Let's talk first about Charles. While Charles is a very sympathetic and hardworking guy (and I know we are all pulling for him), it appears that his contributions would not qualify him as a co-inventor of the work being contemplated for a patent application. His work, albeit skilled and important, did not contribute to the conception of any of the claimed subject matter (the process or the compounds). He appears to have made the compounds he was asked to according to a route that was provided. Although he had to vary certain parameters to optimize certain reactions or separations, those types of changes would appear to be within the normal technical purview of one of ordinary skill in the art.

Now let's talk about Mike. While Mike did not make nearly as many compounds as Charles (8 versus 30) some of his contributions appear to be inventive. For example, Mike came up with an alternative route that allowed entry into the indole compounds that will be part of the claimed invention. Assuming the process of making those indole compounds is claimed in the final patent (remember that claims can change throughout the patent application prosecution), then there is no doubt that Mike would be an inventor of that claimed subject matter. However, even if the process itself is not claimed, Mike still is likely a co-inventor because it appears that he designed

---

[34]It seems Curt's company commemorated the filing of patent applications by awarding each one of the listed inventors on the application a new silver dollar.

at least some of the analogues based on his synthetic scheme. Beyond the novel analogues that he proposed, Mike's contribution may have still risen to the level of inventorship since it appears that Curt's synthetic scheme would have failed to produce the indoles, and thus Mike's synthetic scheme represented an entry into the indole part of the claimed invention. Additional facts to be considered for this latter possibility include the availability of other workable synthetic schemes for making those types of indoles and their likelihood of success. In other words, did Mike's route to the indoles require skill beyond what is ordinary in the art? This latter analysis is more subjective and difficult in such a case because it turns on hypothetical posturing; nevertheless, a good-faith effort to get the right answer is all that is required.

Susan is unlikely to be listed as a co-inventor of this patent. It appears that Susan set up a routine paradigm for testing the chemists' samples. The chemists knew the target and submitted their compounds anticipating that they would be active. Susan did not come up with any novel uses for these compounds that could be separately claimed. In this case, Susan (like Charles) has performed part of the actual reduction to practice but was not involved in the conception of the invention. This is not to say that a biologist could never be an inventor on a patent claiming only compounds and processes for making them. For example, a biologist may have found an unexpected utility in a compound where previously none was known (a compound with no utility cannot be patented). Likewise, a patent application claiming compounds may also claim novel biological uses for those same compounds. Whoever conceived of the novel use would be listed as an inventor of that claimed subject matter.

Next let's look briefly at Curt's contributions. If you have determined that Curt is a co-inventor, give yourself a well-earned pat on the back. By recognizing Curt's contribution, you have appreciated that conception truly is the touchstone of invention. Even though Curt did not make or test a single compound, he is an inventor. Remember, it was Curt who conceived of the compounds that were ultimately made and tested. Moreover, Curt's idea was not simply a pie-in-the-sky wish for a certain result; he provided real molecular targets to be made and tested and provided a scheme for making them. So even though Curt never got physically involved with the work, his mental contributions earn him a well-earned spot at the inventor's table.

Finally, let's move onto our friendly chemistry project management figure, Gus. If you have determined that Gus is not an inventor then you are again congratulated. If you have the guts to explain this to Gus, then you are really congratulated! Explaining to people that they are not co-inventors is one of the more dreaded tasks of a patent agent or lawyer. This takes not only a keen sense for the right words but also real courage. One can see that Gus will not be very happy to receive the news that there will be no silver dollar for his kid, but we really have no choice in the matter. While it might be true that he made the work possible by exercising his managerial authority, there is no conceptual nexus between his efforts and the claimed invention.

CHAPTER 5

Patent Claims

Humpty-Dumpty: "When I use a word, it means just what I choose it to mean-neither more nor less."

Alice: "The question is, whether you can make words mean so many different things."

Humpty-Dumpty: "The question is: which is to be master – that's all."
                              —Lewis Carroll, *Alice in Wonderland, Through the Looking Glass*

5.1   INTRODUCTION TO CLAIM LANGUAGE AND STRUCTURE

The claims in a patent define the "metes and bounds" of the invention because it is the claims themselves that define that which the patentee is granted the right to exclude from others. The claims consist of language on whose interpretation will render the ultimate judgment of whether a claim is valid and whether it is infringed. Unlike the precision and relative universality of numbers and quantities, much of claim language depends on several fields of context; reduced even to its simplest context, a single word can have multiple definitions. In view of the myriad challenges that claim interpretation can present, the courts have developed and continue to refine certain general principles and guidelines to consider when interpreting claim language and hence claim scope. Together we will review some of these principles and guidelines, which will help us understand how to interpret a claim's meaning and hence scope.

Before diving into claim interpretation, let's first review some basic aspects of claims, such as the types of claims, the structure of claims, universal claim transitions, and finally, Markush claiming in chemical patents.

5.2   INDEPENDENT AND DEPENDENT CLAIM TYPES

The claims in a patent define the scope of the invention and therefore set the parameters for the right to exclude others from making, using or selling in the United States or importing into the United States.[1] Every nonprovisional patent application must

[1] 35 U.S.C. 271.

The Chemist's Companion Guide to Patent Law, by Chris P. Miller and Mark J. Evans
Copyright © 2010 John Wiley & Sons, Inc.

What is claimed:

1. A compound having the structure of formula I (*independent claim*):

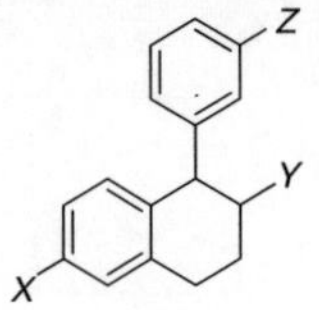

I

where: X is selected from hydrogen, C_{1-3} alkyl, or halogen;
Y is selected from hydrogen, C_{1-6} alkyl, phenyl, or OH; and
Z is selected from hydrogen or halogen.

2. The compound according to claim 1 wherein formula I is (*dependent from claim 1*):

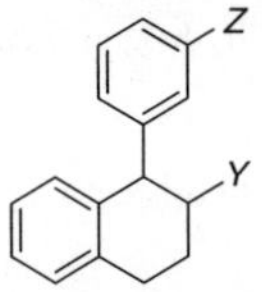

I

where: Y is selected from hydrogen, C_{1-6} alkyl, phenyl, or OH; and
Z is selected from hydrogen or halogen.

3. The compound according to claim 2 wherein formula I is (*dependent from claim 2*):

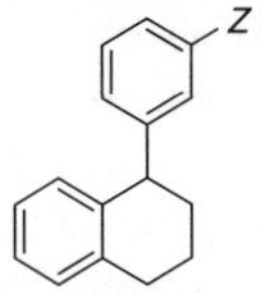

I

where: Z is selected from hydrogen or halogen.

FIGURE 5.1   Independent and dependent claims.

conclude with one or more claims that particularly point out and distinctly claim what the applicant regards as his invention.[2] Claims can be of two types: independent or dependent. Independent claims stand alone, whereas dependent claims rely on the claim they are dependent from.[3] It is important that a dependent claim borrows all of the limitations from the claim it is depending from and adds at least one additional limitation or further restricts a limitation that is already present. To illustrate, a sample claim set is provided in Figure 5.1.

[2]35 U.S.C. §112.
[3]Using correct patent terminology means that one should say that dependent claims depend "from" another claim and not "on." We are not really sure why this convention is used, but if you want to sound really smart and patent savvy, make sure you follow this convention whenever possible.

What is claimed:

1. A compound having the structure of formula I (*independent claim*):

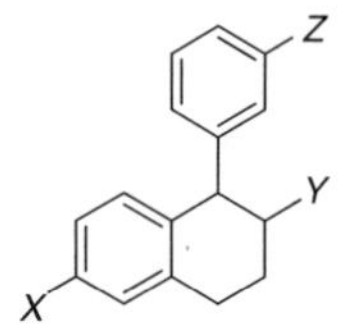

I

where: X is selected from hydrogen, C_{1-3} alkyl, or halogen;
        Y is selected from hydrogen, C_{1-6} alkyl, phenyl, or OH; and
        Z is selected from hydrogen or halogen

2. The compound according to claim 1 where X is hydrogen (*dependent from claim 1*).

3. The compound according to claim 2 where Y is hydrogen (*dependent from claim 2*).

FIGURE 5.2   Alternative claim set.

In this example claim set, claim 1 is independent since it does not depend from any other claim—the claim stands alone. In contrast, you can see that the claims 2 and 3 are each dependent, they incorporate the terms of another claim and then provide further limitation to the claim that they are dependent from. Whereas independent claim 1 allows selecting a compound with a possibility of 3 different groups at 'X,' dependant claim 2 allows only one (hydrogen); so dependent claim 2 is narrower than claim 1 that it depends from. Finally, claim 3 depends from claim 2, meaning that it incorporates all of the definitions from claim 2 and then provides a further limitation. As a result, the compound of claim 3 will have the compound of formula I as defined in claim 1 except that X and Y are both set to hydrogen.

A simpler but equivalent way of presenting these three claims is illustrated in Figure 5.2. The Markush structure is not repeated because it is understood that the dependent claim is importing the entire Markush that it is dependent from, and then further limiting that structure according to the specified language.

Having additional or further narrowing limitations means that the dependent claim is necessarily narrower in scope. A dependent claim that *is in any way* broader than the claim it depends from is not properly dependent. For example, by slightly modifying the claim set just presented, we have converted claim 2 to an improper dependent claim (Figure 5.3). Even though claim 2 is narrower overall, it still is improper for attempting to expand the definition of X to include the group -OH.[4]

---

[4]The claim set can be drafted to avoid this difficulty by redrafting claim 2 as an independent claim and providing the desired Markush definition for X.

1. A compound having the structure of formula I (*independent claim*).

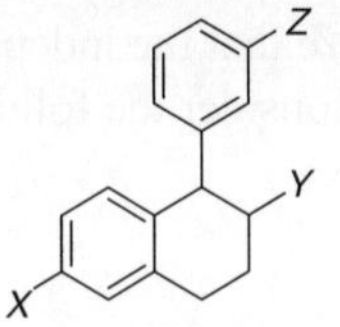

where: X is selected from hydrogen, C_{1-3} alkyl, or halogen;
Y is selected from hydrogen, C_{1-6} alkyl, phenyl, or OH;
Z is selected from hydrogen or halogen

2. The compound according to claim 1 wherein X is -OH (*improperly dependent from claim 1*).

FIGURE 5.3 Claim set with improperly dependent claim.

As a practical matter, understanding the difference between an independent and dependent claim can save time when examining issued patents to understand their claim scope. Specifically, if you are examining an issued patent to determine whether a certain compound, activity, etc. falls within the scope of the patent claims, you need only examine the independent claims. Why? Because the dependent claims cannot be broader in any sense than the independent claims they are dependent from. Accordingly, if you are outside the boundaries of the independent claims, you also will be outside the boundaries of any of the corresponding dependent claims since those claims are completely within the broader boundaries of the independent claims.[5] Likewise, if you are evaluating a set of claims to determine how a piece of prior art might affect the novelty of those claims, you should examine the novelty of the independent claims of interest first. If the independent claims are novel, then the claims that are properly dependent from those claims are presumed novel as well.[6] Notwithstanding, this all assumes that the claims were properly drafted. If the analysis is sufficiently important, one should carefully review the independent claims and review the dependent claims to make sure they are not broader in scope.

You may ask, Why bother to include dependent claims in the first place? If someone is infringing a dependent claim she will also be infringing the independent claim from which it is derived. There are at least a few practical reasons. First, it is often easier for someone in the future to try to have your broadest claims declared invalid for a lack of novelty or for being obvious; broader claims are more likely to read on the prior art because they occupy more turf. Second, it is generally harder to enable one to "make and use" a broader claim because a patent claim needs to be enabled throughout its

---

[5] If you are not in the apartment building, you cannot be in any of the apartments.
[6] The novelty-defeating effect of various types of prior art disclosures on chemical claims will be discussed in depth in Chapter 7.

scope; a broader claim will require more enablement. If a claim is not enabled, then it is invalid because it fails to meet the written description requirement.[7] Finally, the existence of a dependent claim serves to emphasize that the independent claim from which it depends is not so limited. For example, consider the following claims:

1. Compound A or derivatives thereof.
2. The calcium salt of compound A.

The existence of dependent claim 2 points out specifically that claim 1 includes salt forms. This is important because in some instances, claims might be read more narrowly then the drafter intends. The use of dependent claims helps more clearly define the breadth of the claim it is dependent from as well as provide additional, narrower protection.

5.3 CLAIM STRUCTURE

Each claim in an issued U.S. patent application will be numbered, and each claim is exactly one sentence long.[8] The structure of a typical claim contains three parts, which are, in the order of their usual appearance, the *preamble*, the *transition*, and the *claim body* (Figure 5.4 identifies the three parts of the composition claim 1 and the process claim 2).

The preamble to the claim is found at the beginning of a claim, and it will usually start by reciting the type of subject matter (e.g. process, machine, manufacture, or composition of matter) and sometimes will go further by summarizing the invention that is subsequently claimed, such as describing the inventions intended use or properties and possibly even how the invention is distinguished over the prior art (e.g., a compound with improved properties). The primary issue encountered with a claim's preamble is whether part or all of the language of a preamble will be used in limiting the scope of the claim itself. If it does, then the words of the claim's preamble are as important as the rest of the claim since those words will help define the "metes and bounds" of the claim. It is not always easy to know whether the preamble affects the scope of the claim. There is no general litmus test for making this determination, which can be "resolved only on review of the entire . . . patent to gain an understanding of what the inventors actually invented and intended to encompass by the claim."[9] Where the preamble simply states the intended use, asserted benefits or purpose of the claimed subject matter then the preamble will not be found to affect the claim

---

[7]Enablement and the written description requirement will be covered in Chapter 9.

[8]Despite the one-sentence requirement, a generous dollop of commas, semicolons, and colons can easily allow a single claim to run for multiple pages! However, the patent examiner does have an ability to fight back and can reject a claim as prolix (§112) if the scope of the claims is render indefinite due to the excess verbiage.

[9]*Catalina Marketing Int'l. v. Coolsavings.com, Inc.*, 289 F.3d 801, 808 (CAFC 2002) citing *Corning Glass Works v. Sumitomo Electric U.S.A., Inc.*, 868 F.2d 1251, 1257 (CAFC 1989).

What is claimed:

1. A pharmaceutical composition useful for depression (←*preamble*) comprising (←*transition*) a compound of formula I

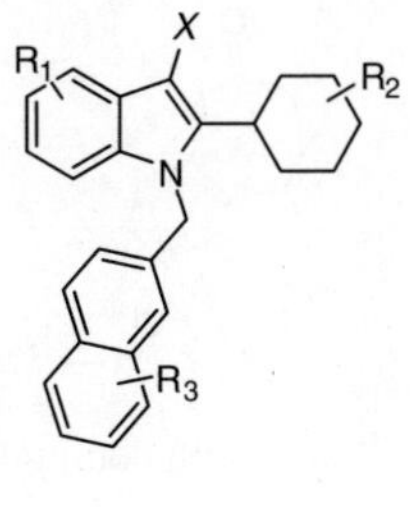

I

or a pharmaceutically acceptable salt thereof, and one or more pharmaceutically acceptable excipients; where R_1 is hydrogen or C_1-C_6 alkyl; R_2 is hydrogen or methyl; and R_3 is C_1-C_6 alkyl, C_1-C_6 alkoxy, or halogen (←*claim body*).

2. A process of preparing a compound of formula I essentially free of impurities (←*preamble*) consisting of (←*transition*) the steps of:
    (a) Reacting a compound of formula II with a compound of formula III in the presence of base

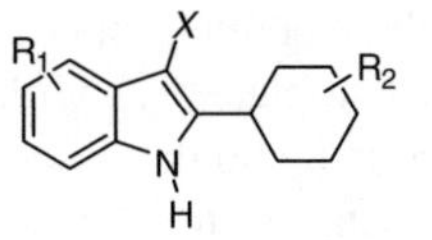

II 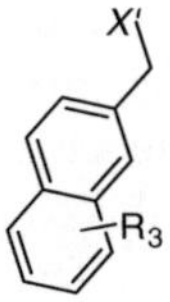   III

    (b) Acidifying the reaction mixture to pH < 7 with aqueous acid;
    (c) Extracting the reaction mixture with an organic solvent having low solubility in the aqueous phase;
    (d) Concentrating the organic solvent to a reduced volume; and
    (e) Chromatographing the residue on silica gel
        where: R_1 is hydrogen or C_1-C_6 alkyl; R_2 is hydrogen or methyl; R_3 is C_1-C_6 alkyl, C_1-C_6 alkoxy, or halogen; X is hydrogen or halogen; and X' is a suitable leaving group (←*claim body*).

FIGURE 5.4 Compound and process claims with claim parts labeled.

scope. However, where the preamble "when read in the context of the entire claim, recites limitations of the claim, or, if the claim preamble is 'necessary to give life, meaning, and vitality' to the claim, then the claim preamble should be construed as if in the balance of the claim."[10]

[10]MPEP 2111.02 citing *Pitney Bowes, Inc. v. Hewlett-Packard Co.*, 182 F.3d 1298, 1305 (CAFC 1999).

We claim:

1. A compound with anxiety-alleviating effects in humans (←*preamble*) having
   (←*transition*) a structure of formula I:

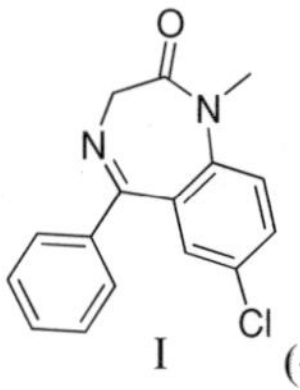

I (←*claim body*)

FIGURE 5.5 Claim with nonlimiting preamble.

For illustrative purposes, an example of a claim with a descriptive preamble is
shown in Figure 5.5. The preamble to the claim tells one that the claim is to a
compound and not, for example, to a method of using the compound for the alleviation
of anxiety. The "anxiety-alleviating effects" of the compound describe a property of
the compound and so the question here is whether that property materially limits the
claim. If it did limit the claim, we would need to read into the claim some further
requirement of the claim, such as a minimum amount of the compound necessary
so that it could fulfill its required property of "anxiety-alleviating effects." However,
the compound being claimed is completely defined by structural formula I, and the
anxiety-alleviating effect is simply one of the inherent properties (and potential uses)
of that compound. In this case, the claim is complete even without the description of
the anxiety-alleviating effects of the compound, and the description of that property
is not "necessary to give life, meaning, and vitality" to the claim. Since the preamble
appears merely to describe one of the properties attributable to the compound and is
not required to make sense of the claim, the phrase *anxiety-alleviating* likely does
not act to limit the claim.

However, the distinction can sometime be very subtle. Consider a patent containing
a very similar claim 1 (Figure 5.6) except that the term *compound* has been changed
to "A pharmaceutical tablet" and the compound of formula I is the "sole active
ingredient" of the pharmaceutical tablet. In this case, a "pharmaceutical tablet" is
what is being claimed instead of only the compound. The question again is whether
the phrase *anxiety-alleviating effects* further limits the claim. In other words, would
a pharmaceutical tablet that did not have anxiety-alleviating effects but contained a
compound of formula I as the sole active ingredient still be covered by the claim?

Without having the patent specification to read we can still manage a little de-
tective work. We do know that the claim requires a pharmaceutical tablet. We also
know that the claim requires a compound of formula I. We further appreciate that a
pharmaceutical tablet is used for the administration to a subject for a pharmaceutical
purpose. In this example, the pharmaceutical tablet being claimed contains the compound of formula I as its sole active ingredient. The purpose of the tablet, as inferred

We claim:

1. A pharmaceutical tablet with anxiety-alleviating effects in humans (←*preamble*) comprising (←*transition*) a compound of formula I as the sole active ingredient:

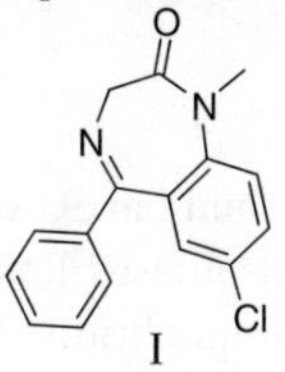

and at least one pharmaceutically acceptable excipient (←*claim body*).

2. The pharmaceutical tablet of claim 1, wherein said tablet contains between 5 mg and 20 mg of the compound of formula I.

3. The pharmaceutical tablet of claim 2, wherein said tablet contains 10 mg of the compound of formula I.

FIGURE 5.6   Claim with limiting preamble.

from the preamble is to have an anxiety-alleviating effect in a human. How much of the compound of formula I needs to be in the tablet? Since the claim is directed to a tablet, and the tablet's purpose is "anxiety-alleviating effects," then the compound of formula I needs to be present in an amount sufficient to alleviate anxiety in a human; anything less would eviscerate the function of the tablet. In this example, the portion of the preamble describing the purpose also would serve as a limitation, at least as to the minimal amount of a compound of formula I that needs to be in the pharmaceutical tablet.

This does not resolve the question of specifically how much of the compound of formula I is required in the pharmaceutical tablet. Normally, the person drafting the claim would have drafted additional dependent claims further narrowing the invention to increase its clarity and limit its scope. For example, claims 2 and 3 give a much clearer picture as to how the pharmaceutical tablet of the previous claim could be further delineated. Notice that claim 1 is independent and claims 2 and 3 are dependent claims since claim 1 can stand alone but claims 2 and 3 borrow all of the limitations from the claim they depend from and then limit the scope further.

Finally, to think further about the importance of determining whether the preamble is limiting, just as a thought experiment let's consider a number of hypothetical questions with regard to the set of claims in Figure 5.6 assuming the phrase "with anxiety-alleviating effects in humans" is indeed construed as a limitation. For the following questions, assume an anxiety-alleviating amount of compound I is at least 5 mg. (1) If another manufacturer sells a tablet containing 1 mg of compound I for the treatment of colds, would the claims be infringed? (2) How about if they sold a tablet containing 6 mg of compound I for the treatment of colds? (3) Finally, how about if they sold a tablet containing 3 mg of compound I for the treatment of colds, but a person had to take two tablets at the same time for the treatment to be effective?[11]

5.4   TRANSITION PHRASES

Transition phrases are limited in number and have a very distinct function for claim interpretation. Structurally, the transition phrases link the claim preamble with the claim body but they serve more than a simple bridge function; they provide specific guidance as to whether to read the claim limitations as being the minimum elements necessary for the invention (open claim) or whether the claim limitations make up the entire invention (closed transition), or somewhere in-between (partially open). Typical transition phrases include *consisting of*, *consisting essentially of*, *comprising*, *having*, *including*, and *containing*.

The transition phrase *consisting of* is a closed transition that indicates that the elements immediate subsequent to the term define the sole elements of the group or subgroup those elements are a part of. Let's consider the very basic composition claim:

1. A composition *consisting of* A and B.

In this claim, the transition *consisting of* is followed by the elements of A and B, and those two elements form the group that the transition phrase is applied to.[12] Since *consisting of* is a closed transition phrase, the elements of the subgroup immediately following the transition are the only allowed elements that satisfy the claim and "excludes any element, step, or ingredient not specified in the claim."[13] However, in a composition claim, the claim is still open to include impurities typically associated with those materials.[14] Another way to read this claim containing "consisting of A and B" is to read it as saying "consisting of A and B and only A and B" and any impurities typically associated with those two materials.

The meaning of the transition phrase *consisting of* can be significant both when one is considering whether one is infringing that claim (freedom to operate), or whether the claim reads on prior art (patentability of the claim). For example, if a prior art

---

[11]For 1, the answer is no because the amount in the tablet is not sufficient to meet the anxiety-limiting requirement. For 2, the answer is presumably yes because the claim is directed to a composition and not a method of treatment. Although the intended indication may limit the amount of the compound of formula I in the tablet, it presumably does not limit the tablet to a certain indication. For 3, the strict answer is likely to be no since the claim is to "*a* tablet" and if the amount in *a* tablet being sold is not sufficient to alleviate anxiety (5 mg) then the claim limitation has not been met. We hope you are beginning to appreciate why every word in a claim is important.

[12]We will see that it is possible for a claim to contain more than one transition phrase. In such a case, the relationship of an earlier transition to a later group of elements is dependent on the form and context of the claim language, including the presence and location of any additional transition phrases.

[13]MPEP 2111.03, citing *In re Gray*, 53 F.2d 520 (CCPA 1931).

[14]MPEP 2111.03, citing *Ex Parte Davis*, 80 USPQ 448, 450 (Board of Appeals, 1948).

composition was described that contained only the material A, then that prior art does not describe exactly the claimed composition since the claimed composition requires A *and* B. Likewise, a prior art composition containing A, B, and C does not describe the exact same thing as the claimed material since the claimed composition requires not only A *and* B, but *only* A and B (and not A, B, and C). Similarly, making, using, or selling a composition consisting of A only, or A, B, and C would not infringe the claim since the claim requires A and B and only A and B, assuming, of course, that C is not an impurity typically associated with A and B. Further, C must be related in some way to the claimed invention and cannot simply be an unrelated item included to avoid infringement of a patent.[15]

On the other end of the transition phrase spectrum are open transition phrases exemplified most commonly by *comprising* and less commonly by *including, containing*, or *characterized by*.[16] Unlike the closed transition *consisting of*, the presence of one of these open transitions means that the listed elements of the claim subsequent to the transition are the minimum elements to satisfy the claim and that additional ingredients, even in very large amounts can be included and the limitations of the claim may still be met. An example of a very basic open claim is:

1. A composition *comprising* A and B.

In this claim, the transition *comprising* is followed by the elements A and B. Since *comprising* is an open transition phrase, the elements following the transition are the minimum elements required to satisfy the claim, and the claim must also be read to include any and all additional ingredients even though unstated.[17] Another way to read this claim containing "comprising A and B" is to read it as saying "containing a minimum of A and B." As we saw in our discussion of the closed phrase *consisting of*, the reading of *comprising* affects the way the claim is read when determining whether one is infringing that claim (freedom to operate) or whether the claim reads on prior art (patentability of the claim). For example, if a prior art composition was described that contained only the material A, then that prior art does not describe the claimed composition since the claimed composition requires A *and* B as a minimum. However, a prior art composition containing A, B, and C would anticipate the claim language since it contains the minimum required elements of A and B. Unlike the potential infringer who wished to design around the previous claim containing the closed transition *consisting of* by adding a little bit of the additional element C to the composition of A and B claimed earlier, the potential infringer would be less successful here, because the open transition means his making, using, or selling the

---

[15]*Norian Corp. v. Stryker Corp.*, 363 F.3d 1321, 1331 (CAFC 2004). In this case the patent at issue (U.S. 6,002,065) covered a set of chemicals in a test kit. Claim 8 used the *consisting of* transition. The infringing kit being sold had the same chemicals plus a spatula. In overturning the District Court, the CAFC stated "While the term 'consisting of' permits no other chemicals in the kit, a spatula is not part of the invention that is described.... Infringement is not avoided by the presence of a spatula, for the spatula has no interaction with the chemicals, and is irrelevant to the invention."

[16]MPEP 2111.03.

[17]MPEP 2111.03, citing *In re Gray*, 53 F.2d 520 (CCPA 1931).

What is claimed:

1. A pharmaceutical composition *comprising* the compound of formula I:

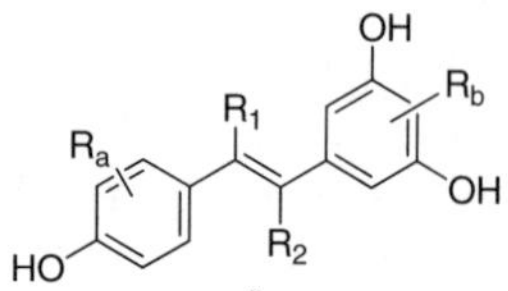

I

where: R_a and R_b are each independently selected from the group *consisting of* hydrogen, C_{1-3} alkyl, and halogen; R_1 and R_2 are each independently selected from the group *consisting of* halogen, phenyl, and C_{1-2} alkyl; and at least one pharmaceutically acceptable excipient.

FIGURE 5.7   Claim with open and closed transitions.

composition containing A and B is sufficient to read on the claim, despite the presence of additional ingredient(s) (like C).

In claim language, it is very common to include a plurality of groups that might be connected by different transition phrases. Unlike mathematics where a commonly established order of operations can be applied, claim language rests closer to real language and as a result, the interpretation of relationships between multiple transition phrases is not always straightforward. Nevertheless, let's briefly consider the claim in Figure 5.7, which uses the phrase *comprising* and twice uses the phrase *consisting of*. Is this an open or closed claim?

You may have noticed the preamble *a pharmaceutical composition* is followed by the transition phrase *comprising* and so, Aha!, the claim is open and therefore requires only the compound of formula I and at least one pharmaceutically acceptable excipient. This is correct, provided we understand that the compound of formula I has a separate set of definitions that cannot be augmented despite the open transition phrase immediately preceding the compound. In other words, this claim may contain compounds or ingredients in addition to a compound of formula I but will not be understood to expand on the definitions of formula I itself. For example, R_a and R_b are each closed sets that consists of hydrogen, C_{1-3} alkyl and halogen, and those groups only. This is likewise true for the definition set of R_1 and R_2. To make this clearer, the claim drafter has placed the language *consisting of* immediately before these Markush subsets, meaning that the group immediately following the transition is a closed group.[18] The key point is that when the phrase *consisting of* appears

[18] In practice, where a chemical structure is claimed according to a Markush format, the variable descriptors at each position of the molecule will almost always (if not always) be read in a closed fashion unless the implication is clearly otherwise. To leave a molecule truly open at any single position might allow for an infinite set of molecules to be generated. Where the compound is an object of the claim, an infinite set of potential molecules is likely to draw a rejection for failure to satisfy the written description and enablement requirements of §112 ¶1, to be discussed in Chapter 9.

in a clause of the body of a claim rather than immediately after the preamble, the *consisting of* limits only the element set listed in that clause.

This begs the question of what effect the transition *comprising* has in this example. We have already seen that we cannot read the claim to allow addition to structural formula I itself. However, the preamble to the claim clearly explains that a pharmaceutical composition is being claimed and not merely a compound. We do not need to get into the exact nature of any claim construction limitations that this preamble might introduce to appreciate that this claim is not limited in scope to the compound of formula I and at least one pharmaceutically acceptable excipient, but would also cover the addition of materials that might be understood to go into a pharmaceutical composition. If the preamble is understood to be a limitation of the claim, then the claim can be read to be broad enough to capture the compound in a pharmaceutical composition but not so broad as to capture the compound in any possible setting, to ensure this interpretation, the patent drafter has carefully added an additional required element, the pharmaceutically acceptable excipient.[19]

Sometimes, the claim drafters wish to draft a claim that is intermediate in scope and look to the transition phrase for help in accomplishing this task. Even though claim scope can be modulated by the judicious use of claim limitations (pharmaceutically acceptable excipients) and controlling preamble language (pharmaceutical composition), it is nevertheless sometimes difficult, if not impossible, to envision the scope and effect all the prior art might have or all the ways that prospective users of your claimed invention might find to design around your claims. The use of *consisting of* may narrow a claim too much, making it harder to enforce against a potential infringer making, using, or selling your composition but including some additional but yet unimportant ingredients. At the other extreme, the use of *comprising* may open the claim up too much, possibly broadening it to include elements that together with the explicitly claimed elements could cause the claim to read upon the prior art or to not be properly enabled for such a broad scope, and in either case adversely affecting patentability. Since an inventor cannot specifically account for each and every trace ingredient that a putative infringer might add to try to avoid infringement of a closed claim, it would be helpful to be able to open up a claim enough to include such inert ingredients without opening the claim to include additional active or otherwise significant ingredients. Conveniently, the closed transition *consisting of* can be broadened to the term *consisting essentially of*, which indicates that claim elements from the subgroup following the transition are open to include additional elements that would not "*materially* affect the *basic* and *novel* characteristics" of the

---

[19]To one of ordinary skill in the pharmaceutical arts, a pharmaceutical composition would be thought of as a mixture containing at least a drug substance and at least one pharmaceutically acceptable excipient (additive). The reason for this is that almost every drug will be given in a tablet, pill, solution, etc., and those drug delivery vehicles all require the use of some excipient—even water can be a pharmaceutical excipient in the case of a drug delivered in solution. Nevertheless, to reduce any claim interpretation issues, it is common practice to include the explicit requirement that the pharmaceutical composition be read to include the claimed compound as well as at least one pharmaceutically acceptable excipient. In this way, the dependency on the preamble is lessened since the additional language acts to more definitively limit the claim.

"... an oxidizing agent selected from the group consisting of potassium permanganate, sodium periodate, potassium dichromate, and manganese dioxide";

"... where: X is fluorine, chlorine, or bromine."

"... where $-Y$ - is a repeating unit selected from  ."

FIGURE 5.8   Examples of Markush groups.

claimed invention.[20] This transition phrase effectively allows the patentee to capture the essence of what he believes is his invention without being rigorously tied down to a precise product that can be readily skated around by a clever infringer, while at the same time not opening the invention up to the extent that it affects the novelty of the claimed composition in view of the prior art.[21] An easy way to remember, if not understand, this transition phrase is that it dictates scope somewhere between the closed transition of *consisting of* and the open transition *comprising* and should be considered where *consisting of* is too narrow and *comprising* too broad.[22]

5.5   MARKUSH CLAIMING IN CHEMICAL PATENTS

By now you have seen numerous generic chemical structures in which a single structure is drawn such that several possible alternatives are provided at different parts of the molecule being described. Each group of possible alternatives is commonly referred to as a Markush group and is best recognized by the format: "X is selected from the group consisting of A, B, and C" or "X is A, B, or C."[23] Some representative Markush groups are illustrated in Figure 5.8.

When a Markush group is used for claiming a chemical invention, the members of the group either need to belong to a common or recognized class or share at least one

---

[20]MPEP2111.03 quoting *In re Herz*, 537 F.2d 549, 551–552 (CCPA 1976), emphasis in the original.

[21]Discerning the precise breadth of a claim having "consisting essentially of" can be challenging as it requires a judgment of what is a basic and novel characteristic of the invention, which in turn likely requires a thorough reading of the specification perhaps as well as an understanding of the prior art. See, for example, MPEP 2111.03, explaining *AK Steel Corp. v. Sollac*, 344 F.3d 1234 (CAFC 2003), where the CAFC found that the applicant's statement in the specification that "silicon contents in the coating metal should not exceed about 0.5% by weight" acted as a limit to the amount of silicon that could be covered by a claim that included the transition "consisting essentially of" before the recitation of the explicit coating contents.

[22]Of course, the elements of the claims themselves are the primary way of controlling the scope of the claim, and the transition phrases can be seen as setting the limits on the inclusiveness or exclusiveness of claim beyond the explicitly cited elements.

[23]MPEP 803.02 quoting from *Ex Parte Markush*, 1925 C.D. 126 (Comm'r Pat. 1925).

1. A compound according to the formula I:

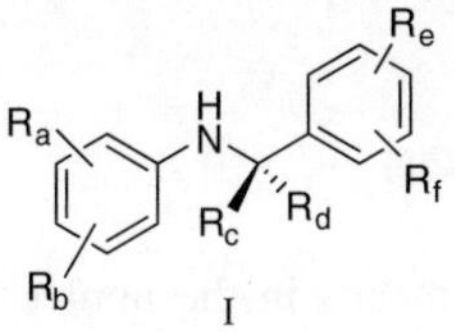

where: R_a, R_b, R_e, and R_f are each independently selected from cholorine, methyl, ethyl, propyl, or methoxy; and Rc and Rd are each independently selected from hydrogen or C_{1-3} alkyl.

FIGURE 5.9   Markush chemical structure claim.

common function. A way to think of the Markush group is that the patent applicant believes that each member of the set can interchangeably serve the function of that position though some may still be preferable to others. The applicant should not include a member in the group unless they believe that group member would provide an operable embodiment of the invention.

The power of Markush claiming is most evident when combinations of Markush groups are all used within the same claim. The number of possible embodiments of the invention multiply in a combinatorial fashion not practically reproduced by drawing all of the embodiments separately. The illustration of a compound claim containing multiple Markush groups is shown in Figure 5.9. Each of the variable descriptors describes a separate and independent set of possibilities, even when they are being chosen from the same group, in this case made evident by the fact that each variable descriptor has a separate designation (R_a, R_b, etc.). To make sure that each variable group cannot be construed as being dependent in any way one from another group, it is common practice to separately state that *each* variable is *independently* selected from the list of possibilities.[24]

Markush groups will frequently contain possibilities that may not be precisely defined. For example, the set of atoms defined by "halogen" in Figure 5.7 is clearly understood. But note that within the Markush group describing the alternative possibilities for R_c and R_d in Figure 5.9, the group "C_{1-3} alkyl" was included. This definition actually allows for a range of possibilities at the position since it also covers a C_1, C_2, or C_3 alkyl. A C_1 alkyl is a methyl group (-CH_3), a C_2 alkyl is an ethyl group (-CH_2CH_3), and a C_3 alkyl is a propyl group (-$CH_2CH_2CH_3$). Since a propyl alkyl group can exist as two isomers, it *might* also include (-$CH(CH_3)_2$),

---

[24]The inclusion of the additional language requiring *each* variable to be *independently* assigned to one of the group choices is probably more critical when two positions have the same designation—for example, when two positions are both designated *R* but the drafter wishes each R to be independently assigned, one from the other, meaning that each R can have the same or a different group member chosen for it. Presumably the *each independently* language should accomplish this purpose; however, it would be safer to simply assign them each a different variable to begin with.

which means an isopropyl group. If both *n*-propyl and iso-propyl groups were meant to be covered in the claim, it would be preferable to specify this within the patent.[25]

5.6   CLAIM CONSTRUCTION

When one asks what a claim means in the nonlegalistic sense, there really is no single answer since it might mean different things to different people. Unlike the universality of a surveyor's platte where the real property being described is a function of its coordinates, patent claims are subject much more to the language choice of the drafter and the interpretations of the reader. Moreover, the multidimensionality of claimable subject matter further complicates an already sometimes vague and shifting landscape. Inherent difficulties aside, the central importance of predictable claim interpretation cannot be understated. Whether one is reading a claim and trying to understand what it actually covers or writing a claim in a way that clearly conveys the desired scope, the importance of common understanding is the key. Since both the reader and writer of claims may have their own particular ideas of what a claim might mean in a given circumstance, the question of what a claim means cannot reside in any given individual's understanding; there must be a common reference point from which all parties can try to orient themselves to. So instead of asking what the inventor intended or what the infringer thought the claim meant, claim terms are given their "ordinary and customary" meaning as understood from the hypothetical vantage point of *a person having ordinary skill in the art* (abbreviated as PHOSITA) at the time of the invention. This is reasonable since patents are typically read by the individuals skilled in the art to which the patent pertains. It is usually a chemist that will read a chemical patent and not a layperson (or a judge), and it is the chemist who will need to understand what the claim means to understand the metes and bounds of the claimed invention.[26] This is especially important for judges in federal courts who try patent cases to keep in mind as they will usually not have the specific technical competence required for understanding the terms of the claims in the context of the technical field of the patent.[27]

---

[25]Whether a C_{1-3} alkyl group includes both *n*-propyl and iso-propyl is a matter of claim construction and how one of ordinary skill in the art would understand such a group to mean. As we will see in the next section, the patentee can define his own terms and if he doesn't do so clearly, it is up to the court, using techniques of claim construction to figure out what is meant by definitional terms in the claims. The safest and best patent practice is for the patentee to provide the definitions in the specification to minimize any ambiguity.

[26]This does not imply that the drafter's intentions are always without force. When the drafter clearly sets out the meaning of claim terms in the patent specification, then those claim terms must be read in accord with the drafter's clear intention. This is not entirely at odds with the preceding discussion since a PHOSITA will take the claim meaning from a thorough reading of the claim terms in view of the entire specification including alternative definitions supplied by the patent drafter. More is said about this later in this section.

[27]You might wonder why the focus on a federal judge when this text is directed to the practicing chemist. In the U.S. legal system, questions of law are determined by judges, and questions of fact are determined by juries (or a judge sitting in as jury), and the determination of claim scope (usually called *claim construction*) is a matter of law. Given the primacy of the claim scope to the questions of claim validity or

It is not surprising that a reasonable place to begin the search for a claim's meaning is the context of the claim itself. Although claims will not often define the terms they include, those term definitions maybe clarified or at least somewhat delineated when put into the context of the rest of the claim language or more broadly, in the language of one or more of the additional claims in the patent. In one recently decided case heard on appeal in *Mars, Inc. v. H.J. Heinz Co.*,[28] claim 1 from U.S. patent 6,312,746 shown below was in dispute.

1. A dual texture pet or animal food product comprising:

   a soft inner component of a dual texture pet or animal food product *containing a mixture of lipid and solid ingredients*, the first component having a water activity, a_w, less than about 0.65 and a total moisture content less than about 15 wt %;

   a cereal based shell component of the dual texture pet or animal food product containing at least one ingredient comprising a carbohydrate, fat, protein or combination thereof, the shell component having a total moisture content less than about 20 wt %;

   wherein the shell component completely surrounds the soft inner component and is formed by the co-extrusion of the soft inner component within the shell component to form one dual component pet or animal food product.

The central issue in this case was the meaning of the claim language "containing a mixture of lipid and solid ingredients." The plaintiffs asserted that the defendant's

infringement, before the trial the judge will conduct what is referred to as a *Markman hearing*, where the claim construction will take place. In much litigation, this is the definitive event because once the claim scope is clarified, the resolution of the pendant issues (novelty, infringement, etc.) typically become more apparent, and often the parties will settle after the claim scope issue is resolved. To assist the judge in understanding the meaning of the invention terms, one or usually both sides will enlist the aid of expert witnesses to explain the technology at hand. It's not surprising that each side's expert witness will agree with the interpretation most favorable to the side paying them. Holding that claim construction is a matter of law is not without its controversy. CAFC Judge Mayer (with Judge Newman) noted that:

> Now more than ever I am convinced of the futility, indeed the absurdity, of this court's persistence in adhering to the falsehood that claim construction is a matter of law devoid of any factual component. Because any attempt to fashion a coherent standard under this regime is pointless, as illustrated by our many failed attempts to do so, I dissent.... While this court may persist in the delusion that claim construction is a purely legal determination, unaffected by underlying facts, it is plainly not the case. Claim construction is, or should be, made in context: a claim should be interpreted both from the perspective of one of ordinary skill in the art and in view of the state of the art at the time of the invention.... These questions which are critical to the correct interpretation of the claim, are inherently factual. They are hotly contested by the parties, not by resort to case law as one would expect for legal issues, but based on testimony and documentary evidence (*Phillips v. AWH Corp.*, 415 F.3d 1303 (CAFC 2005)).

Whether claim construction is a mix of law and fact is an important distinction since a federal court of appeals will review a district court's findings of fact with deference and reverse on a finding of "clear error" only, whereas matters of law are reviewed completely again (de novo) on appeal. Since each appeal on claim construction is reviewed without deference to the district court's finding and claim construction is a very nuanced, complex (and dare we say, often fact based) undertaking, the net result is that many of the district court's claim constructions are modified or reversed on appeal.

[28] *Mars, Inc. v. H.J. Heinz Co.*, 377 F.3d 1369 (CAFC, 2004).

pet food composition infringed claim 1 of their patent, while the defendant's argued that the claim did not cover their pet food product because the claim covered only "lipid and solid ingredients" and that the term ingredients meant that the mixture applied only to the starting components before the actual process of making the final product. The defendant's further argued that since their intermediate product contained a syrup as a starting component, it was outside the literal scope of the claim, because the syrup was neither a lipid nor a solid ingredient. Thus, by focusing solely on the term ingredients, the defendants were trying to make their argument that the claim referred to a fixed point in time, before the actual processing of the ingredients (like the ingredients to a recipe before their actual incorporation in the recipe).[29] The CAFC did not agree, stating

> We conclude that: (1) "ingredients" as used in the phrase "a mixture of lipid and solid ingredients" refers to the components of the inner core at any time after they have been mixed together, and (2) the phrase "containing a mixture" is open-ended. Thus, the claim language, "containing a mixture of lipid and solid ingredients," does not exclude the presence of additional, unnamed ingredients in the inner core mixture that are neither lipids or solids.[30]

In reversing the District Court's decision, the CAFC noted that the term *ingredient* could not be taken out of context since what was being referred to in the claim was already a *mixture* of ingredients. Since the ingredients were already mixed, to some extent they had already been processed, and it would no longer be persuasive to argue that the term was limited to the starting ingredients only. Thus the CAFC held that the term *ingredients* referred to the components at any time after they had been mixed together and included the ingredients as found in the final finished product.

Beyond the immediate context of the claims themselves, claim language should never be taken out of its broader context. In the realm of patents, this broader context includes the entire specification as well as the prosecution history of the patent.[31] While claim terms are given their normal and customary usage, the normal

---

[29] Apparently after processing, the syrup would become part of the solid, and the finished product would be more likely to infringe if the claim was read as being to a product rather than an unprocessed preproduct composition.

[30] *Mars, Inc. v. H.J. Heinz Co., id* at 1377.

[31] *Phillips v. AWH Corp.*, 415 F.3d 1303 (Fed. Cir. 2005). In the normal course of review at the Court of Appeals of the Federal Circuit, appeals from the district courts are typically heard by 3-judge panels. In some instances, different panels will issue apparently conflicting decisions. Often times in such cases, the full Court, consisting of 12 judges will vote and decide by majority to hear and decide the issue (absent personal conflict issues, which may cause 1 or more judges to recuse themselves from a particular case) in what is known in an en banc decision. The holding of the en banc decision will then be established as the precedential rule absent overturning by a U.S. Supreme Court decision or a subsequent en banc ruling. The recently decided *Phillips en banc* decision was just such a decision. Before the *Philips en banc* rehearing, it appeared the court was split in its interpretation of claim language, with some members giving primacy to dictionary or treatise definitions to interpret claim language, whereas others gave primacy to the claim language in the context of the specification. The subject matter of *Phillips* concerned high security walls for use in prisons, where the invention comprised steel-shell panels that could be welded together to form vandalism resistant walls. The panels also contained internal baffles that could help deflect penetrating

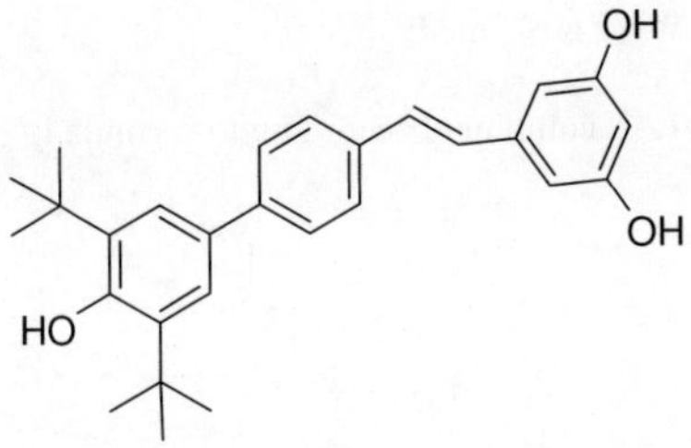

FIGURE 5.10   *Tris*-phenolic food additive.

and customary usage can be overridden by definitions supplied in the patent itself. The federal courts have consistently held that the patentee can act as "his own lexicographer," and this means that patent specifications may provide definitions that vary from a more standard definition, but where the patentee places an unusual meaning to a given term, he must do so "clearly and deliberately."[32] The ability of the patentee to act as his own lexicographer is critical to keep in mind when one is the patentee writing the definitions in the specification. If claim terms are not defined clearly, one may be later surprised to realize that a court has decided a different meaning should be ascribed to those claim terms than what you, the patentee, had intended. Thus it is good practice to define each of the critical claim terms in the patent specification. Likewise, one needs to be equally aware of the primacy of a patentee's definitions when reading the claims of others' patents.

To better appreciate why reading the specification definitions is so important, let's consider the following example. Imagine a chemist has discovered a tris-phenolic compound (Figure 5.10) and is particularly excited about its antioxidant activity, which will be useful as a food additive due to the compound's apparent lack of toxicity.

Hoping to move the compound forward into development, the inventor decided to look through a few patents in the area just to make sure that he wasn't "stepping

---

objects, such as bullets, but the claims were silent as to the angle of the baffles. On its face, it would appear that the baffles could be mounted at a 90-degree angle to the wall, but the specification gave no such examples, and furthermore, such an angle would appear to controvert the ability of the baffles to deflect penetrating objects, which was a stated purpose of the invention. An alleged infringer sold the units with the baffles oriented at 90 degrees, and the plaintiff sued. The *Phillips en banc* court upheld a District Court decision finding no infringement due to a reading of the claims in view of the specification, which indicated that the baffles would not be perpendicularly located to the walls. In *Phillips v. AWH, id.*, the court approvingly quoted *Renishaw PLC v. Marposs Societa' Per Azioni*, 158 F.3d 1243, 1250 (CAFC 1988) for the following: "Ultimately, the interpretation to be given a term can only be determined and confirmed with a full understanding of what the inventors actually invented and intended to envelop with the claim. The construction that stays true to the claim language and most naturally aligns with the patent's description of the invention will be, in the end, the correct construction."

[32] See, *Merck & Co. Inc. v. Teva Pharmaceuticals USA, Inc.* 395 F.3d 1364 (CAFC 2005), quoting *Union Carbide Chems. & Plastics Tech. Corp. v. Shell Oil Co.*, 308 F.3d 1167, 1177 (CAFC 2002): "[the] presumption in favor of the claim term's ordinary meaning is overcome, however, if a different meaning is clearly and deliberately set forth in the intrinsic evidence."

What is claimed:

1. A compound according to formula I:

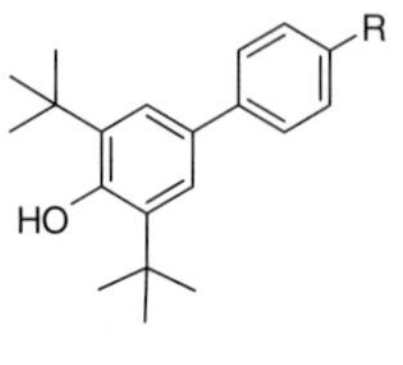

I

where: R = hydrogen or C_{1-6} alkyl.

FIGURE 5.11 Issued claim in patent.

on anybody's toes." One recently issued patent was called to the inventor's attention, but after reading over the closest claim he was comfortable that it was not even close. The relevant claim of concern is reproduced in Figure 5.11.

Looking at the definition of the R-group, the inventor was content to apply a "normal" definition of C_{1-6} alkyl. Confirming his hunch by looking up the definition of *alkyl* in some standard texts, he was confident that his compound of interest was clearly outside the scope of the claimed genus.[33] However, had he taken the time to read the entire patent specification, he would have learned that the patentee as lexicographer had a different definition in mind. In the definitions part of the specification, the following definition for C_{1-6} alkyl was provided:

> C_{1-6} alkyl is a carbon chain of between 1 and 6 carbon atoms in length, wherein said carbon chain may be optionally unsaturated, thus including alkenes and alkynes. Said C_{1-6} alkyl group may be optionally substituted with from 1-3 substituents, wherein said substituents are selected from halogen, hydroxyl, nitro, cyano, and phenyl (wherein said phenyl is optionally substituted with from 1-3 substituents selected independently from the group consisting of halogen, hydroxyl, cyano and nitro).

If the inventor in our hypothetical had carefully read the specification, he would have realized that C_{1-6} alkyl was defined broadly enough that together with the rest of the claimed Markush would encompass the compound he wished to make and use.[34]

---

[33] The Merriam-Webster online dictionary (2007) defines *alkyl* as "having a monovalent organic group and especially one C_nH_{2n+1} (as methyl) derived from an alkane (as methane)." *Wikipedia* notes that "an alkyl is a univalent (or free) radical containing only carbon and hydrogen atoms arranged in a chain. The alkyls form a homologous series with the general formula C_nH_{2n+1}." In normal patent terminology, a *radical* means that a group is connected to something else and thus relies on the group it is connected to fulfill its normal chemical valency.

[34] Working backword from the claim itself, we see that the left-hand portion of the two organic structures are the same. It remains to be seen if the *R* group in the claimed structure is broad enough to encompass the right-hand side (vinylbenzene-1,3-diol) of the molecule. Since R is defined as C_{1-6} alkyl and the

This example, while being of a fairly typical ilk for patents dealing with organic chemistry subject material, is somewhat straightforward in the sense that a fairly clear definition for the claimed Markush was available in the specification. Often, however, the specification does not provide such a clear definition, and one needs to look for clues through the context in which the term is used. In the absence of a clear definition, a consistent association of a term in a particular nexus can reinforce the term's meaning just as clearly as if the term had been set out and separately defined.[35]

Another source of context within the specification that one looks to for clarification of terms are the examples or embodiments detailed in the specification. If we are trying to decide between a narrower or broader reading of a certain claim, the presence of examples that fall outside the scope of the narrower reading but within the scope of a broader reading could persuade us that the inventor intended the broader reading for that term. However, we need to be more cautious when narrowing claim term scope based solely on the examples in an invention. Examples of the invention

---

specification provides that C_{1-6} can be optionally "unsaturated thus including alkenes," the C_2 alkene would be included within the R group definition. The definition doesn't stop there since the alkene can also be optionally substituted to include, among other things, an optionally substituted phenyl group (a benzene radical is commonly referred to as a phenyl group). In this case, the phenyl group can itself be optionally substituted by one to three independently selected groups, including hydroxyl. Putting it all together, the claimed Markush, together with the definition in the specification, appears to cover the inventor's compound.

For this particular example, we focused in on an issue related to freedom to operate. For now, it is important simply to appreciate that critical claim terms may be defined differently in a patent specification than one might typically appreciate and that those definitions, if "clear and deliberate," can change the breadth of a claim in a way that one might not predict from the language of the claim itself.

[35] Even in cases where the nexus is not so abundantly clear, the courts will still look to the rest of the specification to provide the context. For example, in *Ortho-McNeil Pharmaceuticals, Inc. v. Caraco Pharmaceutical Laboratories, Ltd.*, 476 F.3d 1321 (CAFC 2007), the Court construed the range limit for a claimed ratio of the drug tramadol and acetaminophen. The claim in question was directed to a weight ratio of tramadol to acetaminophen of *about* 1:5. The defendant's composition would contain, at its closest ratio, 1:7.5 tramadol to acetaminophen. The Court had to determine whether the defendant's composition of 1:7.5 tramadol to acetaminophen would infringe the claim to "about 1:5." In confirming that the defendant's composition *did not* infringe the claim at issue, the Court upheld the District Court's ruling that the claimed amount "about 1:5" extended no further than 1:3.6 to 1:7.1. The Court noted from one of its previous decisions that the term *about* depends on the "technological facts of the particular case" (quoting *Pall Corp. v. Micron Separations, Inc.* 66 F.3d 1211, 1217 (CAFC 1995)). In supporting the District Court's narrow reading of the claim term, the Court pointed out that the patent specification and many of the claims were directed to ranges of ratios such as "1:1 to about 1:1600." By so doing, the Court held that "one of ordinary skill in the art would understand the inventors intended a range when they claimed one and something more precise when they did not." The Court also noted that broadly interpreting "about 1:5" could "render meaningless another claim's limitation, mainly the 1:1 limitation." In addition, the specification described numerous data points, including ratios of 1:1, 1:3, 1:5, 1:5.7, and 1:15. The Court found that had the patentees found a claimed range desirable, they could have claimed a range, for example, 1:1 to 1:5. As for the 1:3.6 to 1:7.1 range that the District Court interpreted "about 1:5" to mean, the court found no error in the District Court's construction, which came from an expert's testimony about the statistical variability reported in the specification together with "his expertise" that "about 1:5" would not be statistically different from a ratio of 1:7.1 to 1:3.6. Based on this construction and the fact that 1:7.5 was outside the constructed claim range, the Court found no literal infringement of the claim at issue.

do not necessarily define the full scope of an invention, and thus one should avoid "importing limitations from the specification" into the claims."[36] Invention examples are often put in simply to show the practitioner different ways the invention might be used but they should *not be* strictly read to limit the scope of the invention. This is important to understand because one's instincts when reading somebody else's patent (in the freedom-to-operate context) is to narrow the claimed invention at least enough so that what she wishes to do falls outside the scope of the other's claimed invention. One commonly hears exhortations to the effect: "Heck, they didn't even use the solvent that I'm using, there's no way that claim covers my process!" or "I didn't see one example with a phenyl group in that position, that claim can't be any good!."[37] While weak or nonexistent exemplification may be considerations directed to understanding whether an invention satisfies the written description requirement, one should be extra careful before reading limitations from the examples into the claims.

The terms of a claim must also be read in view of the prosecution history of the patent. As we covered in Chapter 2, prosecution history is a record of the correspondence between the patent applicant and the USPTO, which is kept in a file corresponding to that particular application. During the course of that prosecution, the patent applicant (usually through their representative) will make statements in support of the patentability of their claimed invention and will sometimes define certain terms in a way that narrows their definition. Often times such narrowing arguments are used to support a contention that their invention is different from prior art that is being cited against one or more of the claims at issue. Through what can sometimes amount to a "give and take" type exchange between the applicant and the patent office, it is possible to better understand how the inventor and the USPTO understood the invention. However, one must be more cautious when trying to glean terminology from the prosecution history since the record may reflect what amounts to a conversation where definitions are seemingly advanced at one point only to be retracted later.[38] Nevertheless, the patentee will not be allowed to expand definitions in the prosecution history where such expansions are not clearly supported by the patent specification[39] but nor will the court limit a patentee's invention based on prosecution history absent "clear and unmistakable prosecution arguments."[40]

---

[36] *Phillips v. AWH, id.*

[37] We know these types of statements didn't come from you, this sentence is really directed to the other, less-patent savvy chemists out there.

[38] *Phillips v. AWH, id.*

[39] In *Honeywell v. ITT Industrials et al.*, 452 F.3d 1312 (CAFC 2006) the CAFC held that the claim scope was limited to a fuel filter and not all fuel components made of molded material as was argued by the patentee in the prosecution history of the patent. In so holding, the CAFC explained: "Honeywell places too much weight on that statement, as we find it to be ambiguous and possibly inconsistent with the written description. After all, the only fuel component disclosed and claimed in the patent was a fuel filter. In any event, such a broad and vague statement cannot contradict the clear statements in the specification describing the invention more narrowly."

[40] See *SanDisk Corp. v. Memorex Products, Inc.*, 415 F.3d 1278 (CAFC 2005), where the CAFC held that a claim term can be limited where the "patentee makes clear and unmistakable arguments limiting the meaning of a claim term in order to overcome a rejection." The CAFC found that this condition was not satisfied by a prosecution argument that "is subject to more than one reasonable interpretation."

This all might seem quite complicated because, well, it is complicated. This is especially true when claim terms that are not clearly defined in the specification lead to ambiguities. Likewise, claim construction issues can also arise when clear claim language is rendered ambiguous when read in view of the rest of the specification or prosecution history (latent ambiguity). For one's own patents, one can best avoid this issue by being proactive through clearly defining terms in the specification. Whether acting as your own lexicographer by providing clear definitions that may not jive with more art-accepted terminology (and thus clearly repudiating the art-accepted definition), or perhaps even better yet, by providing clear definitions even where those definitions are consonant with more art-accepted terminology, one can do much to avoid many claim construction problems. The alternative can amount to letting those claim terms be defined by federal judges through the uncertain and expensive process of litigation.

CHAPTER 6

Basic Requirements of Patentability: Utility

The woods are lovely, dark and deep,
But I have promises to keep
And miles to go before I sleep
—Robert Frost, "Stopping By Woods on a Snowy Evening"

6.1 THE SIX REQUIREMENTS OF PATENTABILITY

The remaining portions of this text will deal almost exclusively with the requirements of patentability—the substantive law. As we wander through these sometimes dense woods, bumping against a seemingly endless and sometimes bewildering array of terms, rules, exceptions, interpretations, comments, opinions, and so on, it would be all too easy to get lost and forget what we set about to learn in the first place. We should be careful not to lose the forest for the trees in front of us. To help avoid getting lost, the thinnest of outlines is provided in this section; let it be your compass. If you find yourself buried in a section and feeling a little confused (like perhaps you're losing the big picture), please be encouraged to turn back to this section and review the basic requirements of patentability. These are the guiding principles of patent law, all the rest is just nuance and texture.[1] Ready?

a. §**101: Thou Shalt Patent** *Useful* **Inventions Only.** This is the *utility require-ment*. Essentially, there are two parts to the utility requirement. The first part deals with whether the invention falls within statutory subject matter. In effect, not all inventions are created equal, and some fall outside the scope of what can be patented. In order to be statutory subject matter, the claim must be directed to a process, machine, manufacture, or composition of matter. The second part of the utility requirement asks whether the thing to be patented is actually useful. I know what you're thinking: "If the thing isn't useful, why

---

[1]Lots and lots of nuance and texture.

The Chemist's Companion Guide to Patent Law, by Chris P. Miller and Mark J. Evans
Copyright © 2010 John Wiley & Sons, Inc.

would I want to patent it anyway?" Ask that question to the guys who invented the non-stick tape.

b. **§102: Thou Shalt Patent *Novel* Inventions Only.** This requirement is referred to as, surprise, the *novelty requirement* (patent law is tricky, isn't it?). People have this strange tendency to copy each other and then somehow think that it was their idea to begin with. When somebody steals one of your jokes, that's one thing because you probably stole it from somebody else in the first place. But when somebody tries to patent something that isn't new, she is basically trying to steal from everybody; this is really antithetical to the primary purpose of patent law, which is to encourage people to make *new* things.

c. **§103: Thou Shalt Patent *Nonobvious* Inventions Only.** It's a funny thing that happens when people really get their creative juices flowing. They see something they like and then think: "If I change that just a little bit, it will be new and nobody can say that I copied it." Oldest trick in the book, but somewhat surprisingly, patent law doesn't really mind you doing this—just so long as the end result is not an obvious variant of what you started with. In other words, if you're going to copy something and then change it around, you need to make sure that it is changed around enough so that the changes you made would not have been obvious (got it?). But, as we will also learn, if it does appear obvious, the result of your changes must be better than what one would have expected.

d. **§112 ¶1: Thou Shalt Provide Adequate *Written* Description.** The *written description* requirement demands that the patent specification provides a written description of the work sufficient to show that the inventor was in possession of the claimed invention. This means primarily that there has to be enough detail in the patent specification to adequately support the claims. More will be said on what is meant by "adequately support" but for now let's appreciate that through the course of prosecuting a patent application, the originally filed claims often get changed around to meet the patentee's perceived needs as well as the other requirements of patentability (e.g. novelty, nonobviousness). During this course of changing the claims during the patent prosecution, the patent attorneys and agents sometimes get carried away and end up trying to claim something that is not really in their patent application. Let's just say that the patent office frowns on this type of improvisation.

e. **§112 ¶1: Thou Shalt *Enable* Your Invention.** To qualify for a patent, the application, as submitted, must teach one of ordinary skill in the art (the PHOSITA) how to make and use the claimed invention without undue experimentation; this is commonly referred to as the *enablement requirement*, and it is part of the patentee's grand bargain with the public, the one where the inventor receives a valuable right in return for enabling the public in how to make and use the invention for which the patent is granted. We just learned that the whole purpose in getting a patent is so that we could exclude others from making and using our claimed invention. But in order to get the patent, we have to teach them how to make and use the claimed invention. Ironic, yes?

Utility
Novelty
Nonobviousness   ————   **UNN-(WEB)!**
⎛Written Description⎞ ————
⎜Enablement   ⎟
⎝Best Mode   ⎠

FIGURE 6.1   Patentability requirement mnemonic.

f. **§112 ¶1: Thou Shalt Teach the *Best Mode*.** The *best mode requirement* means that that the patentee must not only teach one of ordinary skill in the art how to make and use the claimed invention but the patentee must also disclose the best mode for the claimed invention. It is not acceptable to play hide the ball in a patent application. If the *patentee(s)* believe(s) certain examples or embodiments of the claimed invention are the best way of practicing the claimed invention then they have a duty to disclose those examples or embodiments or they risk rendering their patent nonenforceable (they still get to keep the certificate as a souvenir though!).

Unfortunately, that does not leave much in the way of a good mnemonic but if you find this sort of thing helpful, take a look at Figure 6.1. The last three requirements are grouped together (WEB) because they all relate to the level of disclosure required in the patent application.[2] Here and in subsequent chapters, we will delve into each of these patent commands in more detail.

6.2   STATUTORY SUBJECT MATTER OF THE UTILITY REQUIREMENT

The "first door which must be opened on the difficult path to patentability" is the utility requirement.[3] The utility requirement in U.S. patent law traces its origins to the Article 1, Section 8 of the U.S. Constitution, which states: "To promote the progress of science and *useful* arts, by securing for limited times to authors and inventors the exclusive right to their respective writings and discoveries" (emphasis added). The utility requirement has been legislatively codified in 35 USC §101:

> Whoever invents or discovers any new and *useful* process, machine, manufacture, or composition of matter, or any new and useful improvement thereof, may obtain a patent therefore, subject to the conditions and requirements of this title (emphasis added).

There are two basic components to the utility requirement as included in §101. The first concerns itself with statutory subject matter, meaning the types of inventions or discoveries that are patentable, which are limited to a "process, machine, manufacture

---

[2]The written description, enablement and best mode requirements all fall under the first paragraph of the same statute, 35 U.S.C. §112 ¶1.
[3]*In re Bergy*, 596 F.2d 952, 960 (CCPA 1979)).

or composition of matter." The second component is the requirement that the invention falling within the defined statutory subject matter categories be "useful," where the definition of useful relates to both the quantitative level of usefulness as well as the credibility of the proof of that usefulness.

Statutory subject matter that qualifies for a patent has to be a "new and useful process, machine, manufacture, or composition of matter or any new and useful improvement thereof." The courts have generally found that the types of discoveries that are not patentable subject material are abstract ideas, laws of nature, and natural phenomena; in effect, the invention or discovery must be reducible to a tangible form.[4] However, one might still be able to patent practical applications of any of these three categories. For example, even though Newton discovered gravity, he would not have been entitled to a patent on gravity itself, since gravity is a law of nature. However, a hydroelectric dam that generates electricity from gravity action on water might have been patentable. Likewise, one cannot patent a product of nature (mineral, natural organic product, protein) but one might be able to patent a manipulation of that product of nature such as patenting an isolated form, a method of using the product, etc.

Understanding the point where a discovery crosses over from being the simple recognition of a scientific law to the practical application of that law is not always easy. For example, imagine a researcher were to "discover" that the level of folate and cobalamin (two vitamins) in a person was inversely related to that same person's homocysteine levels such that a high homocysteine level would indicate a folate and cobalamin deficiency. As you can readily discern, such a discovery is not a process, machine, manufacture, or composition of matter but rather a natural relationship, akin to a natural law, and a claim to the relationship itself would not be patentable subject matter. However, think of how this discovery might be rewritten in a way that it could broadly protect the discovery without actually requiring any additional inventive input. In particular, imagine that the following claim was included in a patent application:

1. A method for detecting a deficiency of cobalamin or folate in warm-blooded animals comprising the steps of:

   assaying a body fluid for an elevated level of total homocysteine; and correlating an elevated level of homocysteine in said body fluid with a deficiency of cobalamin or folate.

Assume that monitoring homocysteine levels was not new at the time of the filing but that the combination of assaying the homocysteine level and correlating that level with a folate or cobalamin deficiency is new. As you can see, this claim now describes a process since it no longer stakes its claim in the natural law itself. By converting the natural law to a process that merely uses the principle of the natural law, the patentees gave their discovery the broadest possible scope without claiming

---

[4]MPEP 2106 IV. See also *Mackay Radio & Telegraph Co. v. Radio Corp. of America*, 306 U.S. 86, 94 (1939), which states "While a scientific truth, or the mathematical expression of it, in not patentable invention, a novel and useful structure created with the aid of knowledge of scientific truth may be."

the law itself. With this claim, anybody who assays an individual's body fluid and correlates that result with a cobalamin or folate deficiency will infringe this claim. Notice that the claim does not require anything more. Do you think that a federal court with patent jurisdiction would find such a claim to be patentable subject matter since it is literally a process, or do you believe that the court would find that the patent applicants have simply rephrased a law of nature in such a way that it merely resembles a process and is actually aimed at capturing the full scope and value of the natural law? As it turns out, we may never know for sure![5]

As you might appreciate from the forgoing, the dividing line between a claim to a natural law or relationship and a process using that relationship can at times be very thin, gray and sometimes wavering.[6] For those practitioners of the chemical arts that

---

[5]The facts from the "hypothetical" case recited above were taken directly from a recent case that was appealed all the way to the U.S. Supreme Court (126 S.Ct. 2921 (2006)), which initially agreed to hear the case for the issue of whether the patent claim reproduced above (claim 13 from US 4,940,658) was patentable subject matter. Instead of hearing the case and issuing a decision on that issue, the Supreme Court ultimately discharged the case without a hearing on the issue, finding that the issue had not been properly raised in the lower federal court (CAFC) hearing (*Metabolite Labs., Inc. v. Lab Corp. of Am. Holdings*, 370 F.3d 1354 (CAFC 2004)) and thus it was not appropriate subject matter for appeal. Nevertheless, three of the Supreme Court's justices (Breyer, Stevens and Souter) issued a dissent from the decision to not hear the case on procedural grounds, and it is clear the dissenters would have found the claim invalid. The following quote from the three-judge dissent summarizes the policy grounds for not allowing certain subject matter to be patentable:

> The justification for the principle does not lie in any claim that "laws of nature" are obvious, or that their discovery is easy, or that they are not useful. To the contrary, research into such matters may be costly and time-consuming; monetary incentives may matter; and the fruits of those incentives and that research may prove of great benefit to the human race. Rather, the reason for the exclusion is that sometimes too much patent protection can impede rather than "promote the Progress of Science and useful Arts," the constitutional objective of patent and copyright protection. The problem arises from the fact that patents do not only encourage research by providing monetary incentives for invention. Sometimes their presence can discourage research by impeding the free exchange of information, for example by forcing researchers to avoid the use of potentially patented ideas, by leading them to conduct costly and time-consuming searches of existing or pending patents, by requiring complex licensing arrangements, and by raising the costs of using the patented information, sometimes prohibitively so. Patent law seeks to avoid the dangers of overprotection just as surely as it seeks to avoid the diminished incentive to invent that underprotection can threaten. One way in which patent law seeks to sail between these opposing and risky shoals is through rules that bring certain types of invention and discovery within the scope of patentability while excluding others (126 S.Ct. 2921, 2922–2923).

[6]In a recent but unrelated opinion, the CAFC provided some clarification in regard to where the line between a strictly mental process begins and a patentable process takes place. In the decision of *In re Stephen W. Comiskey*, (CAFC 2009), the CAFC took a case on appeal from the USPTO regarding a "business method" patent. In the action on appeal, the patentability of several claims directed toward methods of conducting arbitration disputes were analyzed by the CAFC in regard to whether they were in a statutory subject matter, i.e. was the process described one of the useful arts? The broadest claims in the application described processes that were independent of any tangible device such as a machine or computer to manipulate the data input, presumably it could all be done by mental processes. The application contained additional claims that required that the data input be manipulated or in some way be handled by a "computer or other machine." The CAFC distinguished between the two types of claims, rejecting those

rely on the more theoretical aspects of the chemical arts (e.g., computer/molecular modelers), the statutory subject guidelines of the utility requirement need to be considered carefully so that the claimed "invention" represents more than an algorithm or the recognition of a relationship. For the majority of us more applied practitioners of the chemical arts, the discovery and patenting of basic scientific principles is not likely to be a common occurrence and thus will not likely affect our day-to-day work in the industry.

6.3 WHAT MAKES A CHEMICAL INVENTION USEFUL?

Now, let's turn our attention to the second component of the utility requirement, which requires that the claimed subject matter must be useful. As one court opined: "a simple everyday word [such as *useful*] can be pregnant with ambiguity when applied to the facts of life."[7] The contours of the utility requirement can be defined by two principles. First, the claimed invention must have "specific" and "substantial" utility and must provide sufficient information to make its use "readily apparent to those familiar with the technological field of the invention."[8] Second, the alleged utility must be credible.

Great ... so what does this all mean? First, one can think of the requirements of specific and substantial as being one, where the combination means that the claimed invention is applicable to a real-world utility. Probably the easiest way to discern whether an invention has the requisite specificity and substantiality is to look to whether the claimed invention in question can be of immediate use or whether it's utility would require further work or analysis to verify. This means that the utility cannot be speculative. Firming up this definition as it might be applied to chemistry related inventions, some situations *lacking* a specific and substantial utility are listed:[9]

- Basic research, including such things as analyzing or studying the properties of a compound or material.
- A method of treating an unspecified disease or condition. Likewise, a compound alleged to be useful in treating unspecified disorders or possessing useful biological properties, without more, is unlikely to meet the test.

outright that did not require any sort of tangible (machine, computer, communication device) involvement from those that did. In particular, the CAFC found that where a mental process was involved as the sole activity of the claim then the claim was not patentable. However, a claim might qualify as containing statutory subject matter under 35 U.S.C. §101, if a mental process is *part* of a claim where that mental process also involved one or more of additional statutory subjects (i.e., a machine, a manufacture, or composition). Returning to *Laboratory Corp. of America Holdings v. Metabolite Laboratories, Inc.*, could you make the argument that more than mental processes are involved?

[7] *Brenner v. Manson* 383 U.S. 519, 529 (1966).

[8] MPEP, 2107.01 citing *Brenner v. Manson*, 383 U.S. 519 (1966) and *In re Ziegler*, 992 F.2d 1197 (CAFC 1993).

[9] MPEP 2107.01, I, B.

1. A compound having the formula

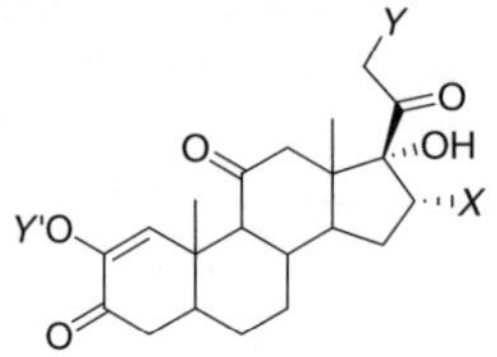

wherein X is selected from the group consisting of hydrogen and methyl, Y (is selected from the group consisting) of hydrogen and an acyloxy radical of a lower organic carboxylic acid having 1 to 7 carbon atoms, and Y is an acyl radical of a lower organic carboxylic acid having 1 to 7 carbon atoms.

FIGURE 6.2 Compound claim from *In re Joly*.

- An intermediate product for use in making a final product, where the final product itself has no specific, substantial, and credible utility.
- A method for making a material where the material has no specific, substantial or credible utility.
- A method of assaying or identifying a material where the material itself has no specific, substantial or credible utility.

To better elucidate the utility requirement as it relates to the chemical arts, let's look at a federal court case dealing with the issue of when claims in a patent application to a novel chemical moiety and a process for making that moiety lack a specific and substantial utility. The case of *In re Joly*[10] arose out of a patent application for a group of steroid precursor compounds. The first compound claim of the patent application on appeal is reproduced in Figure 6.2 and is considered illustrative for purposes of the remaining discussion. Additional composition of matter claims 2 and 3 were directed to specific compounds falling within the genus of claim 1, and claims 8 through 12 were directed to processes for making the claimed compounds. In the applicant's specification, he explained the utility of the compounds of the invention as follows: "The products obtained by the invention, the lower organic carboxylic esters of enols in the 2-position of Δ 1,2-steroids, are particularly useful as intermediates in the preparation of steroids, especially steroids having a ketone group in the 2-position, such as 21-acetoxy-pregnane-17α-ol,2,3,11,20-tetraone and 16α-methyl-pregnane-17α-ol-2,3,11,20-tetraone by acid hydrolysis and customary separation steps." The USPTO examiner rejected these claims, stating "the applicants have stated only that the (claimed) final products may be converted to the corresponding 2-keto compounds, which compounds have no known utility."[11]

---

[10] *In re Joly*, 376 F.2d 906 (CCPA 1967).
[11] *In re Joly* at 1161.

The applicant appealed the rejected claims to the USPTO Board of Appeals, which affirmed the examiner's rejections:

> The portions of this specification ... show that the claimed compounds can be used to prepare the corresponding 2,3-keto compounds. Having arrived at this point, what has been accomplished? Appellants do not assert that the latter compounds have known utility ... such conversion does not constitute a disclosure of utility for the claimed compounds because there is no disclosed utility nor any indication of a known utility for the 2,3-keto derivatives. A useless product does not become useful by virtue of conversion into another useless product.[12]

The applicants subsequently appealed the rejection to the U.S. Court of Customs and Patent Appeals (CCPA).[13] In affirming the Board's rejection, the CCPA refuted the applicant's argument that the 2,3-diketosteroids that could be prepared from the claimed compounds, using the claimed processes, were "closely related" in chemical structure to compounds of known usefulness, the glucocorticoids prednisone and cortisone. The CCPA found that there was no evidence in the record that the claimed compounds possessed activity in common with the allegedly closely related compounds prednisone and cortisone. Since the rejection of all of the claims was upheld *including* the process claims 8 through 12, it is useful to read the CCPA's relational logic between the two types of claims (compound and process for making compound) as they apply to the utility requirement. Quoting from a sister case decided at the same time, the majority stated:[14]

> [T]he conclusion is inescapable that, just as the practical utility of the compound produced by a chemical process "is an essential element" in establishing patentability of the process, (Brenner v. Manson) 383 U.S. 519 (86 S.Ct. 1033), so the practical utility of the compound, or compounds, produced from a chemical "intermediate", the "starting material" in such a process, is an essential element in establishing patentability of that intermediate. It seems clear that, if a process for producing a product of only conjectural use is not itself "useful" within 101, it cannot be said that the starting materials for such a process—i.e., the presently claimed intermediates—are "useful."

From this opinion, we not only can derive the principle that the claimed compounds themselves are not patentable where their utility is only speculative, but also that the process for converting those intermediates to their final products is also not patentable, where the final products have speculative utility only.[15]

The second aspect of the useful portion of the utility requirement is that the utility asserted in the patent application must be credible. Normally, when we think of

---

[12]*In re Joly, id.*

[13]Why would the applicants care so much about a set of allegedly useless compounds? Note that the CCPA is the predecessor court to the CAFC.

[14]A very similar issue and decision were addressed at roughly the same time in *In re Kirk*, 376 F.2d 936 (CCPA 1967).

[15]The requirement of a showing of a specific and substantive utility is quite often an issue in the biological arts where sequences, tags, genes, etc. are routinely isolated, synthesized, or otherwise identified but often have speculative use only.

credibility in the legal context our thoughts automatically go to a courtroom scene, with a perspiring witness being questioned on the witness stand before a judge and jury—every facial tic or downward glance of the eyes considered as evidence of deceit. In the patent prosecution process, there is no judge or jury in the normal sense but there is a standard by which the application is judged and the jury, if you will, is our ever present touchstone: the hypothetical person having ordinary skill in the art. Since this is an objective standard, we know that the measure of credibility is not the apparent veracity of the applicant's own belief (that would be a subjective standard) but whether one of ordinary skill in the art would find the asserted utility credible in view of the teachings in the specification itself, arguments by the applicant together with what is known in that particular art at the time the patent application was filed.

Naturally, the burden is much greater on the applicant where the utility being asserted is incredible by its very nature (e.g., "solar powered, anti-gravity, pull your-self up and fly by your own bootstrap boots"). In such an instance, the evidence put forward by the applicant must be sufficient to satisfy the hypothetical jury of persons having ordinary skill in the art that the claimed invention is likely to demonstrate sufficient utility. Please appreciate, however, that the claimed invention need not be safe, it need not be the best, it need not even be very good at all.

For chemists working in the pharmaceutical industry, this requirement often raises the question of the credibility of a claim directed to the treatment of a human disease. It is often the case that an invention is alleged to have a certain utility whereby that utility is proved indirectly by showing some effect on an experimental parameter that is believed to relate to a particular outcome in animals, especially humans.[16] For example, imagine a patent application that claims compounds allegedly useful for the treatment of acquired immune deficiency syndrome (AIDS). It is unlikely that such an application will contain data from a clinical trial. Rather the claim may be supported by disclosing results from in vitro assays where the compounds of the invention demonstrate activity, such as inhibition of various viral proteins produced by the human immunodeficiency virus (HIV), which causes AIDS in humans.[17] In such a case, an appreciation of the mechanism whereby the claimed compounds function, through their interaction with certain components of HIV, is helpful for establishing the ultimate utility of treating HIV infection in humans.

As a practical matter, the ability to establish utility indirectly is an important aspect for credibly asserting utility, especially for chemical inventions that are de-signed ultimately for use in humans. As just mentioned, a company or organization probably will not have human data related to the use of their particular compound or composition before their filing of a patent application containing those compounds or

---

[16]It is not incumbent on the patentee to demonstrate how the invention works. In other words, the patentee does not have to indicate a mechanism by which the claimed invention proves useful but simply needs to establish credible evidence that it is useful. Quoting *Fromson v. Advance Offset Plate, Inc.*, 720 F.2d 1565, 1570 (CAFC 1983), "It is axiomatic that an inventor need not comprehend the scientific principles on which the practical effectiveness of his invention rests." However, it is sometimes useful to convince an examiner of the utility of the invention where its working principles are understood. A credible mechanism can sometimes be helpful where empirical data are less believable.

[17]Assays or experiments that are done without dosing a living animal are typically referred to as *in vitro* whereas experiments performed in a living animal are typically referred to as *in vivo*.

compositions. This is logical when one considers that (1) most chemical inventions having pharmaceutical utility never even make it to human testing and (2) once a promising compound or composition is identified, it is usually desirable to file a patent application as quickly as is consistent with preparing a well-drafted and well-supported application. Given the large investment of time, money, and resources required for running human clinical trials, it is not surprising that an organization would want to ensure that the patents are filed and even preferably granted before embarking on such an expensive proposition.[18] Fortunately for those companies making such investments, the courts have found that utility can be established by in vitro data alone, but this does not mean in vitro data will always be sufficient. The following statement is an instructive insight to the CAFC's perspective on the sufficiency of in vitro data for establishing utility:

> We perceive no insurmountable difficulty, under appropriate circumstances, in finding that the first link in the screening chain, *in vitro* testing, may establish a practical utility for the compound in question. Successful *in vitro* testing will marshal resources and direct the expenditure of effort to further *in vivo* testing of the most potent compounds, thereby providing an immediate benefit to the public, analogous to the benefit provided by the showing of an *in vivo* utility.[19]

The types of data required to credibly support an alleged utility will depend on whether, and to what extent, the claimed invention and supporting disclosure comport with established scientific principles. Applying this to chemical inventions asserting pharmacological utility, the courts have found that there need only be a "reasonable correlation" between the demonstrated activity of the compounds and the asserted utility.[20] The reasonable correlation can be established by relying on statistically relevant data, argument or reasoning, or teachings of relevance in the art (journals, etc.). Furthermore, evidence of structural similarity to compounds having known activity may also be taken into consideration.[21] From this, we can discern that the utility determination will be a factual inquiry sensitive to several variables.

---

[18]The court addressed this point in *In re Brana*, 51 F.3d 1560 (CAFC 1995):

> Usefulness in patent law, and in particular in the context of pharmaceutical inventions, necessarily includes the expectation of further research and development. The stage at which an invention in this field becomes useful is well before it is ready to be administered to humans. Were we to require Phase II testing in order to prove utility, the associated costs would prevent many companies from obtaining patent protection on promising new inventions, thereby eliminating an incentive to pursue, through research and development, potential cures in many crucial areas such as the treatment of cancer.

[19]*Cross v. Iizuka*, 753 F.2d 1040, 1051 (CAFC, 1985).

[20]MPEP 2107.03 I, citing *Cross v. Iizuka, id.*; *In re Jolles*, 628 F.2d 1322 (CCPA 1980); and *Nelson v. Bowler*, 626 F.2d 853 (CCPA 1980).

[21]This may seem at odds with *In re Joly, id.*, but one should appreciate that in that case it appeared that structural similarity was the sole basis for the utility argument rather than in addition to other information or data. In other cases, such as *In re Jolles, id.*, a close structural relationship in addition to a shared pharmacological activity with known useful compounds were sufficient for patentability. One should use caution in this regard, however, since such arguments may also be used against an applicant or patentee with a charge that the claimed compounds are obvious in view of the prior art.

As a practical matter, the USPTO examiners are instructed to find a utility assertion credible if it would be believable to one of ordinary skill in the art in the pertinent subject matter based on the totality of the evidence and reasoning provided.[22] A utility assertion is found *not* credible only if the logic underlying the assertion is not consistent with the facts used to support the assertion.[23] The USPTO bears the burden of making the initial showing that a utility is not credible, and examiners are cautioned against rejecting an application for lack of credible utility.[24] Where such rejections have been upheld, they have been typically upheld only when an applicant either asserts no utility in the application or when an assertion of utility would be "incredible in view of contemporary knowledge and where nothing offered by the candidate would counter what the contemporary knowledge might otherwise suggest."[25]

While it is not a typical chemical case, the following finding of no utility due to an incredible assertion is instructive. Professors Dash and Keefe (Dash) appealed to the CAFC to challenge the USPTO's final rejection of claims 4–11 of their patent application for alleged lack of utility and enablement.[26] Dash and Keefe's patent application disclosed an electric cell that generated heat energy from a palladium cathode and inert anode. The cell contained H_2SO_4 and D_2O and was alleged by the applicants in their patent application to produce heat energy, hydrogen, deuterium, oxygen gases, and possibly heavy water through the recombination of deuterium and oxygen. The USPTO argued that the heat energy was used to mean "excess" heat energy and that the invention was directed to achieving cold fusion. The USPTO further produced a significant amount of scientific literature casting doubt on the ability of electrolytic cells to produce cold fusion. The CAFC upheld the rejected claims on appeal, finding that the examiner needed only to "establish that a person of ordinary skill in the art would reasonably doubt the asserted utility" and that the examiner fulfilled that duty "based on the number and quality of cited references that debunked claims of cold fusion." Dash's attempted rebuttals were found lacking by the CAFC. For example, Dash alleged that a comparison of their D_2O cell with a H_2O cell demonstrated that their D_2O cell generated more heat. However, the USPTO examiner asserted that such an increase heat was predicated on known differences in the properties of D_2O and H_2O, such as their different heats of absorption. The examiner's response was found persuasive by the court and likewise for the additional responses offered by the examiner to the numerous other evidence and arguments introduced by the applicants.

---

[22] This is known as the Langer test for utility and is based on the court statement: "As a matter of Patent Office practice, a specification which contains a disclosure of utility which corresponds in scope to the subject matter sought to be patented *must* be taken as sufficient to satisfy the utility requirement of §101 for the entire claimed subject matter *unless* there is a reason for one skilled in the art to question the objective truth of the statement of utility or its scope" (emphasis in the original.) *In re Langer*, 503 F.2d 1380, 1391 (CCPA 1974).
[23] MPEP 2107.02.
[24] MPEP, *id*.
[25] MPEP, *id*.
[26] *In Re John Dash and Patrick S. Keefe* (CAFC 2004).

So let's briefly summarize what we know about utility. The utility of the invention must be statutory subject matter, meaning that it must describe a process, a composition of matter, a machine, or a manufacture. It must be tangible; a purely mental process, scientific law, or mathematical relationship will not do. Further, the utility must be substantial and specific, meaning that the utility must be of immediate and real value and cannot be speculative. Finally, the utility asserted in the patent application must be credible. Fulfill these criteria, and you are on your way to a patent. But don't get too excited yet—as we have seen, the utility rejection hurdle is actually rather low.

CHAPTER 7

Basic Requirements of Patentability: Novelty

In every matter that relates to invention ... we are borrowers.
—Wendell Philips, "The Lost Arts"

By now, you probably have gotten the idea that you can't patent something if it isn't new. This should come as no surprise because we already have recognized that a patent is a limited grant to the patentee, but in return the patentee must have given something in return. If that something the patentee gives to the public is not new, then the erstwhile patentee will be getting something for nothing. Even worse, the patentee (notice we didn't use the word *inventor*) would be taking back from the public what already belonged to the public. Therefore the novelty inquiry is among the first inquiries the prospective inventor should make before embarking on a particular field of research. To appreciate whether something is novel, we need to first understand that novelty is determined in view of what we refer to collectively as the "prior art" (Chapter 3). When we determine whether an invention is novel over the prior art, it is important to appreciate that the invention we are concerned with is the *claimed* invention. It is the claim language (as properly constructed) that defines the invention, and thus the specific claim language used is a very important starting point for determining not only the scope of the invention but also whether it is novel.[1]

---

[1]Claim construction was discussed in Chapter 5, but if you happen to need a quick refresher, here is the Cliff's Notes version. The claim language is paramount for determining the "metes and bounds" of the invention. While it is the words of the claim itself that determine the invention's scope, those words must be read in view of the patent specification and the correspondence generated between the patent attorney prosecuting the case and the USPTO (this is called the *patent prosecution history*). Claim language terms are given their normal and customary usage, unless the patentee has defined the terms otherwise, either, in the prosecution of the patent or the patent specification itself. In this regard, the patentee may be her own lexicographer for purposes of defining their invention. In patent litigation, claim construction is performed first in what is known as a Markman hearing in the federal district court hearing the case. Claim construction is considered a matter of law and thus is the responsibility of the judge overseeing the litigation before it. Once the claims are construed, the parties can than proceed onto additional motions or straight into the trial phase, where either a jury or a judge sitting as fact finder will hear the rest of the case.

The Chemist's Companion Guide to Patent Law, by Chris P. Miller and Mark J. Evans
Copyright © 2010 John Wiley & Sons, Inc.

7.1   REQUIREMENTS OF THE PRIOR ART TO DEFEAT NOVELTY

In Chapter 3, we reviewed what types of information constituted prior art but we did not describe how that prior art is analyzed in view of a later claimed invention. Once a reference is qualified as prior art under one or more of the appropriate 102 sections, the next step is to analyze the properly qualified prior art to determine whether it affects the novelty of the claims we are examining. For the purpose of determining novelty (and also nonobviousness, as we'll see later), *each and every* element and limitation in the claim language must be considered. The reason that each and every claim element and limitation must be considered is that in order for the prior art to render not novel (i.e. to *anticipate*) the claimed invention, the prior art must contain *each* and *every element* and *each and every limitation* of the later claimed invention that is being scrutinized for novelty. In addition, each and every element and limitation must be found in a *single* reference.[2] This is of great significance because it means that novelty cannot be defeated by picking one feature from one reference and another feature from a different reference and then combining them to anticipate the later claimed invention.

Beyond the one reference requirement for novelty determinations, in order for a prior art reference to anticipate a later claimed invention, the reference must present each and every element and limitation being claimed in a manner that does not require one to piece the components of the reference together. For that single reference to anticipate the later claimed invention, the claimed invention must allow one to be able to *at once envisage* the claimed invention from the disclosure. In other words, the elements or limitations of the claim language that are allegedly found in the prior art reference(s) cannot be presented in a way that they would require one to guess at which and how the pieces need to be put together to arrive at the claimed invention— the prior art reference cannot require combining or ordering elements that are not already combined or ordered in the required fashion.

Finally, in order for the prior art reference to anticipate a later claimed invention, the prior art reference must be enabled. This requirement stipulates that one of ordinary skill in the art must be able to make and use the anticipating teachings in the prior art reference without resorting to undue experimentation. Something described in the prior art is not really described if one of ordinary skill cannot actually put the described thing together. The linchpin of the analysis turns on whether the allegedly

---

[2]Although each and every element must be contained within a single reference, that reference may incorporate other references, and what is contained in those incorporated references may also be considered for novelty purposes. Incorporated references must be *specifically* incorporated, a situation typically encountered only in a patent specification. Other supporting references can be used, however, to demonstrate the reference was enabled. For example, one reference might disclose a composition but not provide a method of preparing it. The reference might still be enabled if one can show that the composition could have been prepared by one of ordinary skill in the art without undue experimentation. Such evidence can be provided by another reference or combination of references that indicate that such a composition could have been made. In addition, other references maybe used to explain the meaning of a term or to demonstrate that a claim element not disclosed in the primary reference is nevertheless present inherently (more will be said about inherent elements later in this section).

novelty-defeating disclosure in the prior art reference could have been made without undue experimentation. A critical difference between the prior art enablement requirement and the prior art anticipation analysis is that the prior art reference can be enabled by teachings and knowledge outside the four squares of its disclosure. For example, if one reference teaches the structure of a compound that you wish to later claim and another reference or combination of references teaches how to make that compound, then the reference is enabled.[3] Prior art for U.S. patents are presumed enabled, including not only the material in the claims but the entire patent specification as well.[4] Outside of the category of prior art consisting of U.S. patents, the question of whether prior art is enabled or not is very fact specific and, at the same time, subjective, meaning that easy and fast conclusions should be avoided, especially where large investments of time and/or money are being contemplated.

7.2 ANTICIPATION IN CHEMICAL PATENTS

To help understand the concept of anticipation, let's consider the following example. A chemist has found that a particular combination of pigments and additives results in

---

[3] A logical extension of this reasoning is that it is not necessary that the prior art reference actually made the thing (e.g., compound) being described but rather the production of the thing needs only to be enabled in the art. *In re Donohue*, 766 F.2d 531, 533 (CAFC 1985). It is interesting that although the prior art must be enabled such that one of ordinary skill in the art could make and use the later claimed invention in order for that later claimed invention to be anticipated, the use portion of the requirement does not have to be credible. In contrast, a credible use must be present to satisfy the patent utility and the patent enablement requirement. In an interference decision by the USPTO on appeal to the CAFC in *Rasmusson v. Smithkline Beecham Corp.*, 413 F.3d 1318 (CAFC 2005), one party's prior art patent application taught how to make and use a 5-α-reductase inhibitor (finasteride) for the prevention of prostate cancer, though at the time that application was filed, the utility was deemed not credible for purposes of getting a patent because evidence of efficacy was lacking. However, the prior art patent application was deemed to be sufficiently enabled as a §102 prior art reference against a later filed patent application to a different company. This brings up the interesting possibility that a party could describe in an earlier filed patent application an incredible machine such as a cold fusion device that is not deemed to be credible at the time of filing, meaning it is not patentable. If science were to at a later date confirm the principles of the machine such that it was credible, it would be too late to save the earlier application. Moreover, a later filed patent application to the same machine could have the earlier filed patent application used against it as prior art. One does not want to be too far ahead of her times it seems.

[4] *Amgen Inc. v. Hoechst Marion Roussel*, 314 F.3d 1313 (CAFC 2003). The logic of presuming enablement is in part due to the fact that an issued U.S. patent has been examined by a USPTO examiner who must examine the patent for enablement along with the rest of the patentability requirements. However, the logic of presuming all of the disclosure in the patent to be enabled has been criticized on the grounds that not all of the patent specification has in fact been examined and that if there is to be a presumption of enablement, it should be limited to the claims only. Perhaps a devious minded person with lots of time and money on her hands could computer generate millions upon millions of plausible chemical structures and load them into multiple patent applications together with one compound that actually meets all of the patentability burden in each patent application. The applicant could then claim that enabled compound and get a patent issued on that compound and have the rest of the paper structures become enabled prior art (remember that the patent application will publish and become prior art). Please don't do this.

a paint with an especially attractive and durable coating. In a U.S. patent application, the chemist claims the composition as follows:

We claim:

1. A composition consisting of:
   a) pigment A;
   b) pigment B;
   c) additive F;
   d) additive G; and
   e) additive H.

Several prior art references with effective dates before our chemist's invention are available for examination against the invention described in claim 1 above. The pertinent details of each reference are:

Reference 1. Describes the preparation of pigments A and B and additives G and H and also describes a composition consisting of pigments A and B together with additives G and H.

Reference 2. Describes the preparation of pigment B and additives F, G, and H and also describes a composition consisting of pigments B and additives F, G, and H.

Reference 3. The reference describes the preparation of pigments A, B, and C and additives F, G, and H. The reference also describes a composition consisting of pigments A, B, and C and additives F, G, and H.

Reference 4. Describes a composition consisting of pigments A and B and additives F, G, and H but does not describe or reference the preparation of the pigments or additives.

Do any of these references affect the novelty of the claim in the patent application?

The first step in the analysis of this problem is to engage in a little claim construction. To begin with, the claim preamble is directed toward a "composition," a generic term that tells us in this case that a material is being claimed and nothing more. If you recognized that the next two words in the claim *consisting of* is a closed transition and affects the scope of the claim, then congratulations, you have applied your learning well! Since the claim is closed to further ingredients, we know that prior art compositions that describe less and/or more elements than are required by the claim will not anticipate the claim. The claim as now constructed should read as: A composition containing pigments A and B, and additives F, G, and H and those ingredients only.[5] Now that we have constructed the claim, we can proceed to the prior art examples and see how they might affect the novelty of the claimed invention.

---

[5] Every comparison of a claim with a prior art document does not per se require that the claim be written out as we might construe it, but more often one simply takes mental note of the language used in the claim and that is sufficient, particularly for relatively simple claims.

**TABLE 7.1   Comparison of Pigments and Additives in the
Claimed Composition and in the Prior Art References**

	Claimed Composition	Reference 1	2	3	4
Pigment A	×	×		×	×
Pigment B	×	×	×	×	×
Pigment C				×	
Additive F	×		×	×	×
Additive G	×	×	×	×	×
Additive H	×	×	×	×	×

Let's next compare each of the possible prior art references to the claimed composition as shown in Table 7.1.

A quick scan down the table is sufficient to demonstrate that additive F is missing in prior art reference 1, so it does not anticipate the later claimed composition; it does not contain each and every element of the composition our chemist wishes to claim. Prior art reference 2 describes the preparation of pigments B and additives F, G, and H as well as their combination into a composition, but it also fails to disclose a composition that contains all of the claimed elements because it does not disclose pigment A as part of the composition. But what about the fact that prior art reference 1 does disclose pigment A, can the combination of pigment A from that reference cure the defect in prior art reference 2? The answer is unequivocally no. Novelty cannot be defeated by combining references to provide missing elements. Unlike the first two prior art references, the composition of prior art reference 3 does contain each and every element of the claimed composition, so does it anticipate the claim? Recall the complete definition of anticipation requires that not only must the prior art reference contain each and every *element* of the later claimed invention but it also must contain each and every *limitation*. Our claim construction exercise revealed that the claimed composition required the listed elements and those elements only, so the limitation is that no additional ingredients be part of the claimed composition. Prior art reference 3 does contain every element of the claim, but it contains the additional element C so it does not contain every limitation of the claim and, therefore, does not anticipate the later claimed invention. We can see the importance of the transition phrase because if it had been *comprising* instead of *consisting of*, the claim would have been anticipated by reference 3.

Prior art reference 4 contains each and every element (and only each and every element) of the claimed composition and so appears to anticipate the later claimed invention. If we've read carefully, however, we realize that prior art reference 4 does not describe the preparation of any of the required elements. A prior art reference cannot anticipate that which is not enabled. Is prior art reference 4 enabled? Notice that the question is directed to whether the reference is enabled and not whether the reference is *by itself* enabled. Even though reference 4 does not enable the preparation of the required pigments and additives, those preparations are within the skill of one of ordinary skill of the art since the prior art itself provides that information. In particular, references 1 and 2 together or reference 3 by itself describe the preparation of all of the pigments and additives. So reference 4 does indeed anticipate the claimed invention.[6]

In the example just discussed, the comparisons made were relatively straightforward in that each reference disclosed a composition, so it was just a matter of comparing the prior art compositions to the claim as constructed. Sometimes the prior art is not so accommodating in the sense that prior art references often disclose a lot of information and a single prior art reference may contain each and every element of a later claimed invention, but it may not do so in a way that anticipates a later claimed invention. For example, consider the following prior art reference in view of the earlier claimed invention.

Prior art reference 5 describes the preparation of numerous pigments (A, B, C, D, and E) and additives (F, G, H, I, and J) as well as their combination into several compositions, though none of the compositions is identical to the claimed composition. Further description in prior art reference 5 explains that additional compositions can be "derived by combining the various possible sub-combinations of the listed pigments with any of the various possible sub-combinations of the listed additives." Since the claimed composition is actually a subcombination of pigments (A and B) together with a subcombination of additives (F, G, and H), it might be argued that prior art reference 5 anticipates the later claimed invention. However, this argument will fail; prior art reference 5 does not render the claimed invention nonnovel. Why not? What is missing is the required level of specificity. The anticipating reference not only must contain the requisite elements but must also combine those elements with the requisite level of specificity. It is not sufficient for this prior art reference to list a group of various ingredients, give a few examples and then suggest additional combinations as possibilities. The prior art reference needs to provide both an element-for-element parity and must combine those elements in such a way that there is no question that the prior art is describing the later claimed invention. A composition invention consisting of several ingredients is more than a listing of separate ingredients; the invention is the specific combination of those ingredients. Many inventions are combinations of known ingredients or parts but their novelty depends on the way those known ingredients or parts are combined. In this regard, it is appropriate to think of novel inventions that contain combinations of known ingredients or parts to be similar to a written novel that contains combinations of known words and/or letters, such a novel is still "novel" despite the fact that all of its letters and words are already known.

Since patent novelty is a somewhat abstract concept in the absence of context, it is helpful to consider another example. Gas liquid chromatography instruments

---

[6]An anticipatory prior art reference must be enabled as of the date of the applicant's invention. In this example, all of the references were in existence before the chemist's discovery. However, patent prosecution can sometimes take a long time so that a prior art reference that was not enabled when the invention was discovered may be enabled by the time the claims get examined, this is more likely in fast-moving biotech areas. For purposes of the claim examination in this context, the date of the invention is the date that needs to be considered.

can be effective ways for the separation of component mixtures and typically work via vaporization of a sample and injection into a heated column. The sample is then carried through the column by a flow of inert gas functioning as a mobile phase. The column itself typically contains a liquid stationary phase adsorbed onto a solid support located within the column. Molecules are retained by the column according to their particular affinity for the stationary phase and their relative volatility, and different molecules will elute from the column at different time points, depending primarily on these variables. After exiting the column, the sample is processed through a detector and then displayed via readout. An analytical chemist has "invented" the subject of the following claim:

We claim:

1. A gas chromatograph comprising:
   a) a sample injection port;
   b) an 8-foot column contained within an oven;
   c) a detector; and
   d) a LCD readout.

In the course of examining the patent application, the patent examiner at the USPTO cites three prior art references with reference dates before the chemist's invention date, and he alleges that each of them anticipate the claimed gas chromatograph. The key features of the three references that the examiner has cited are explained as follows:

Reference A. Discloses and enables a gas chromatograph, which shows a sample injector port, a column of 8 feet in length contained within an oven, a detector, and a graphical chart readout.

Reference B. Discloses and enables a gas chromatograph, which shows a sample injection port, a column of undisclosed length contained within an oven, a detector, and an LCD readout.

Reference C. Discloses and enables a gas chromatograph, which shows a sample injection port, one column of 8 foot length, preceded by a separate filtration column of 2 foot length (both columns are located inside an oven), a detector, and a LCD readout.

Can you guess which, if any, of these three prior art references would anticipate the claim to the gas chromatograph given in claim 1?

First, we again engage in some claim construction, the major observation being that the claim transition is open (comprising) and thus the listed parts are the minimal elements necessary to fulfill the claim language.

Starting with reference A, we begin our analysis by looking at each of the required elements in the claim language, and then determine whether each and every element of that claim language has been met. As it turns out, each and every element is not met by the prior art reference A. The reason? The claim requires a LCD readout. If you look carefully, reference A discloses all of the elements except that it does not contain an LCD readout. It contains a graphical readout. Thus reference A does not anticipate the claimed gas chromatograph.

Moving onto reference B, we see that it does not anticipate the claim in question, since it misses one of the limitations. It is important to note that, while reference B discloses the inclusion of a column, it does not disclose the column length. The length of the column is a limitation in the claim language. Remember that each and every claim limitation must be met by the prior art document in order for the claim to be anticipated. In this case, one of the limitations has not been met.

As for reference C, it contains all of the elements of the claim, but it also contains some additional elements. In particular, the reference provides an 8-foot column but it also has a 2-foot filtration column. Since the required elements of the claim are preceded by the term *comprising*, we recognize that the claim is open-ended and thus may include additional elements to those listed. Thus reference C anticipates the claim to the gas chromatograph. What if instead of *comprising*, the claim to the gas chromatograph had used the term *consisting of*? In that case, reference C would not have anticipated the claimed gas chromatograph since *consisting of* is close-ended and thus does not allow for additional, unlisted elements.

What if the patent examiner alleged that references A and B together anticipated the claimed gas chromatograph? For example, the examiner might point out that reference A provides all of the required elements except the limitation of column length. Reference B provides for an 8-foot column length. By combining reference A and reference B, one can match element for element the claimed gas chromatograph. However, we should know by now that such a rejection is invalid for anticipation since each and every element must be found in a *single* reference.[7]

Although the technical definitions for novelty that we have discussed are important to understand, they don't necessarily help us in all of our specific novelty inquiries. As a practical matter, it is sometimes difficult to determine whether a prior art disclosure actually anticipates a later claimed invention. One novelty test that has been found to be generally useful is stated as follows: "That which if literally infringes if it comes after, anticipates if it comes before."[8] To apply this test, one needs to take the allegedly anticipating reference and imagine that instead of being prior art, it occurs after the claimed invention. Then one looks at the pertinent elements of the reference to see if they would result in a literal infringement of the

---

[7] As we will see in the next section, obviousness may be found from a combination of references.

[8] The term *literal infringement* means that the infringing activity falls directly within the literal scope of the claim. This is in contrast to the doctrine of equivalents, which allows for claim infringement even when the infringing activity does not fall within the literal scope. Generally, the doctrine of equivalents is subject to significant limitations but *might* apply if *each element* of the alleged infringing practice performs substantially the same function, in substantially the same way, to achieve substantially the same result as each element of the claim in question.

claim in question. If the prior art disclosure does provide an embodiment that would infringe, then it would anticipate the claim in question. Since literal infringement requires that each and every element and/or limitation of the claim must be met by the infringing item, the patent novelty/infringement test is tautological in its application but useful nevertheless, as we will see in many of the examples and problems to come.[9]

While patent applicants cannot control the prior art, they (together with their agent or attorney) can control the language used in the claiming of the invention. Almost always, an invention can be described in a number of different ways. Dimensionally speaking, these different ways of claiming the invention typically vary in terms of the breadth of the claims. Claims can vary in scope from broad to narrow and each has potential advantages and disadvantages. A broad claim will contain the least limitations and will thus occupy a large amount of intellectual turf. A broad claim has the advantage of providing a wide berth of protection and thus is more likely to keep others far away from the inventor's staked out turf. However, a broad claim has the disadvantage of being more difficult to support from an enablement perspective. This is true since a claim needs support in the specification adequate to enable one of ordinary skill in the art to make and use the claimed invention throughout the claim scope. Thus broader claim scope requires greater support in the specification.[10] An additional complication with broader claim scope and particularly relevant to this chapter is that a broad claim, since it occupies more "turf," is more likely to run afoul of the prior art, and thus it is possible that the something in the prior art will act to render the broad claim not novel.

Conversely, a claim that is narrow in scope will allow others more room to maneuver outside of the narrower claim scope. However, narrower descriptions of the invention are often more readily enabled since they don't require broad support in the patent specification, and furthermore, the invention is less likely to be anticipated by the prior art since the narrower claim, by occupying less territory, is less likely to overlap with the prior art. So what is an inventor to do? Should the inventors go for broke and claim their invention as broadly as possible but at the same time increase their risk of patent invalidity? Or should they instead take a conservative approach and claim narrowly? Well, as it turns out, the skillful applicant might be able to have her cake and eat it too! A U.S. patent can issue with an unlimited number of claims and thus a chemical invention can be claimed from broad to narrow, all within the

---

[9] You can even try this test out on the preceding problem. If reference A had come after the claimed gas chromatograph, you can quickly ascertain that is would not infringe the gas chromatograph claim since it does not contain a digital readout device, as the claim requires. Reference B, if it came after, would not be proved to infringe the claimed gas chromatograph since the required column length has not been met (a patentee must prove infringement). Finally, reference C would infringe if it came after since it does contain each and every element of the claimed gas chromatograph. The inclusion of the additional precolumn would not affect the analysis since the claim term *comprising* means that the claim will still be infringed by the inclusion of additional elements that come after the term *comprising*.

[10] Enablement for purposes of patenting is covered in Chapter 9.

confines of a single patent application.[11] As a result, when a given piece of prior art is being considered for whether it anticipates a later claimed invention, that piece of prior art must be compared to each of the relevant claims of the later claimed invention. It is very often the case that a piece of prior art can anticipate one or more of the broader claims of a later claimed patent or application, while not anticipating one of the narrower claims.

As an example, consider a chemist at a pharmaceutical company who has synthesized a number of related compounds that are capable of inducing sleep in a mammalian species when administered via various routes, including oral and subcutaneous routes. In addition, she believes that the compounds will be effective in the treatment of bacterial infections since the compound inhibits several gram positive and gram-negative bacterial strains. She has not done a thorough search but she believes that the compounds are novel. Her company wishes to get patent protection for the compounds because they believe that they might be useful as drugs and would like to further pursue their development. For purposes of simplification and illustration, the compound structures are represented as: A, B, C, D, and E. The claims to the invention are as follows:

We claim:

1. A compound, or pharmaceutically acceptable salt thereof, that is selected from the group consisting of A, B, C, D, and E.
2. The compound of claim 1, wherein said compound is A.
3. A compound according to claim 1, wherein said compound is isolated and at least 95% pure.
4. A compound according to claim 1, and at least one pharmaceutically acceptable excipient.
5. A method of treating bacterial infections in a mammal, comprising the administration of a composition according to claim 4 to a mammal in need thereof.

---

[11] As is so often the case, there is a catch. Although one is unlimited in terms of the number of claims that can be presented for examination in a given patent, there is an examination fee based on the number of claims presented in the prosecution of the application. Thus the basic regular application examination fee in fiscal year 2009 ($385 for a small entity and $770 for a large entity, not including a separate search fee) allows no more than 3 independent claims and 20 claims in total. Additional independent claims can be presented at a cost of $43 (small entity) or $86 (large entity) and an additional charge of $9 (small entity) or $18 (large entity) for each claim (of any type) presented over 20. While the fees can add up for cases with a large number of claims, they are typically only a fraction of patent prosecution and maintenance costs. Keep in mind that a patent application that contains distinct inventions will typically be divided by the examiner in a restriction requirement by which the inventions have to be pursued in separate patent applications. So while it is true to say that there is no claim limit in issued patents, practical considerations provide their own controls. The USPTO has proposed numerous rule changes that would, among other things, severely limit the number of claims in a patent application as well the number of continuing applications or requests for continued examination. The rules are currently being litigated, and it is not presently clear what the final rule changes (if any) will look like.

6. A method of inducing sleep in a mammal comprising the administration of a composition according to claim 4 to a mammal in need thereof.

7. The method of either claim 5 or 6, wherein said mammal is a human.

8. The method of claim 7, wherein said administration is oral.

9. The method of claim 8, wherein the pharmaceutical composition to be administered comprises:

   a) from 5 to 50% of compound A, B, C, D or E;

   b) from 1 to 10% microcrystalline cellulose; and

   c) from 10 to 50% starch.

In the first office action by the USPTO, the examiner has cited several references:

Reference A. Discloses the identification of compound B as a by-product of an industrial scale chemical synthesis. The compound was unambiguously identified but was not isolated from the mixture it was in. No utility for the compound was discussed.

Reference B. Discloses the preparation of analytically pure compound A. No utility for the compound was discussed.

Reference C. Discloses the dissolution of essentially pure compound C in an aqueous vehicle, and the administration to laboratory rats to determine its effect on the sleep latency time in rats. The compound decreased sleep latency (time to sleep onset) by 22 minutes. A separate reference is cited that shows that compounds like compound C can be readily prepared by synthetic methods common to one of ordinary skill in the art. This reference is cited to show that compound C is enabled.

Now, let's separately examine each of these references for their effect on the listed claims in regard to anticipation under §102.

Reference A discloses the discovery and identification of an industrial by-product that has the same structure as compound B. Claim 1 encompasses a compound of structure A, B, C, D, *or* E (or a salt thereof). Claim 1 separately claims each of the listed compounds meaning that an anticipation of any of those compounds means the entire claim is anticipated.[12] Notice that there is no additional limitation on any of the five compounds of claim 1. The claim does not require that the compounds have a specified purity or be present in a minimal amount. Thus the prior art disclosure of compound B contains each and every element of compound B of claim 1, since claim 1 requires only the enabled disclosure of one of its listed compounds and nothing else. However, there might still be a question regarding whether reference A

---

[12] A claim is either anticipated or not. In this case, claim 1 is separately claiming five different compounds. If the claim to any one of those compounds is anticipated then the entire claim is anticipated. If this does not seem right, then ask yourself what would happen if claim 1 were issued and an infringer wished to make, sell, use, or import the compound of formula B. Would they be infringing the claim? The answer is clearly yes; making, using, selling, or importing A, B, C, D, *or* E would infringe the claim. We know this from the tautology between infringement and anticipation.

was properly enabled. Would one of ordinary skill in the art be able to prepare the compound B? Although questions of enablement are fact-based inquiries, we should appreciate that the level of enablement required needs to be consistent with what the reference discloses. In this example, the prior art reference can anticipate claim 1 without providing a method of isolating the pure material because the claim contains no such limitation. Recall also the prior art reference need not contain a utility. So the examiner will issue a §102 rejection of claim 1.

What about claim 2? Clearly the disclosure of the compound B in reference 1 does not anticipate the separate claim to compound A. This helps explain the advantage of separately claiming A as is done in claim 2. Imagine that out of the five compounds, compound A was specially preferred for some reason. It might be advantageous to claim A separately, that way if one of the "pack" of compounds fell (as happened in claim 1 due to reference A), the most desirable member might still survive since it is claimed separately. Note that claim 2 is a dependent claim in regard to claim 1, which we have just said will be rejected by the examiner. In its present form, claim 2 could issue except for its dependence on a rejected claim. In this scenario, the examiner will "object" to claim 2. The patent applicant can resolve this objection in multiple ways. For example, claim 2 could be made an independent claim. Alternatively, claim 1 could be modified so as to longer be rejected by removing compound B.[13]

As you scan through the remaining claims (3–9), take notice of the way the invention is sliced into different embodiments. Various compositions containing the compounds as well as methods of using the compounds of the invention are described. Although claim 1 may have been anticipated by reference A, the remaining claims are not.

Reference B discloses the preparation of compound A. Notice that the prior art reference B anticipates claim 1 since it discloses compound A, and it anticipates claim 2 as well. Due to the disclosure in the reference that the compound was prepared "analytically pure," claim 3 is presumably anticipated as well. So claims 1, 2, and 3 will all be anticipated and rejected under §102. However, note that claim 4 requires the addition of a pharmaceutically acceptable excipient and that reference B probably does not disclose such a composition, and thus reference B does not anticipate claim 4. Likewise, claims 5–9 all include limitations directing the claims to methods of inducing sleep and/or treating bacterial infections and will not be anticipated by reference B.

Reference C discloses compound C, and that disclosure is *probably* enabled since a separate reference describes the preparation of similar compounds. This assumes that the level of ordinary skill in the art is sufficient to allow one to bridge the gap between what is disclosed in the separate reference and what would be required to prepare compound C. Assuming enablement, claim 1 is anticipated by reference C. Claim 2 is not anticipated because that claim is directed to compound A. Claim 3 is anticipated if we assume that *essentially pure* means >95% pure; the ultimate determination would

---

[13] MPEP 706.01 Can you determine which of the claims in this scenario are directly or indirectly dependent on claim 1 and thus will have the same problem if claim 1 is rejected?

likely require a thorough reading of the specification to determine whether and how the procedures used to isolate the compound might yield such a purified material. Claim 4 describes the inclusion of a compound of claim 1 and a pharmaceutically acceptable excipient. Is water a pharmaceutically acceptable excipient? The answer is probably yes, particularly when one considers that water is used in many pharmaceutical vehicles, including in the prior art reference being cited. However, as we have already learned, the patent applicant can be "his own lexicographer" and so before jumping to any quick conclusions, one would be well advised to carefully read the patent specification to better understand how the applicant might have defined the term. Presumably, reference C would impact upon claim 6 because the compound was used to induce sleep in an a mammal; however, the thoughtful patent drafter further narrowed the claim 6 by dependent claim 7 that limits the type of mammal to a human. So reference C will anticipate claims 1, 3, maybe 4, and 6. The remaining claims may need to be rewritten to be in proper format but they would not be anticipated by reference C.

7.3 ANTICIPATION OF A CLAIMED GENUS BY A SPECIES FALLING WITHIN THAT GENUS

In the course of chemical research and discovery, it is often determined that it is not just a single compound that can perform the objectives of the invention but that a common activity exists throughout a genus of compounds.[14] In cases in which there are only a few compounds in the family or genus, it is easy enough to specifically list each of them in a claim. In cases where the family or genus of compounds contains large amounts of compounds, the exercise of specifically listing each member of the genus quickly goes from merely tiring to nearly impossible. The most commonly employed solution to this problem is the use of Markush structures. As an example of a Markush format as applied to a set of organic molecules is shown in Figure 7.1, in which nine naphthalene structures are shown as well as a Markush representation of those nine structures. As you can see even from this very small set, the Markush representation is much more economical in terms of both the time it would take to individually draw each structure as well as the space it would require to display all of the individual structures.

Recall that in this chapter, we are discussing a basic tenant of patentability—That is, for an invention to be patentable, it must be novel. Since the Markush format is a very common way to claim chemical genera,[15] it is critical that the practicing chemist appreciate the effect that disclosures in the prior art have upon the patentability of their claimed Markush structure. For this reason it is quite reasonable to ask, What

---

[14]*Genus* refers to a class or group of objects sharing at least one common property. Typically, in chemical inventions, a claimed genus will share at least one common structural element or common utility, or both. Each member of the genus is called a *species*.

[15]*Genera* is plural form of genus.

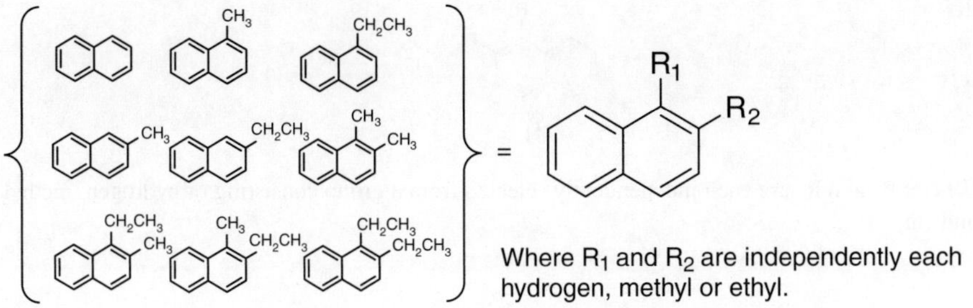

FIGURE 7.1   Naphthalene Markush.

sort of disclosure in the prior art would render my claimed genus not novel (i.e., would anticipate my genus)? As we saw before, in order for something to be anticipated, the prior art must, in a single document, disclose each and every element of the chemist's claimed invention.

Let's consider a patent application that contains a claim including the Markush group we have drawn in Figure 7.1. The claim containing the naphthalene Markush is shown in Figure 7.2. A prior art search reveals the following disclosures:

Prior art disclosure 1 discloses and enables the preparation of the genus of compounds represented by the structural formula II in Figure 7.3.

Prior art disclosure 2 discloses and enables the preparation of the genus of compounds represented by the structural formula III in Figure 7.3.

Prior art disclosure 3 discloses and enables the preparation of the compound represented by the structural formula IV in Figure 7.3.

We claim:

1. A compound having the formula (I)

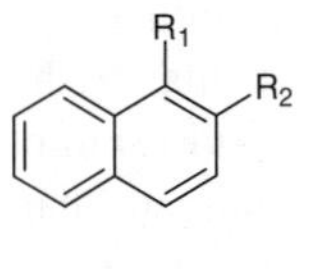

(I)

Where: R_1 and R_2 are each independently selected from a group consisting of hydrogen, methyl and ethyl.

FIGURE 7.2   Claim to naphthalene Markush.

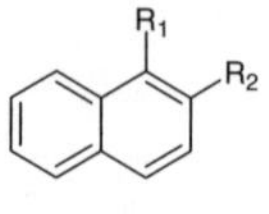

(I)

Where: R_1 and R_2 are each independently selected from a group consisting of hydrogen, methyl and ethyl.

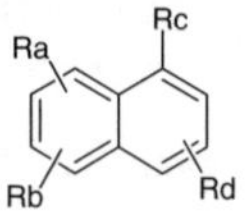

(II)

R_a, R_b, and R_d are each independently selected from the group consisting of hydrogen, fluorine, chlorine, methyl, or phenyl; and R_c is *n*-propyl.

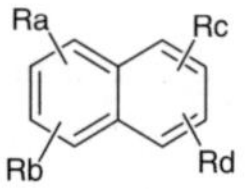

(III)

R_a, R_b, R_c and R_d are each independently selected from the group consisting of hydrogen, fluorine, chlorine, methyl, or phenyl.

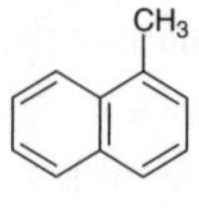

(IV)

FIGURE 7.3 Claimed genus (I) from Figure 7.2 and prior art genera and species.

Can you figure out if any of these disclosures will anticipate the claimed genus of naphthalenes of formula I?

First, let's analyze the prior art disclosure 1 (compound formula II). The Markush group described by formula II is similar to the Markush group defined by our claimed structure I. For example, they both contain core-naphthalene rings with various optional combinations of substituents attached to the ring. Careful examination of the possible substituents defined by the prior art disclosure 1 indicate that it must always contain an *n*-propyl group at position 1 of the naphthalene ring. In contrast, the claimed genus of formula I *cannot* have an *n*-propyl group at the 1-position (the definition for R_1 in formula I is hydrogen, methyl, or ethyl). If one were to draw a Venn diagram of these two described genera, one would have the situation as shown in Figure 7.4. Where a disclosed genus in the prior art does not overlap with the genus one wishes to claim, without more, that prior art genus *will not* anticipate the claimed genus.

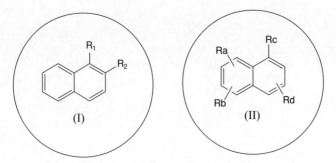

FIGURE 7.4   Venn diagram relationship between claimed genus (I) and prior art genus (II).

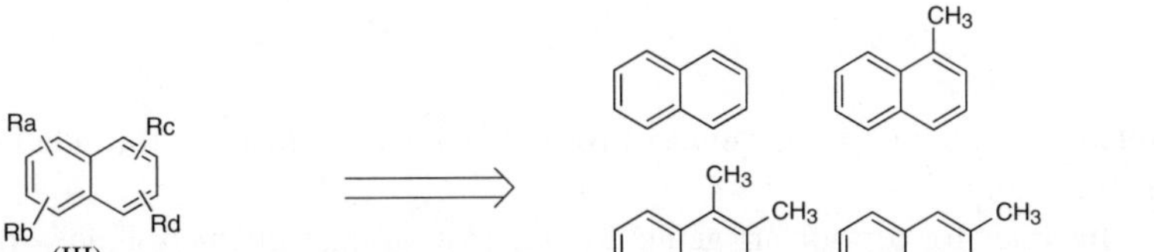

R_a, R_b, R_c, and R_d are each independently selected
from the group consisting of hydrogen, fluorine, chlorine,
methyl, or phenyl.

Four hypothetical compounds from
(III) that fall within (I)
From definition of (III)
When R_a and R_b are hydrogen
When R_c and R_d are hydrogen or methyl

FIGURE 7.5   Prior art genus (III) with particularly selected variables.

The case of prior art disclosure 2 is more complex since there is an overlap between the prior art genus represented by formula III and the claimed genus of formula I. With the right picking and choosing of allowable substituents in formula III, we can come up with hypothetical examples that fall within the claim scope of genus I.[16] As shown in Figure 7.5, if we set R_a and R_b = hydrogen, and R_c and R_d to hydrogen or methyl, where R_c and R_d are positioned at the -1 and -2 position on the naphthalene ring, this combination will produce a subset of four hypothetical compounds, where each of those four compounds fall within the scope of the claimed genus of formula I.

The four hypothetical compounds produced by specifically setting and placing R_c and R_d are just a few possibilities out of a much larger number of compounds that could have been constructed from the genus. A Venn diagram demonstrating the

---

[16]The word *hypothetical* is used because the fact pattern dictates that none of the four compounds was specifically exemplified in prior art disclosure 2. Thus, to find such a compound, we have to construct it from the variable choices provided for formula (III). There is no evidence that such examples have been actually prepared or specifically identified.

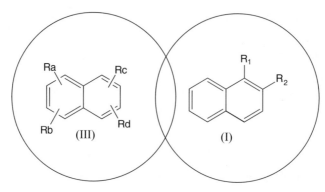

FIGURE 7.6   Venn diagram overlay of prior art genus (III) with claimed genus (I).

relationship between the genus of formula III and the claimed formula I is shown in Figure 7.6.

In order for a prior art genus to anticipate a later claimed genus, that prior art genus must place that the later claimed genus into the possession of the public, allowing one to *at once envisage* the later claimed genus. In the present case, prior art disclosure 2 provides a description of a genus that contains thousands of hypothetical compounds. Out of those thousands of compounds, only four hypothetical members fall within the scope of the claimed genus defined by formula I. To arrive at those four hypothetical examples falling within the claimed Markush I, one needs to pick and choose from various variables, carefully setting each one to a specified value and to a specific position. This is far from allowing one to at once envisage the later claimed Markush. While it may have not taken much effort for us to determine that some members of prior art reference 2 fall within the claimed Markush structure, this is with the benefit of *hindsight* (i.e., we have the claim to work backward from). However, putting ourselves into the position of the inventor of formula I *at the time of the invention*, the question is whether the disclosure of formula III would have allowed him to at once envisage the claimed Markush I. The answer is no because the claimed genus of formula I is but one of an infinite set of overlapping hypothetical Markush structures available to the inventor at the time of her invention. Referring back to the first example in this chapter describing the pigments and additives, we can appreciate that the situation with these two Markush groups is very similar to the prior art reference 5 of that example. In the prior art reference 5, one needed to mix and match pigments and additives to arrive at the later claimed composition; there was no composition actually disclosed that matched the later claimed composition. In the current example, the same thing is required since variables and positions must be determined and set in order to match the reference to the later claimed invention; the requisite level of specificity is not present. Anticipation requires that the thing being claimed must have already been identically disclosed. As such and absent more, the mere disclosure of a broadly overlapping genus will be insufficient to anticipate the later disclosed and claimed genus.

Finally, let's consider prior art disclosure 3. For those of you who already figured out that prior art disclosure 3 would be an anticipatory disclosure, congratulations! It is an axiom of patent law that *the prior art disclosure of a* single *species falling within a later claimed genus will anticipate the claim to that genus.*[17] Notice that the prior art disclosure of 1-methylnaphthalene requires absolutely no picking and choosing of variables to arrive at its structure and that one can at once envisage its structure. It is also helpful to note that because 1-methylnaphthalene falls within the claimed generic structure, its structure matches exactly one of the possible structures falling within the claimed Markush group of formula I (Figure 7.1).[18] Remember that we can think of the Markush claim as separately claiming each of the individual species falling within its range of possibilities, so we could rewrite the naphthalene genus claim 1 as shown in Figure 7.7.

As we mentioned before, the claim is either anticipated or not. If one is claiming the equivalent of nine objects (compounds in this case) separately, the anticipation of one of those objects anticipates the claim. It should now be clear that the prior art compound formula IV is one of the nine choices (Figure 7.8).

Finally, if we are not already convinced that the prior art species IV anticipates the later claimed genus I, we can apply the anticipation/infringement test. If we assume that the claimed genus I had issued into a patent and the prior art compound of formula IV was made after the claim genus was issued, we can agree that the making or selling of that compound in the United States would infringe the claim

---

[17]Few rules in chemical patent law are as absolute, so when we do have one, it is nice to be able to boldly state it.

[18]MEPE 2131.02 quoting *In re Slayter*: "A generic claim cannot be allowed to an applicant if the prior art discloses a species falling within the claimed genus." *In re Slayter*, 276 F.2d 408, 411, 125 USPQ 345, 347 (CCPA 1960). See also MPEP 2131.02 for the explanation of why the number of individual species in the prior art reference is not a consideration with respect to whether the prior art reference anticipates the later claimed genus. In *Ex parte A*, 17 USPQ2d 1716 (Bd. Pat. App. & Inter. 1990), the applicant's claimed compound was specifically named in a reference that also disclosed 45 other compounds. The board found that the extensive listing of compounds in the prior art reference did not affect the fact that their claimed compound was still specifically taught. The board analogized the situation to one in which the compound was specifically taught in a pharmacological reference such as the *Merck Index*, saying: "the tenth edition of the *Merck Index* lists ten thousand compounds. In our view, each and every one of those compounds is 'described' as that term is used in 35 U.S.C. § 102(a), in that publication." Clearly, at its extremes, this logic could present a conundrum. From the extant case law, we can surmise that a prior art disclosure of a genus as a Markush structure may not have the same effect as a prior art disclosure of an individual listing of each of the species contained within the Markush. Putting aside any prior art enablement questions, one can envision a computer program that systematically generates huge numbers of specific chemical structures and then places those structures (or chemical names) on a website. While this would be the functional equivalent of a very broad Markush, its effect as anticipating prior art against later attempts to claim a specific compound may very well be different. Although this would seem to be a rather illogical outcome, it appears to be consistent with current federal patent law jurisprudence. For example, quoting the CAFC in *In re Gleave* (CAFC 2009): "For the purposes of whether they are anticipatory, lists and genera are often treated differently under our case law. Compare *Perricone v. Medicis Pharm. Corp.*, 432 F.3d 1368, 1376 (Fed. Cir. 2005) (rejecting "the notion that [a compound] cannot anticipate because it appears without special emphasis in a longer list") with *Atofina v. Great Lakes Chem. Corp.*, 441 F.3d 991, 999 (Fed. Cir. 2006) ("It is well established that the disclosure of a genus in the prior art is not necessarily a disclosure of every species that is a member of that genus").

We claim:

1. A compound selected from one of the following:

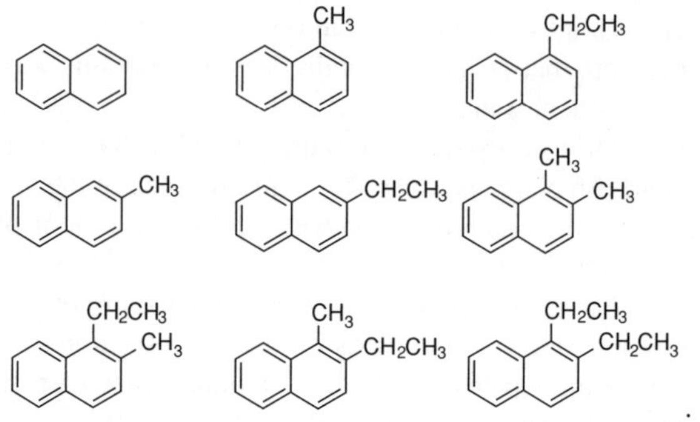

FIGURE 7.7   Claiming genus formula (I) as nine separate species.

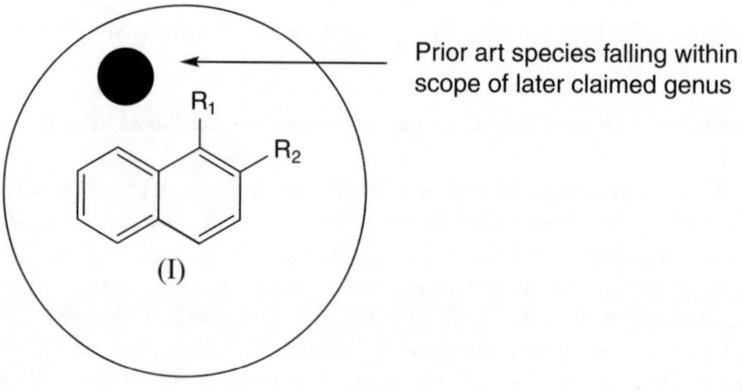

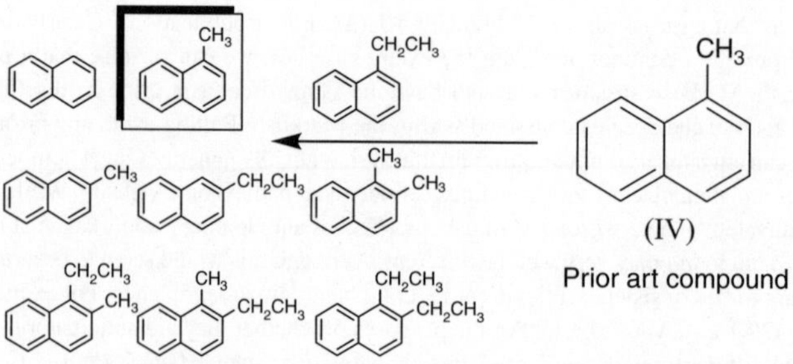

Prior art species falling within scope of later claimed genus

Prior art compound

FIGURE 7.8   Anticipation of genus by prior art compound.

to the genus. It clearly falls within its claim scope. Accordingly, the making of that compound before the filing of the genus claim that covers the same compound will anticipate that compound. Therefore, the test is satisfied. The single species anticipates the genus.

7.4 ANTICIPATION OF A SPECIES CLAIM BY A PRIOR ART GENUS

From the preceding section, we know that a prior art genus that does not overlap a later claimed genus will not anticipate that claimed genus. We also know that a prior art genus that partially overlaps a later claimed genus but that does not contain actual examples falling within the later claimed genus will not anticipate the later claimed genus. Finally we also know that a prior art disclosure of a species falling within the scope of a later claimed genus will anticipate that later claimed genus. A situation we did not discuss is that which occurs when the prior art discloses a genus of compounds and the practicing chemist wishes to claim a single species that happens to fall within that genus. Will the claim to that species be novel in view of the prior art disclosure of the genus?

Imagine you are a chemist who has been laboring to find a compound that is capable of inhibiting a very important pathway in a human disease state. After many years of hard work and false leads, you find a compound—compound 4 in Figure 7.9—that appears to possess all of the necessary attributes. However, your information scientist reports that there are prior art references that disclose a total of three different Markush structures, each of which encompasses your compound as shown in Figure 7.9. The prior art reference(s) that contain(s) these three genera provide(s) the general methods of making the compounds, and the preparation is enabled for one skilled in the art.[19] A few specific examples have been made that fall within genus 1 (and hence genera 2 and 3 as well) of the prior art, but your exact compound 4 has not been specifically disclosed.[20] The question is now whether compound 4 is anticipated and rendered nonpatentable per se in view of any or all of the three prior art Markush disclosures.[21]

Take a few moments to look at the three prior art Markush structures and see if you can determine which Markush structure represents the broadest disclosure? We hope you've determined that the genus represented by formula 3 is the broadest disclosure since it contains within its reach many thousands of hypothetical compounds.[22] The next broadest disclosure is the genus represented by formula 2, which contains hundreds of hypothetical compounds, followed by the narrowest disclosure represented by formula 1, containing only 14 hypothetical compounds. Since we are interested in

---

[19]For the purposes of this analysis, it does not matter whether the particular Markush groups all come from the same reference or different references.

[20]If your exact compound had been disclosed (and enabled), it would be anticipated.

[21]Remember we are referring to the patenting of the compound *itself*. Even if the compound were anticipated, it still might be patentable as a method of treatment using the compound or in a formulation or even in a dose administration schedule.

[22]Recall that the compounds of the genus that are not actually made or otherwise individually exemplified are hypothetical.

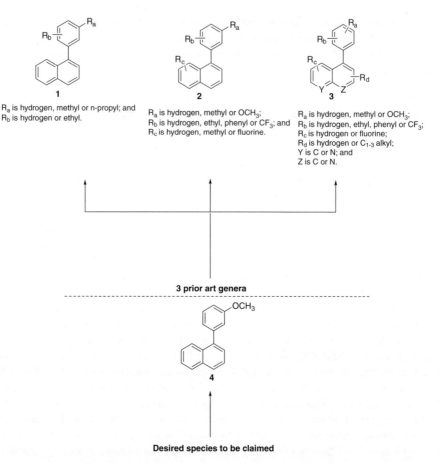

FIGURE 7.9 Prior art genera and species to be claimed.

determining the anticipatory effect (if any) of these three prior art Markush structures on our claimed species (compound), we might find it useful to engage in a little thought exercise to make better sense of the rule of law that will eventually follow. In this experiment, we will exaggerate the scope of the prior art Markush structures with our goal being to determine whether a narrower or broader encompassing scope in the prior art is more likely to be anticipatory. To begin, let's imagine that the intellectual property space occupied by the broadest Markush, structure 3, was grabbed on each end and stretched so large that instead of containing thousands of hypothetical compounds, it now contains millions or even billions of hypothetical compounds.[23] Now let's take the smallest Markush, structure 1, and shrink it down to a single species

---

[23] Of course an actual broader Markush would not be obtained by physically stretching its presence on the page, but rather by adding additional variables on the structure and/or allowing each variable set to contain more possible members.

point, that species point being our claimed compound 4. With these exaggerated Markush embodiments in hand, let's think back to what anticipation means—the ability of a prior art reference to allow one to "at once envisage" the later claimed invention. In our case, the claimed invention is the single compound represented by formula 4, and so the question at hand is which hypothetical (stretched or shrunk) Markush will allow us to at once envisage our claimed compound having the structure of formula 4? Let's start with the very broad (hypothetically stretched) Markush first. If we wished to derive our claimed compound from its many variables, we would have to limit each variable group to a single choice such that we would get just the right combination. It would be like trying to open a huge combination padlock without any clues. Certainly one could not at once envision our compound from the millions or billions of hypothetical compounds contained within this broad disclosure and even more certainly, such a broad disclosure does not identify any single compound sufficiently enough to satisfy the requisite specificity needed for chemical anticipation. In contrast, the Markush that had been shrunk down to a single species clearly anticipates our claimed structure because it would allow us to at once envisage our claimed structure from its disclosure since ... it is our claimed structure! Between these two extremes is a molecular Grand Canyon of uncertainty. The primary issue can then be presented as: At what point, if any, can a prior art genus become small enough to anticipate a later claimed species falling within the scope of that genus?

For guidance, we begin by looking to see if there are any federal court decisions likely to be controlling for our situation.[24] In the case of *In re Petering*,[25] a patentee was attempting to claim a particular compound in view of a prior art reference that contained a generic class of 20 hypothetical compounds, the members of which were very similar in structure and all of which would be expected to possess the same properties.[26] In its analysis, the court noted that even though the prior art reference

---

[24]The term *controlling* is one of legal art and signifies that not every previous court decision must serve as precedent. By way of nonlimiting example, an earlier court decision that was properly overruled by a higher court on that issue would not serve as a controlling decision for that issue (though the higher court's decision may be). Likewise, some decisions by the CAFC are clearly labeled nonprecedential. Another noncontrolling circumstance is when a court's discussion in a case opinion is not necessary for the final outcome of the case. The legal term for nonprecedential discussion in a court opinion is referred to as *dicta* and might be persuasive, but it is not controlling legal precedent that must be followed by future courts. Litigants will argue that cases being cited by opponents are not controlling since the specific facts of their dispute take them outside the realm of the rule of the cases cited by their opponent. They might argue that their opponent is relying on dicta and not the rule of law from the case. Instead, they will argue the implementation of a rule from a case that will lead to the result they want and argue that particular case or cases is controlling. Thus in some respect, the term *controlling* is more a conclusion to be argued for. Much of patent law derives from court decisions that are written to address the specific controversy before the court. While general guiding principles maybe gleaned from these previous court decisions, these guiding principles may often conflict, and so the result for a new case is often not easy to predict successfully. The more your situation resembles the specific facts of a previous court decision, the more likely your result will be like the decision in that particular case. However, since no two cases ever have identical facts, there is always room for alternate interpretations.

[25]*In re Petering*, 301 F.2d 676 (CCPA 1962).

[26]In the case of *Petering*, the class of 20 compounds was not actually present as a single Markush structure, but rather was the product of the court's construction of a genus from the entire teaching of the prior art

did not specifically disclose the compound that the applicant later attempted to claim, the court found that such a narrow disclosure (genus of 20 compounds) of related compounds sharing the same core would allow one to at once envisage the later claimed compound falling within that generic scope. The court's characterization of the relationship between the prior art disclosure and the later claimed compound clearly foreshadowed their ultimate conclusion; the narrow prior art genus anticipated the later claimed compound. In so doing, the *Petering* court established the important precedent that forms the basis for our own hypothetical analysis.

We can begin our analysis with Markush 3, which represents the broadest set of compounds. While the exact number of compounds has not been calculated, it contains at least many thousands of hypothetical compounds. Furthermore, there is not even a constant core, as the variables Y and Z allow for C or N, and thus the core can be naphthalenes, quinolines, and 1,8-naphthyridines as well. This is not a tight set of compounds as in *Petering*, where there were only 20 possible compounds sharing a common core. Thus it would appear that the holding in *Petering* would not be controlling for our situation as it relates to genus 3, and thus genus 3 would likely not anticipate our later claimed species compound 4.

Let's move to the other end of the spectrum and analyze prior art genus 1. This Markush contains 14 hypothetical members, all of which share a common naphthalene core. The variability of the R_a is not great, containing the possibility of hydrogen, methyl and *n*-propyl only. Given these facts, it is quite likely that *Petering* would be controlling and Markush 1 would anticipate the later claimed species compound 4 and thus render it nonpatentable.

How about the genus of formula 2? Let's pull out the crystal ball and take a look. Darn, the crystal ball has a lot of dust on the glass and we can't see very much! The reason for the cloudy crystal ball is that there is very little case law beyond *Petering* to light our way. We do know that the prior art Markush represented by formula 2 does not contain nearly as many hypothetical compounds as the Markush of formula 3, but it still contains at least several hundred hypothetical compounds. Moreover, even though the core does not vary from a naphthalene, it does have fairly widely varying substitution on the pendent phenyl, since R_b can range from hydrogen to trifluoromethyl to a phenyl group. Given the fact that there are still many more compounds in formula 2 then what was found in the prior art references discussed in *Petering*, together with the relative disparate nature of the substituents (especially for R_b), it is plausible/reasonable to expect that the Markush of formula 2, absent more, would not anticipate the later claimed compound 4.[27]

Summarizing, our actual determination based on the case law precedent of *Petering* paralleled our general prediction based on the hypothetical stretching and shrinking we did to the same prior art genera disclosures. As we saw, the broader the disclosure,

---

reference including preferred embodiments and actual examples. Had the 20 compounds existed as a single Markush, the final decision would likely been the same since a single Markush structure could provide an even clearer set of compounds from which one could at once envisage the later claimed species.

[27] See MPEP 2131.02 for further discussion on the treatment of prior art genus disclosure on later claimed species falling within the prior art genus.

the less likely it was to anticipate and vice versa. Thus we saw that the broadest disclosure represented by genus 3 did not anticipate our later claimed species whereas the narrow genus 1 did anticipate. The genus of formula 2 occupies the middle ground and may not anticipate our claimed species. Having just invested our time learning the law of *Petering* and anticipation of species by a narrow prior art genus, we might logically ask in what context this situation is most likely to arise. In other words, why is such a seemingly very specialized situation worthy of such a detailed discussion?

One important situation where anticipation of a species by a prior art genus (*Petering*-type situation) can occur is in the case of a selection invention.[28] A selection invention happens when a later claimed invention falls within the boundary of an earlier invention (like the species falling within a genus). The reason why selection inventions often occur is that one may wish to improve on an earlier disclosed genus. More specifically, a chemist may discover interesting activity in a series of compounds and wish to claim that series of compounds through the use of a Markush structure. After the patent application filing, the chemist or the organization she works for may continue working within the genus by making additional analogs. In this later work, it often happens that some surprising or interesting activity is found within a narrower subgenus or a species falling within the original genus. This surprising activity might be in the form of a discovery that a change of one functional group to another functional group results in significantly improved properties or activity. For example, it may be found that changing a nitro group to an amino group may convert the compound from an agonist to an antagonist. The advantage to filing a subsequent application with claims to the subgenus or species falling within the earlier claim is that this will both provide more specific coverage for the subgenus or species and also might provide a new, *later*, filing date, (meaning the new subgenus or species can be covered by a whole new 20-year patent term, assuming the later claimed species is not considered obvious in view of the prior art genus).

Alternatively, a series of compounds may be made and claimed in a Markush format, and further work is dropped on that series of compounds. In the meantime, many of the compounds may have been deposited into the organization's compound library where they are plated for screening. Many years down the road, one or more of the compounds may hit on a new target, and work is taken up on the old series of compounds. In the process of optimizing the old compounds to the new use, new species will be discovered, and it is possible that some of these new species of compounds might fall within the boundaries of one of the previously claimed or described Markush structures. Although the species is new in the sense that it has never been made or specifically described, for it to be patentable, it must be novel according to the legal requirements of §102. And as we saw earlier, it is possible that a compound can be found not novel despite the fact that it has not been specifically identified in the prior art disclosure. This can occur when the prior art genus is relatively narrow with a common core and closely related groups for allowed substituents. *Petering*-type anticipation is very important in this venue because once it is determined that the earlier disclosed genus anticipates the later claimed species,

---

[28] Selection inventions are sometimes also referred to as *improvement inventions*.

that species can no longer be patented as a chemical species (although patents directed to methods, dosages, formulations, etc. might still be possible). Thus *Petering* can conceivably have significant ramifications for the practicing chemist and is always worth keeping in mind.

7.5   ANTICIPATION OF A RANGE BY A PRIOR ART SPECIES FALLING WITHIN THAT RANGE

Sometimes it is necessary to claim chemical compositions that consist of more than one material. Such compositions can be thought of as mixtures and have practical utility across the spectrum of the chemical arts. For example, alloys are typically mixtures of metals that can possess desirable properties very different from what any given component might have. Other examples include pharmaceutical formulations, which typically contain a mixture of one or more drug substance together with one or more pharmaceutical excipients, chosen in such a way as to maximize the desired pharmaceutical properties of the drug substances. Chemical mixtures are typically claimed by representing different components of the mixtures together with a percentage range for that component in the mixture. For example, consider two composition claims for a titanium-based alloy:

> We claim:
>
> 1. A titanium metal base alloy consisting essentially by weight of about 0.6% to 0.9% nickel, 0.2% to 0.4% molybdenum, up to 0.2% maximum iron, balance titanium, said alloy being characterized by good corrosion resistance in hot brine environments.
> 2. A titanium base alloy as set forth in claim 1 having up to 0.1% iron, balance titanium.

As you can see, the claimed alloy can contain up to four different metals, wherein at least three metals must be present (nickel, molybdenum, and titanium), and one metal can be present in up to a limited amount (iron, up to 0.2%).[29] The second claim is dependent from claim 2, having a narrower scope since the iron is still an optional ingredient but it is now limited to a narrower range. Accordingly, it is possible to infringe claim 1 without infringing claim 2, but it is not possible to infringe claim 2 without infringing claim 1. Just as for the compounds discussed in the previous section, a claim to a mixture must be novel to be patentable. In order to understand what makes a range novel or not novel, we'll examine the case *Titanium Metals Corp. v. Banner*,[30] which involved the exact titanium alloy claims just presented.

The USPTO rejected the claims, alleging a lack of novelty under sections §102(a) and §102 (b), citing a short, highly technical Russian article published in 1970, several years before the Titanium Metals application filing date of July 25, 1975. There was no dispute between the parties that the Russian publication could serve as both a §102(a) reference and a §102(b) reference. The Russian publication disclosed a

---

[29]Elements that may or may not be present are sometimes referred to as optional elements.
[30]*Titanium Metals Corp. v. Banner*, 778 F.2d 775 (CAFC 1985).

number of graphs that showed various data points for ternary metal alloys consisting of combinations of titanium, nickel, and molybdenum. From the graphs, it could be discerned that the Russian reference disclosed a metal alloy consisting of 0.25% molybdenum and 0.75% nickel, the balance being titanium, which, as the CAFC noted, falls within the ranges of claim 1 and 2 of Titanium Metals' application in suit.[31] Applying the patentability/infringement test, the CAFC noted that "there can be no question that claims 1 and 2 read on it and would be infringed by anybody making, using or selling it" (the *it* being the alloy disclosed in the Russian publication). This relates as an exact analogy to that which we saw previously for the effect of a prior art species falling within a later claimed Markush structure. In fact, one can think of the prior art mixture disclosed in the Russian publication in the *Titanium Metals* case as a species falling within the claimed ranges of the applicant's invention.

An interesting and important aside to the *Titanium Metals* case is that the CAFC rejected the plaintiff's argument that their claimed alloys were patentable because the Russian publication did not disclose the many uses and advantages of the particular alloy disclosed in the plaintiff's application, such as corrosion resistance to brine. These arguments were to no avail since Titanium Metals' claim to the alloy simply claims a metal alloy having a certain percent composition, with no limitation to a specific use. Any metal falling within the range claimed by the plaintiffs is presumed to have such properties since such properties are inherent to the composition itself.[32] In this vein, the CAFC pointed out that one cannot take something old and make it new for purposes of patenting simply by discovering heretofore undisclosed properties of the old material. Finally, the CAFC did not accept Titanium Metals' argument that the disclosure of the alloy in the Russian publication was not enabled. Although the publication did not disclose how the particular anticipating alloys were made, the court noted the plaintiff's specification did not describe the preparation of the alloy either; if the court accepted that the prior art was not enabled then Titanium Metals' claimed alloys would not be enabled either. In addition, the court noted that applicant's own expert stated that one of ordinary skill in the art would have known how to prepare the alloys disclosed in the Russian publication by any one of at least three ways.

7.6   INHERENT ANTICIPATION

A prior art reference will anticipate a later claimed invention where that prior art reference contains each and every element and limitation of the claimed invention. However, there are situations where the prior art does not specifically *disclose* an element or limitation but will still be treated, for purposes of defeating the novelty of a later claimed invention as if it had. Elements that are not disclosed explicitly but that are *always and necessarily present* in the prior art reference are deemed inherent

---

[31] Implicit in applicant's claims is the fact that iron is allowed in a quantity from 0% to 0.2% (claim 1) and 0% to 0.1% (claim 2). While the court did not specifically address this point, the Russian reference did not disclose any iron content in their composition and thus is assumed to be 0%, which falls within the claimed composition range for iron as well.

[32] Inherent elements and anticipation will be discussed more in the next section of this chapter.

elements. Inherent elements cannot be a matter of possibilities or probabilities; they must be inherent to the disclosure and such inherency must be demonstrable. As it has been sometimes expressed, the inherent limitation or element must necessarily flow from the practice of the prior art disclosure. In the case of inherent anticipation, simply practicing the prior art will manifest the claimed invention, even though the prior art does not specifically describe all of the elements or limitations necessary to, or resulting from the practice of the art. If this sounds confusing based on all that was said about anticipation before, please don't feel bad because confusion seems to be inherent to this doctrine![33]

To better understand this oftentimes nebulous and occasionally controversial doctrine, it is instructive to look at some examples that should help more fully illuminate some of the contours and contexts of inherent anticipation. As an example, consider a chemist who has been working very diligently in the area of trying to discover new compounds suitable for the production of polymers that can in turn be used to prepare films, fibers, and molded parts.[34] In particular, he believes that $2,2',6,6'$-tetramethylbiphenyls (TMBP), which is $4,4'$ disubstituted with CO_2H, would be a good agent for polymerization. The chemist measured several of the properties of the compound and claimed the compound, as shown in Figure 7.10.

A prior art reference was cited that disclosed, among other compounds, the exact compound of formula I. However, the reference made no mention that the compound was suitable for the production of polymers. Given the failure of the prior art to teach the suitability for polymer production limitation, does the prior art reference disclose each and every element and limitation as required in order for it to be an anticipating reference? The short answer is yes. The reason for this is that a molecule and its properties are *presumed* to be inseparable.[35] Thus the suitability of the compound

---

[33] It could be that what is so *inherently* difficult about the doctrine of anticipation by inherency is that anticipation itself requires that something be part of the prior art, and inherent things, by their very nature are not disclosed (otherwise they would be explicit). Part of the patentee's grand bargain with the public is that he give the public something new (and nonobvious and enabled) in exchange for his limited term of right to exclude others. However, if something is in the art but it is not apparent, it might not really be doing much good to anybody until it is identified. Thus it might seem that if one had invented something that was already there but undiscovered, one had provided the public with that thing by virtue of having discovered and disclosed it. Another difficulty arises when one considers the desirability of having inventions that we are confident are not anticipated by the prior art. Oftentimes in the chemical arts, a great deal of money will be invested in the development of a compound or material and that investment requires a reasonable amount of certainty that what is being patented is not anticipated inherently by something in the prior art. While proper searching is a must under any circumstances, one can imagine the difficulty of searching not only for things that are specifically disclosed in the art but also for those things that are not disclosed but are inherently a part of the art. This is seldom possible. This seeming difficulty can often be fixed by claiming the invention in as many ways as possible. Thus, for example, although the invention may lack novelty if claimed in one form (as it existed in the prior art), it is generally possible to take advantage of the inventor's additional work with the disclosed art. For example, naturally occurring proteins cannot be claimed as composition claims with no further limitations since the proteins already exist (are not new and moreover are products of nature); however, they still might be claimed as isolated proteins, or recombinant proteins, or methods of producing a recombinant protein with that structure, etc.

[34] This example is loosely based on the facts of an actual case, *In re Donohue* 766 F.2d 531 (CAFC 1985). Some of the information has been modified to bring inherent anticipation to the fore.

[35] *In re Spada*, 911 F.2d 705, 709 (CAFC 1990).

What is claimed is:

1. A compound with the formula (I):

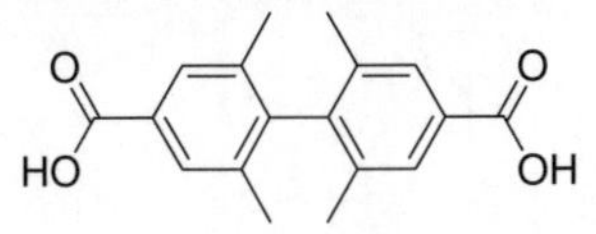

(I)

wherein said compound of formula (I) is suitable for the production of polymers.

FIGURE 7.10   Claimed polymerization agent.

of formula I for polymer production is inherent to the structure itself. Therefore, the compound of formula I disclosed in the prior art reference will be assumed to have the same polymer suitability as the compound claimed.[36]

A second example demonstrates just how diverse the prior art can be. Although everything under the sun made by man might be suitable subject matter for a U.S. patent, suitable subject matter for prior art is even broader. In fact, prior art includes not only everything made by man but made inside of man as well. Schering-Plough Corporation is the patent assignee of patent U.S. 4,659,716 (the '716 patent), which contains claims to descarboethoxyloratadine (DCL), which is the descarboethoxy human *metabolite* of loratidine (an antihistamine marketed as Claritin®). Loratadine is the subject of a different patent, U.S. 4,282,233 (the '233 patent), also held by Schering-Plough. The structures of the two compounds are shown in Figure 7.11. The '233 patent is available as prior art against the '716 patent (the '233 patent issued on August 4, 1981, over 1 year before the earliest filing date claimed for the '716 patent, February 15, 1984).

Several generic drug makers wished to market a generic drug version of loratidine, and filed abbreviated new drug applications (ANDAs) with the FDA. Since the '233 patent covering loratidine would be expired by the time the generic products came to market, there was no issue of infringement of that patent. However,

---

[36]Oftentimes, chemical reactivity and solubility may be contingent on more than just the identity of the compound itself. The particular physical state that a compound is in can also affect its solubility and thus kinetic reactivity. For example, an amorphous solid material may dissolve more quickly than a crystalline material and thus may have different reactivity rates in a given solvent. Likewise, in the case of drugs, compounds with different crystalline states (polymorphs) often have different dissolution rates that may in turn lead to different absorption characteristics that might ultimately affect a drug's pharmacokinetics and biological activity. However, in the absence of evidence otherwise, the chemical structural presumption of inherency will apply but might be rebutted should the patentee be able to show that the presumed inherent element fails to adequately match the claimed limitation. In the instant case, had the patent applicant found that a particular crystalline form was suitable for polymerization while the others were not, the "suitable for polymerization" might be given force should the patentee be able to demonstrate that he enabled such a form and that practicing the prior art would not be predicted to yield such a form. In such a situation, the examiner might require him to explicitly limit the claim to the crystalline form possessing the differentiating property.

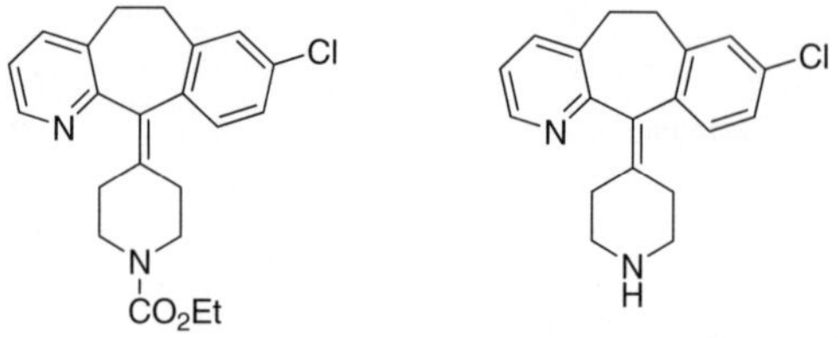

Loratadine ('233 patent)   Descarboethoxyloratadine ('716 patent)

FIGURE 7.11 Structure of loratidine and descaroethoxyloratidine.

We claim:

1. A compound having the structural formula

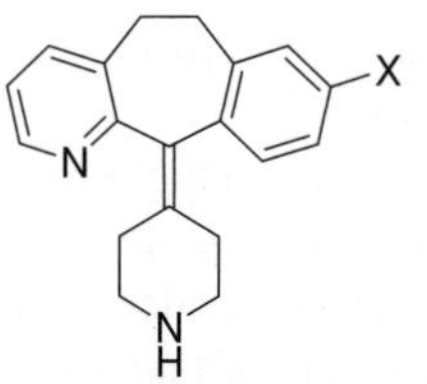

or a pharmaceutically acceptable salt thereof, wherein X represents Cl or F.

3. A compound having the structural formula

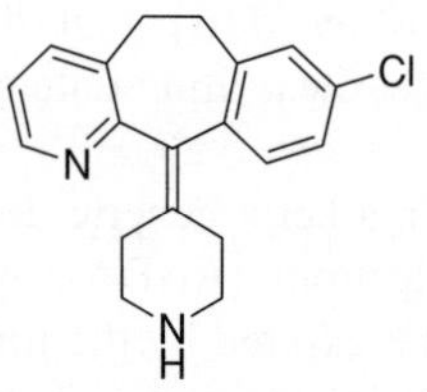

or a pharmaceutically acceptable salt thereof.

FIGURE 7.12 Asserted claims from '716 patent.

Schering-Plough's later '716 patent would still be in force, and so Schering-Plough sued for infringement of claims 1 and 3 of the '716 patent (shown in Figure 7.12)[37] .

Although the generic companies were interested in manufacturing loratadine, the federal court hearing the case construed claims 1 and 3 to cover DCL in any form (both parties agreed to this construction). This is because claims 1 and 3 simply do not provide any limitation as to the purity of DCL nor the quantity required for infringement of the claim. However, you may have noticed that the structure

---

[37] *Schering Corp. v. Geneva Pharmaceuticals et al.*, No. 02-1540 (CAFC 2003).

of loratidine does not fall within the scope of the '716 patent and so what is the basis for infringement? As it turns out, it was well established by the time of the infringement suit that the administration of loratidine to a human subject necessarily results in the formation of DCL, the subject of claims 1 and 3 of Schering-Plough's '716 patent.[38] This means that the pharmaceutical use of loratidine, the subject of the Schering-Plough's '233 patent, would generate the structure corresponding to DCL, the subject matter of the '716 patent. It would appear that Schering-Plough has come up with a very clever way to effectively extend the useful lifespan of their loratidine patent by covering its human metabolites in a later patent.

However, the federal court poured cold water on this idea because they found that claims 1 and 3 of the Schering-Plough '716 patent were invalid because they lacked novelty over Schering-Plough's own disclosure in its own prior art '233 patent. In particular, the court found that the '233 patent disclosed loratidine and methods of using loratidine, including administration to humans. According to the court's reasoning, even though DCL was not specifically disclosed in the '233 patent, DCL would always and necessarily be formed when loratidine was administered according to the methods taught in the invention. In other words, DCL necessarily flows from the teaching of the '233 patent itself. Further, the court noted that it did not matter whether the metabolite had actually been formed in the '233 patent, it was simply sufficient that an enabling disclosure had been provided for its production. Since this disclosure was made more than 1 year prior to the '716 filing date, claims 1 and 3 of '716 patent were anticipated under §102(b). Finally, the court noted that a metabolite could be claimed 1) in a pure form, 2) as a pharmaceutical composition, or 3) in a method of use claim. In fact, additional claims in the '716 patent covering pharmaceutical compositions and methods of treating allergic reactions by administering DCL were not anticipated by the '233 patent. Can you see though why this did not help Schering-Plough in the infringement case?[39]

---

[38]This part of the case is easy since the judge so constructed the claims and the party's agreed to that construction. You might wonder how it is that the company that produces the noninfringing substance (loratidine) needs to be worried about what happens to that compound once somebody else ingests it. It might seem logical to assume that it is the person ingesting the loratidine and converting it into the infringing substance that is the infringer. In fact, such a person would also be an infringer, but it is hard to imagine a scenario where Schering-Plough would have found it in their best interest to hunt down people ingesting generic loratidine and sue them for infringement. Instead, Schering-Plough presumably sued the generic companies under the theory of contributory (or possibly induced) infringement, whereby the product the generic drug companies would be selling could only be used in a way that would result in the user infringing their patent.

[39]Schering-Plough actually markets DCL as Clarinex®. However, the alternative claiming forms in the '716 patent would not have helped Schering-Plough to prevail in this infringement suit because none of these alternative claims to DCL would have been infringed by a generic company's marketing of loratidine. A claim to a pharmaceutical composition containing DCL would not be infringed by the metabolic production of DCL in a human's body. Likewise, the method of use comprising the administration of DCL would not be infringed since it is not DCL that is being administered. Finally, a claim to a pure form of DCL would not have been infringed since the DCL that is metabolically produced would not be in a pure form but would be contained within whatever the physiological milieu it exists in after its metabolic production.

CHAPTER 8

Basic Requirements of Patentability: Nonobviousness

The more original a discovery, the more obvious it seems afterword.

—Arthur Koestler

8.1 THE BASIS FOR THE NONOBVIOUSNESS REQUIREMENT

As you read about the novelty requirement as described in the previous chapter, you might have thought to yourself, "Patenting a chemical invention is easier than I thought, I never knew just changing one little thing was enough to get a patent granted." In fact, with the possible exception of inherent anticipation, you probably can appreciate that a thorough appraisal of the prior art should allow one to carefully design or claim one's invention in such a way so as to avoid the literal, element-by-element requirement of novelty.

However, patentability is neither so narrow as the novelty requirement alone would dictate, nor is its inquiry so relatively clear-cut. If we allow our own intuition to guide us in determining what an invention is or even what it should be, we might all come up with somewhat different definitions. However, a common theme would likely run through our collective responses—inventiveness should require something more than simply being novel under the strict definitions outlined in the prior chapter. This common theme of requiring more than absolute novelty is reflected in patent law through the nonobviousness requirement.

From the original Patent Act of 1790, more than 150 years of American patent law jurisprudence passed before a congressional act codified an inventiveness standard that required more than novelty and utility.[1] Known officially as the 1952 Patent Act

---

[1] In his spare time, Thomas Jefferson was our country's first "patent commissioner," who administered the 1790 Patent Act and later contributed to the 1793 Patent Act. Jefferson was also an accomplished inventor with a colonist's inveterate aversion to monopolies. Jefferson was reportedly leery of patent grants and did not believe in any natural right to an invention. However, he did see value in the patent system as an inducement to bringing forth new knowledge. In Jefferson's view, patents should be granted only

---

The Chemist's Companion Guide to Patent Law, by Chris P. Miller and Mark J. Evans
Copyright © 2010 John Wiley & Sons, Inc.

(35 U.S.C. §103) but more generally referred to as the nonobviousness requirement, it lays down the framework for federal court nonobviousness jurisprudence through the last 55 years and will serve as the starting point for our present discussion. The key part of the nonobviousness requirement is codified in 35 U.S.C. §103(a):[2]

(a) A patent may not be obtained though the invention is not identically disclosed or described as set forth in section 102 of this title, if the differences between the subject matter sought to be patented and the prior art are such that the subject matter as a whole would have been obvious at the time the invention was made to a person having ordinary skill in the art to which such subject matter pertains. Patentability should not be negatived by the matter in which the invention was made.

Note that 35 U.S.C. §103(a) lays out the overarching framework of the nonobviousness requirement.[3] To facilitate a detailed discussion of obviousness as it relates to chemical inventions, we first need to break down this section to better appreciate the finer topological contours of its landscape.

8.2   UNDERSTANDING §103(a)

Obviousness determination relies on §102 prior art so employment of the prior art in an obviousness determination (§103) will possess the same general characteristics as use of the prior art in novelty determination (§102). Thus an obviousness rejection based on §102(a) art can be overcome if the applicant is able to show that he invented the subject matter before the publication or art in question (just as a novelty rejection based on §102(a) can be antedated). Likewise, an obviousness rejection based on art

for discoveries that were more than small details, obvious improvements, or frivolous ideas. In spite of Jefferson's leanings, it was not until 1851 in *Hotchkiss v. Greenwood*, 51 U.S. 248 (1850) that the U.S. Supreme Court formulated a general condition of patentability with a clear statement of inventiveness reaching beyond novelty. In *Hotchkiss*, the Court held that the substitution of porcelain or clay for wood or metal in doorknobs was not sufficient to merit a patent. In its criticism of such an "invention," the Court held:

Unless more ingenuity and skill . . . were required . . . than were possessed by an ordinary mechanic acquainted with the business, there was an absence of that degree of skill and ingenuity which constitute essential elements of every invention. In other words, the improvement is the work of the skilful mechanic, not that of an inventor (p. 267 of the opinion).

The holding in *Hotchkiss* laid the groundwork for what ultimately became the 1952 Patent Act, where the obviousness requirement was codified in statute by Congress. Subsequent federal courts have further defined the obviousness requirement to bring us to where the law is now and to what will be the subject of our further discussion.

[2]§103(b) refers only to certain biotechnological processes. In brief, the section provides that biological processes that are used to produce novel and nonobvious compositions of matter will also be considered nonobvious provided certain conditions are met; §103(c) was discussed in Section 3.4.

[3]The European patent equivalent to the nonobviousness requirement in the United States is referred to as the *inventive step requirement*. Although the terminology is different, the overall effect is generally the same.

falling under §102(b) prior art *could not* be antedated by evidence of prior invention because, as we have already discussed, art published more than 1 year before applicant's filing date is a statutory bar, and prior invention by the applicant is no longer helpful to the applicant. There are two exceptions to the use of §102 prior art for §103 obviousness analysis. The first is that the §102 prior art used must be what is called *analogous art* (defined and discussed later). The second is the narrow but important exception provided for in 35 U.S.C. §103(c), by which obviousness rejections under §102(e), (f), and (g) can be overcome if at the time the claimed invention was made it was subject to an assignment or owned by the same entity or was the subject of a valid research collaboration. This was discussed in Section 3.4.

Reading further into the statute instructs one that a patent *will not* be granted if the differences between what one intends to patent (one's claimed invention) and the prior art are such that the *subject matter as a whole* would have been obvious. The phrase *subject matter as a whole* refers to the *entire* subject matter of the claim. This means that an interpreting body cannot pick the elements of a claim apart and find support in the art for each of the claimed elements without considering how the elements themselves combine into the whole to make up the invention. This is especially important when one considers that inventions are often combinations of disparate elements, wherein each element by itself is known in the art, but it is the particular combination that makes up the invention. This rule also prevents an examiner or court from focusing on certain obvious steps or elements of the claim while ignoring other steps that are nonobvious. Since infringement of a claim requires that an infringer make, use, or sell a device or process that includes each and every element of the claimed language (or its equivalent under the doctrine of equivalents), it would be unfair to deny patentability by ignoring one or more of the required claim elements when comparing to the prior art. This principle is well demonstrated in *In re Hirao*[4] in which the sole claim at issue was directed to the following three-step process:

1. A process for preparing foods and drinks sweetened mildly, and protected against discoloration, Strecker's reaction, and moisture absorption, which comprises:

    adding α-1,6-glucosidase and β-amylase, under such conditions and in a quantity sufficient to produce straight chain amylose, to enzymatically liquefied starch which consists essentially of amylopectin thereby producing straight-chain amylose;

    subjecting the resulting amylose to the action of β-amylase and purifying and drying to obtain high purity maltose in crystalline powder form of 90–95% maltose; and then

    adding said high purity crystalline maltose powder to foods and drinks as the essential added sweetener.

This case was on appeal from the USPTO, which denied patentability, asserting obviousness in view of two prior art references. The first prior art reference described how to make high-purity maltose, and the second prior art reference described using

---

[4]*In re Hirao*, 535 F.2d 67 (CCPA 1976).

high-purity maltose as a sweetener. The USPTO agreed that the first two steps of the process were nonobvious over the prior art method of making purified crystalline maltose powder, while the third step of simply adding the purified crystalline maltose powder to food and drinks was an obvious step. In its decision, the CCPA phrased the single issue as whether appellants' three-step process is obvious, the first two steps being unobvious but forming a known product, and the third step being the use of this known product in an obvious way and concluded "that due to the admitted unobviousness of the first two steps of the claimed combination of steps the subject matter *as a whole* would not have been obvious to one of ordinary skill in the art at the time the invention was made."[5]

Reading further in §103(a), we see that the obviousness inquiry must take part from the perspective of a "person having ordinary skill in the art" (a PHOSITA). This is an objective standard, meaning that the fact finder (judge or jury) responsible for making the obviousness inquiry must figuratively put themselves in the shoes of the "person having ordinary skill in the art."[6] The standard is really a form of legalized fiction (have you ever met a real person named PHOSITA?) that has been instituted to maintain objectivity throughout the obviousness inquiry. In this vein, the PHOSITA is not considered a layman, a judge, a jury, one skilled in remote arts, or a genius in the art at hand.[7] As this standard is applied to the chemical arts, it is generally recognized that the level of skill in the chemical arts is high, but not so high that a person having ordinary skill in the chemical arts would require exceptional insight. Factors that the federal circuit court have established thus may be considered for what is ordinary skill in the art include the educational level of the inventor, type of problems encountered in the area, prior art solutions to those problems, rapidity with which innovations are made, sophistication of the technology and education level of workers in the area.[8]

An important consideration is that a person having ordinary skill in the art is charged with the knowledge of all prior art "to which the subject matter of their

---

[5]The claim went onto issue as claim 1 in US 4,001,435.

[6]Patent examiners at the USPTO are usually highly educated in their respective fields, with most examiners in the chemical, pharmaceutical, and biotechnology areas holding a Ph.D. in their respective discipline. Accordingly, the examiners should be able to understand this standard from a technical standpoint and, we hope, be able to apply it in a meaningful way. However, patent litigation is initiated in a federal district court, where the obviousness inquiry will be undertaken by a jury and/or judge (jury or judge makes factual conclusions, obviousness conclusions from these facts are made by the judge as a matter of law), neither of which are likely to understand the chemical arts very well. Of course, this is where expert testimony might assist by helping the judge or jury by explaining her side's view of whether something would be obvious from the standpoint of a PHOSITA. However, depending on which side the expert represents, one is likely to get a very different opinion. One might expect that a plaintiff looking to enforce a patent would argue for its validity from an obviousness attack by asserting that the skill in the art would not have been high enough to make the invention obvious, whereas the defendant of an infringement action might look to invalidate the patent due to obviousness and will allege that the invention was obvious to one of ordinary skill in the art because the skill in the art is high.

[7]*Environmental Designs, Ltd. v. Union Oil Co.*, 713 F.2d 693 (Fed. Cir. 1983), cert denied, 464 U.S. 1043 (1984).

[8]*Id.* at 696.

invention pertains," but not every teaching in every art (in other words, a PHOSITA in chemistry will know *everything* that has ever been published in the chemical arts but will not be expected to know aeronautical engineering). Thus it might not be appropriate to apply working knowledge or techniques known to a person of ordinary skill in one art to those of another art.[9] This is known as the analogous art requirement and is a distinction from the novelty requirement, where there is no such limitation; though as we will see, this distinction has not made much of a difference. Furthermore, it is impermissible to apply the teachings of the patent application itself to indicate the ordinary skill in the art. This rule works to discourage applying hindsight gleaned from the applicant's disclosure itself against the applicant when considering the nonobviousness of the claimed invention in question.

Two cases will help explain the analogous art requirement as it pertains to chemical inventions.[10] In *Ex parte Bland* the applicant for patent argued that references cited by the USPTO to support an obviousness rejection were nonanalogous art and thus should not be cited against applicant.[11] The applicant's claimed invention related to the adsorption of propionic acid onto expanded perlite or verxite (a purified vermiculite) having a particular particle size range, where those compositions were useful for the inhibition of bacterial and fungal growth in a foodstuff while being nontoxic to the animals. The cited art that the applicant objected to included (1) a reference where propionic acid had been adsorbed onto silica for use as a food preservative; (2) a reference stating that the food additive verxite could safely be used in animal feed; (3) a reference that stated that verxite was a carrier for such agricultural products as fungicides and fumigants; (4) a reference disclosing a composition containing propionic acid and a suitable adsorbent such as perlite (the reference did not specify expanded perlite or the same particle size as applicant), and (5) the bacteriostatic and fungostatic use of such a composition for application to corn fodder and other agricultural products.[12] The Board disagreed with the applicant's position that the cited prior art was nonanalogous, indicating that all the prior art references were concerned with absorbing biologically active materials onto carriers. The Board further noted that the teachings in each of the references would have been pertinent to each other as well as to the problem solved by applicant's invention.

Likewise, in *In re Mlot-Fijalkowski*, the CCPA upheld the Board's rejection of the applicant's claimed invention as unpatentable for obviousness.[13] The applicant's claimed invention involved the use of penetrating dyes for the detection of surface imperfections, where the penetrating dye included a substance that undergoes a color change when treated with a developer. Two of the cited prior art references (Skelly and Vincent) described the use of color formers or dye intermediates with clay developer particles, but did not disclose the use of such dyes for the imagining of surface

---

[9]"A person of ordinary skill is also a person of ordinary creativity, not an automaton." *KSR v Teleflex* 550 U.S. 398 (2007).

[10]The USPTO analysis of analogous and nonanalogous can be found in MPEP 2141.01(a).

[11]*Ex parte Bland*, 3 USPQ2d 1103 (Bd. Pat. App. & Inter. 1986).

[12]Can you understand why none of the cited references as described would *anticipate* the applicant's claimed invention?

[13]*In re Mlot-Fijalkowski* 676 F.2d 666 (CCPA 1982).

imperfections. An additional reference (Sockman) described the use of a colored dye for the detection of surface imperfections. The Board noted in upholding the rejection of the patent application for obviousness in view of the cited art that "the concepts fairly contained in the examiner's references... would suggest to one skilled in the art... (the) substitution of the color former-activator systems of either Skelly or Vincent into the processes of Sockman." The applicant objected to the use of the Skelly and Vincent dye references and asserted that they were nonanalogous art. In particular, the applicant noted that the uses of the dyes in the Skelly and Vincent references were not related to the use that the applicant claimed for his dye but were focused on the duplicating paper art. The CCPA rejected this logic and instead pointed out that the applicant misplaced the focus onto the process the dye was going to be used for, rather than on the substitution of one dye for another. The CCPA agreed with the applicant's contention that the problem confronted by the applicant was the "enhancement and immobilization of dye penetrant indications" but characterized the problem as one of dye chemistry, and accordingly, one's search for a solution would not be limited to the field of dye penetrant inspection but would include the dye arts in general. As these two representative cases indicate, analogous art has been broadly construed, and it is not likely one will succeed in overcoming an obviousness rejection or challenge on the basis of the cited art being nonanalogous.

Finally we turn to the second sentence of §103(a): "Patentability should not be negatived by the matter in which the invention was made." Between the time the U.S. Supreme Court first required something akin to a nonobviousness requirement and the time Congress codified that requirement in 35 U.S.C. §103 approximately one century later, it appeared to Congress that the judiciary had begun to impose a subjective test that looked at the mental state of the inventor rather than the traits of the invention itself. In particular, the Supreme Court instructed that to be patentable a device must reveal "the flash of creative genius, not merely the skill of the calling."[14] Due to this language, Congress asserted the last sentence of §103(a) to make it clear that the patentability of the invention would be judged by the invention itself and not the state of mind of the inventor. Thus presumably it would not matter (one way or the other) whether the inventor had a flash of genius or just simply a brief and dim glimmer of routine recognition; the invention stands or falls on its own merits and not the state of mind or the actual process by which the invention was discovered.

8.3 GRAHAM FACTORS ANALYSIS OF OBVIOUSNESS

All of this preliminary material sets the stage for the crux of our present inquiry, What is obvious? This is a key inquiry for patent law and one of the most crucial yet challenging concepts to understand because its meaning goes to the core of the definition of inventiveness. We have seen and reviewed novelty and should well understand its predicates. Unlike novelty, however, obviousness is a much more open-ended inquiry. Whereas novelty instructed us that we were limited to a single reference, obviousness can be based on information separately incorporated from

---

[14]*Cuno Engineering Corp. v. Automatic Devices Corp.*, 314 U.S. 84 (1941).

numerous references. Obviousness does not require an element-for-element order match with the prior art and moreover, missing elements or limitations can sometimes be provided by the knowledge or typical practice of a PHOSITA (such as, for example, through the routine manipulation of experimental variables). In the remainder of this chapter, we will review numerous examples, hypothetical and real, where the obviousness inquiry is carried out. Even though many apparent rules will be laid down, the ultimate satisfaction of an all-encompassing criterion of obviousness will remain forever elusive since obviousness at its very heart requires a value judgment. It requires the fact finder to reach a conclusion that remains subjective in nature, no matter how much we might try to pretend otherwise. Undoubtedly, there will be arguments for and against the obviousness of an invention, and the ultimate decision will turn on how high the party performing the analysis decides to set the inventiveness bar. This in turn will be a function not only of the personal values of the one reaching the ultimate conclusion but the policy environment that person is operating in. Since patent policy is ultimately public policy, one might expect the inertia of case precedent to constantly adjust to the shifting tides of public policy.

After Congress passed the 1952 Patent Act creating 35 U.S.C. §103, it remained for the federal courts to interpret its meanings and color in the details through obviousness issues arising in patent cases. In 1966, two cases dealing with the interpretation and application of the statute reached the Supreme Court, in which they were consolidated and decided in *Graham v. John Deere*.[15] While the subject of the two cases were the mechanical arts (in *Graham v. John Deere*, specifically, the issue was the shock-absorber system for a plow shank[16]) and not directly applicable to the chemical arts, the Court's broader holding has formed the touchstone interpretation of 35 U.S.C §103 and is applied to obviousness inquiries for all cases, including those in the chemical arts. Although an obviousness inquiry does not always end at *Graham*, it will always begin with it. The Supreme Court in *Graham* set out a four-step inquiry that forms the beginning framework for any obviousness inquiry:

1. Determine the scope and content of the prior art.
2. Ascertain the differences between the claimed invention and the prior art.
3. Assess the level of skill in the art.
4. Evaluate evidence of secondary considerations.

---

[15] *Graham v. John Deere*, 383 U.S. 1 (1966).

[16] The subject matter of the first case in *Graham v. John Deere* concerned a patent that the plaintiff Graham wished to assert against an alleged infringer, John Deere. The patent described and claimed an improvement for a plow that was characterized by the Court as a combination of old mechanical elements involving a device designed to absorb shock from plow shanks as they plow through rocky soil that works to prevent damage to the plow. The Court held the patent invalid due to obviousness in view of a prior patent to Graham as well as a prior use clamp device. The second case concerned the validity of a patent describing and claiming a plastic finger-activated sprayer with a hold-down lid used as a built-in dispenser for containers packaging liquids, especially useful for household insecticides. This patent was also found invalid, in view of three prior art patents. In its discussion of these two cases, it appeared that the 1966 Court believed that the patent office was not maintaining a high enough standard of inventiveness in its patent grants and attributed this, at least in part, to a very large workload.

We shall discuss each of these in turn (note that the sequence of examining these four factors may be different in various cases, but each factor will be considered).

The first requirement of determining the scope and content of the prior art makes logical sense since it is the prior art that will be a primary consideration in finding whether a claimed invention is obvious or not. As we have already discussed in our analysis of §103, the scope and content of the prior art is analogous to the prior art as laid out in §102, with the exception that the art must be analogous art. We also discussed that §103(c) provides an important exception to §102(e) art arising from the same entity or from a research collaboration. Furthermore, it is possible that an applicant's own characterization of references as prior art can make those references prior art even where they might not have been considered so otherwise.[17] Ascertaining the scope and content of the prior art means evaluating all of the prior art and not just the prior art that supports a finding of obviousness. When an inventor makes his discovery, he may have been confronted with prior art that suggested to do what he ultimately did, but there may also have been conflicting art which pointed away from what he ultimately did. Art that suggests against doing what the patent applicant ultimately did is referred to as art that "teaches away," and such contradictory prior art can be persuasive evidence of inventorship. In a related way, the predictability of the art must be considered as well. The results of modifying highly unpredictable prior art are themselves highly unpredictable; *an obvious variation from the prior art requires at least a reasonable expectation of success.*

The second *Graham* step, ascertaining the differences between the claims to the invention of interest and the prior art, is central to the obviousness determination. This process requires interpreting the claimed invention *as a whole* as well as the prior art *as a whole* when making the comparison (recall our discussion of *In re Hirao*, in which one obvious element in a claim did not render the entire claim obvious). The claims are not read in a vacuum but find their meaning in the context of how one of ordinary skill in the art would understand the claims in view of the broader teachings of the specification.[18] Dictionaries and treatises can be consulted as secondary sources but are deemed to be inherently less reliable than the specification of the patent himself.[19] Once the claims are properly constructed, it is the job of the patent examiner (or federal court judges) to compare the properly interpreted claimed invention as a whole to the prior art as a whole. An invention contained within a claim cannot be reduced to an abridged version of the invention that a particular fact-finder thinks is the key component.[20]

---

[17] The USPTO follows the Graham four-step process in its examination of patents for obviousness. See MPEP 904.02 for a description of the USPTO search guidelines. Note that for some arts, such as the chemical arts, the USPTO recommends that its examiners, who likely hold advanced degrees, still use the services of USPTO personnel specifically trained in specialized tools for searching the chemical arts.

[18] *Phillips v AWH Corp.*, 415 F.3d 1303 (Fed. Cir. 2005).

[19] The patent prosecution history is also considered to be part of the intrinsic record and is publicly available for applications that have published.

[20] See for example, *Panduit Corp. v. Dennison Mfg. Co.* 810 F.2d 1561 (CAFC 1987), in which it was found that the District Court improperly distilled claims down to a one-word solution.

Understanding the invention as a whole also means placing the invention into the broader context of what the invention accomplishes in relation to what the art as a whole teaches. For example, consideration of whether the invention discovers the source or cause of a particular problem and whether the problem itself was known in the prior art are part of understanding the invention as a whole. Thus an obvious solution to a nonobvious source of a problem can lead to a patentable invention provided that the solution to the problem was not the same solution used for a similar problem known in the art. To better understand this principle, let's look at a hypothetical problem.

Imagine that a particular commercial chemical process used to make an important intermediate in the fragrance business is subject to irreproducible yields, resulting in occasional significant loss of materials as well as increased effort and expense in isolating the product. After considerable effort, including modifying the catalyst, varying solvents, reaction times, and temperatures, the chemist realizes that the problem may not be in the catalyst but rather the starting material itself. Recognizing that the process to make the starting material includes an alkanethiol reactant, he wonders whether trace amounts of the alkanethiol are being carried over and poisoning the catalyst in the process. To investigate this possibility, the chemist washes a solution of the starting material with an aqueous sodium bicarbonate solution containing dilute hydrogen peroxide, and it is found that the alkanethiols can be removed without degrading the starting material and that the process of using the prewashed starting material results in a significantly more robust reaction and a higher-quality product.[21] Since the previous process used to prepare the fragrance intermediate is already known in the literature, the chemist's organization files a patent application claiming the process including the oxidative wash step to remove the alkanethiol impurities.

It is well known in the literature of analogous arts that washing materials containing alkanethiols with basic aqueous peroxide solutions oxidizes organic thiols to sulfonic acids, which are subsequently extracted into the aqueous solution as their conjugate base. As was mentioned previously, the process absent the washing step was already known as well. Given the prior art familiarity with both of these processes, an overzealous patent examiner might argue that a new process combining these two prior art processes together therefore is also obvious. However, this would be a mistake because the patent examiner needs to keep in mind that which we just learned: It is the claimed invention as a whole that needs to be examined, including the problem that is solved by the claimed invention. The inventor was confronted with an irreproducible reaction and identified the problem as trace amounts of alkanethiols in the starting material. The primary locus of inventive activity was the discovery of the problem. Once he discovered the source of the problem, the solution *may* have been obvious, but that misses the point. The claimed invention as a whole would not have been obvious because it would not have been obvious to wash the prior art starting material with sodium bicarbonate and hydrogen peroxide because the prior art did not recognize alkanethiol as an impurity that needed to be removed. Whereas

---

[21] Alkanethiols have additional issues, including the fact that they tend to smell pretty awful.

the solution might seem obvious, the problem was identified from what was likely a very large number of candidates. The reality is that in many cases, problematic steps in chemical processes are never brought to such a clean and successful conclusion (paper chemistry is always easier than the real thing).

The third step of the Graham inquiry requires determining the level of skill that one of ordinary skill in the art would possess. As we learned earlier in this chapter, one of ordinary skill in the art is charged with the knowledge of the prior art that relates to the field of the invention, but the level of skill in the art relates more directly to the ability of one of ordinary skill in the art to apply that knowledge to solve problems. The higher the level of skill in the art, the greater the insight an ordinary practitioner is going to be assumed to have. Things obvious to one who understands chemical structure–activity relations through the eyes of an experienced chemist would not be obvious to one with no more than a high school level of chemistry knowledge. The baseline determination for the level of ordinary skill in the art would be an important exercise, except that one can expect a relatively universal standard of a high level of ordinary skill in the art across the chemical disciplines. As a practical matter, the average chemist is college-educated and possesses several years of practical experience.

The fourth step of the Graham inquiry requires that any obviousness determination must also consider secondary evidence of nonobviousness. This is a very important consideration where chemical inventions are concerned, and we will see its practical application in due course, but for now let's consider the overarching import of what the fourth step of Graham entails. By referring to "secondary" evidence of nonobviousness, the Graham Court did not mean that such evidence is of secondary importance but rather that such evidence is not necessarily evident from the patent application itself or the prior art. These secondary considerations of nonobviousness relate to the invention itself and include such items as unexpected results, commercial success, long-felt need, copying by others, licensing, skepticism of experts, and independent development; they will be explored further later in the chapter.

8.4   FOCUSING THE OBVIOUSNESS INQUIRY: PRIMA FACIE OBVIOUSNESS AND THE CHEMICAL INVENTION

From our review of the statutory language of §103 as well as the application of that statute as spelled out in *Graham v. John Deere*, we have seen that obviousness is determined by an inquiry that initially takes into account the scope and content of the prior art, the differences between the claimed invention and the prior art, the level of skill of one of ordinary skill in the art, and any secondary evidence of nonobviousness. In a patent application examination at the USPTO, in order for an examiner to make an obviousness rejection of the invention, the examiner must present a prima facie case of obviousness. *Prima facie* is Latin for "on the face of it," and it means that the patent examiner must put forth sufficient written argument or evidence in view of the appropriate legal precedence such that the claimed invention, as described in the specification and in view of the prior art, is legally nonpatentable for obviousness.[22] This does not mean that the invention will ultimately be determined to be nonpatentable for obviousness, but rather that the examiner has established sufficient arguments to place the burden of rebuttal onto the applicant. The rebuttal itself may take the form of legal arguments by the applicant, evidence of secondary considerations, or a combination of arguments and evidence. Since secondary evidence of nonobviousness is often not evident from the patent application itself, the examiner's prima facie case will typically focus on the apparent features of the claimed invention in view of the prior art; this is essentially the first three steps of the Graham inquiry. It is very often in the rebuttal to the examiner's assertion that the patent applicant will assert secondary evidence of nonobviousness (the fourth step of the Graham inquiry). Although the obviousness inquiry is ultimately a single determination, collectively taking into account the primary arguments and secondary evidence, it is most helpful to break the two types of analyses up into separate parts.

As was directly stated earlier, the obviousness inquiry cannot avoid a great element of subjectivity. However, this is not to say that the subjectivity is desired, it is not. Inventors need to know with the maximal amount of certainty, in advance, whether what they perceive to be an invention is worthy of a strong and enforceable patent. Something so valuable (and costly) as an enforceable patent right should not be subject to the luck of the draw regarding which examiner, jury or judge is assigned to evaluate whether the claimed subject matter is worth a patent or not. Perhaps just as important, those who would declare an invention obvious in view of the prior art have the advantage of hindsight gleaned from the patent application itself. Just as the answer to a brain teaser or riddle can seems so obvious once we have been given the answer, so too can an inventive solution seem obvious once we've seen the disclosure of the inventor's patent application. However, unlike the person looking at the patent application who slaps herself on the forehead and says "Of course, now I see—he just combined A with B and sprinkled in a little C—how obvious," the patent applicant does not have his own solution in front of him at the time of the invention. That is the thing he needed to figure out. In recognition of at least these two difficulties (hindsight bias, objectivity), the federal courts have established and refined a useful analytical framework for assisting those conducting the obviousness inquiry. This analytical framework provides that in order for a claimed invention to be obvious, the examiner or judge(s) should be able to point specifically to some teaching, suggestion, or motivation (TSM test)—either in the references cited or in the knowledge generally available to one of ordinary skill in the art—or to the nature of the problem itself, or to combine references or modify the references in such a way as to arrive at applicant's claimed invention. Furthermore, the modification or

---

[22] As the CAFC has put it, "[R]ejections on obviousness grounds cannot be sustained by mere conclusory statements; instead, there must be some articulated reasoning with some rational underpinning to support the legal conclusion of obviousness" *In re Kahn*, 441 F.3d 977, 988 (CAFC 2006). This rationale is frequently cited by the Board to reverse an examiner's obviousness rejection, (e.g. *Ex parte Erkey et al.*, Appeal 20071375 and *Ex parte Crawford et al.*, Appeal 20062429). For examples of acceptable prima facie cases of obviousness, see MPEP 2143.

combination of prior art references must represent a reasonable chance of success, and of course, the prior art references must teach, suggest, or motivate one to arrive at all of the claim elements or limitations; the claim as a whole must be obvious.[23] However, as a very recent Supreme Court opinion on the subject made clear, the TSM test should not be rigidly applied but is rather one useful inquiry among many that can be used to evaluate whether an invention is obvious.[24]

The TSM test is a very useful construct because it asks that the examiner provide objective evidence to support his rejection. This requirement is especially important in many chemical cases where the obviousness of a chemical structure is being asserted. In order for a claim to a chemical structure or composition to be obvious, the examiner must be able to allege not only structural similarity of the chemical structure or composition but also where the prior art provides the teaching, suggestion, or motivation to make the suggested change or combination or, if not in the prior art, why one of ordinary skill would have been motivated to combine or modify the prior art in the absence of a specific teaching in the prior art itself.

---

[23] MPEP 2142.

[24] In *KSR v. Teleflex, Inc.*, 550 U.S. 398 (2007), the Supreme Court (in a rare unanimous decision) reversed a CAFC decision that found that a patentee's invention was not obvious as a matter of law (in so doing, the CAFC had reversed a District Court's summary judgment that the patentee's claimed electronic accelerator that incorporated an electronic sensor was obvious in view of a combination of prior art accelerator pedals). The Supreme Court reasoned that the CAFC had too rigidly applied the TSM test. In particular, the Supreme Court was especially critical of the CAFC's apparent requirement that the prior art itself must provide the specific teaching or suggestion to combine prior art teachings or references in the way that the patentee had done. Rather, the Supreme Court noted:

> As our precedent makes clear, however, the analysis need not seek out precise teachings directed to the specific subject matter of the challenged claim, for a court can take account of the inferences and creative steps that a person of ordinary skill in the art would employ. . . . A person of ordinary skill is also a person of ordinary creativity, not an automaton.

To many, the *KSR* case may have elevated the bar for invention at the expense of objectivity and predictability. Not only must the party reviewing the invention put themselves in the position of one having ordinary skill in the art but they must also assume ordinary creativity in the art as well, suggesting we are traveling back in time to the "flash of creative genius" implied previously in *Cuno*. However, this reading probably overstates KSR's importance to the chemical arts for multiple reasons. First, federal court jurisprudence has always allowed that the motivation to combine or optimize the prior art could be derived from the nature of the problem to be solved, including allowing for routine optimization. As we review many of the Court decisions that have been made where the invention consisted of manipulation of known, results-affecting variables, prima facie obviousness will be found more easily; *KSR* does not change this. Second, the subject matter of much of chemical patent law is by its very nature often less predictable than that of mechanical cases. The workmanlike adjustment of routine variables with expected results hardly applies to much of chemistry, where the results obtained are less predictable and more empirical. As the CAFC recently noted in one post-*KSR* chemical case: "To the extent an art is unpredictable, as the chemical arts often are, *KSR*'s focus on these 'identified, predictable solutions' may present a difficult hurdle because potential solutions are less likely to be genuinely predictable" See *Eisai v. Dr. Reddy's Laboratories,* No. 2007-1397 (CAFC 2008). Finally, despite the temptation to reduce all experimentation to allegedly routine processing, one needs to remain focused on the invention being claimed. If the *process* used in discovering the invention is deemed to be a routine *process*, then a claim to that *process* of discovery will be affected but not the discovery itself. Let's recall, "Patentability should not be negatived by the matter in which the invention was made" (35 USC §103(a)).

As we will see in many of the case discussions to follow, the TSM test can be quite useful but that *motivation* as a legal term of art is still a tricky one. Its strictures sometimes appear elastic, stretching in such ways that previous forms cannot always be recognized. To appreciate how federal courts have applied the test to specific chemical cases, we'll first stipulate to the characterization that the motivation requirement is a patentee-friendly policy. It puts the burden on the party alleging obviousness to not only describe the features that make the prior art disclosure(s) similar to applicant's claimed invention but also asks him to explain why somebody of ordinary skill in the art would have wanted to make the change. The difficulty of this task to a large part rests on how broadly the courts want to define motivation. Defined most broadly, one might readily find motivation everywhere, by simply saying something to the effect: "One of ordinary skill in the art would be motivated to modify the prior art references to arrive at applicant's claimed invention because it is 'obvious to try' variations and combinations of things that are already known." On the narrow end of the spectrum, one might hear: "Show me where the prior art motivates one to make this specific modification or this specific combination—'obvious to try' cannot be the standard." The actual position taken by any federal court or patent examiner is going to be a reflection of the specific facts before her as well as, to some extent, her own reference point. To better appreciate how this very important test is applied, several examples from federal court opinions on the subject will be presented. Even when not discussed specifically, aspects of the TSM test will still be evident in many of the cases that have been highlighted for other purposes.

8.5 APPLICATION OF THE TSM TEST TO THE CHEMICAL ARTS

The decision in *In re Albrecht* results from an appeal from the USPTO to the CCPA where the applicants' eight claims directed to bis-esters and bis-amides of carbazoles were rejected.[25] Claim 1 of the invention, directed to a Markush genus of compounds is reproduced in its entirety from the opinion and shown in Figure 8.1.

The additional rejected claims, 2–8, were directed to additional, presumably narrower embodiments of claim 1. In rejecting claims 1–8 for alleged obviousness, the USPTO examiner (and majority of the USPTO Board that upheld the rejection) relied on a single piece of prior art from the chemical literature.[26] The cited art related to a study of the local anesthetic properties of certain carbazoles, dibenzofurans, and dibenzothiophene derivatives. There were no bis-basic esters of carbazoles disclosed (only mono-basic esters), but there was one bis-basic ester of a dibenzofuran that was disclosed, having the structure shown in Figure 8.2.

The bis-basic ester of the dibenzofuran from the cited art is structurally similar to the Markush genus disclosed in Albrecht's application but does not fall within it since the core structure is a dibenzofuran and that of Albrecht is a carbazole.[27] This

---

[25]*In re Albrecht*, 514 F.2d 1389 (CCPA 1975).
[26]R.R. Burtner et al. *J. Am. Chem. Soc.* 62 (1940), 527–532.
[27]They are the same core structures except the dibenzofuran has an oxygen in place of the N-R_3 of the claimed carbazole.

1. A compound of the formula,

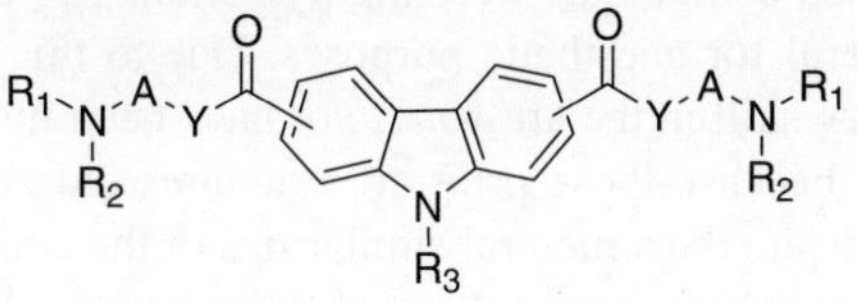

where:

(A)   each of R_1 and R_2 is hydrogen, (lower) alkyl, cycloalkyl of 3 to 6 ring carbon atoms, alkenyl of 3 to 6 carbon atoms having the vinyl unsaturation in other than the 1-position of the alkenyl group, or each set of R_1 and R_2 taken together with the nitrogen atom to which they are attached is pyrrolidino, piperidino, N-(lower)-alkylpiperazino, or morpholino;

(B)   each A is alkylene of 2 to about 8 carbon atoms and separates its adjacent Y and amino nitrogen by an alkylene chain of at least 2 carbon atoms;

(C)   each Y is oxygen, or N-R wherein R is hydrogen, methyl or ethyl; and

(D) R_3 is hydrogen or (lower) primary or secondary alkyl, or an acid addition salt thereof.

FIGURE 8.1   Claim 1 from *In re Albrecht*.

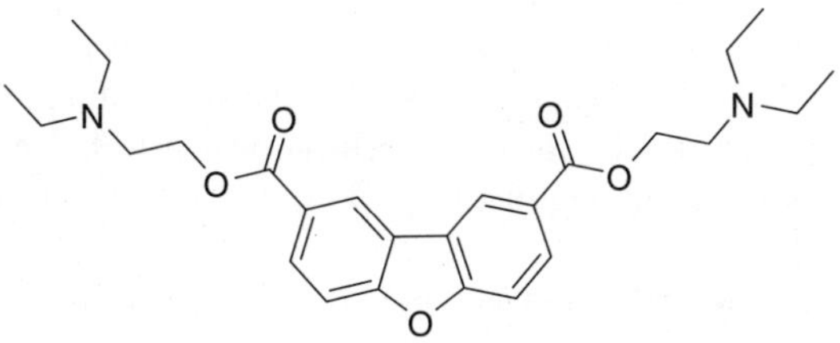

FIGURE 8.2   Prior art dibenzofuran.

is an important point because if the core structure of the prior art example shown in Figure 8.2 was a carbazole instead of a dibenzofuran, the compound would fall within Albrecht's genus and thus would have anticipated the genus, and it would have been game over for Albrecht, at least for the compounds of claim 1.[28]

In rejecting claims 1–8 of Albrecht's application, the examiner noted that certain mono-basic esters of carbazole were reported to have "powerful" local anesthetic activity, whereas the corresponding dibenzofuran compounds were less effective. The further disclosure of the bis-basic dibenzofuran as having some minimal anesthetic activity was sufficient, the examiner alleged, to motivate a PHOSITA to make the bis-basic esters of the carbazole compounds like those claimed in Albrecht's application, since the prediction would be that the bis-basic ester of the carbazole would be a more effective anesthetic than the disclosed bis-basic dibenzofuran. It would, therefore, be obvious to make these compounds, and the claims should be rejected.

---

[28] See Section 7.3 for a discussion of anticipation of a genus by a species.

The applicant appealed and in its opinion, the CCPA note the cited reference also indicated that the disclosed compounds were highly irritating to the skin and could not be considered as useful for anesthetic purposes. Due to this feature, the court found that one of ordinary skill in the art *would not* have been motivated to modify the prior art compounds because those prior art structures did not have an overall interesting activity. So despite the structural similarities of the compounds of claims 1–8 of the Albrecht et al. application and those of the prior art reference, the CCPA reversed the USPTO's rejection. In driving home its logic for reversal, the CCPA quoted from its own earlier decision:

> How can there be obviousness of structure, or particularly of the subject matter as a whole, when no apparent purpose or result is to be achieved, no reason or motivation to be satisfied, upon modifying the reference compounds structure? Where the prior art reference neither discloses nor suggests a utility for certain described compounds, why should it be said that a reference makes obvious to one of ordinary skill in the art an isomer, homolog or analog of related structure, when that mythical, but intensely practical, person knows of no "practical" reason to make the reference compounds, much less any structurally related compounds?[29]

The compounds from the prior art reference cited against the Albrecht application possessed utility in that they were active as local anesthetics, but they lacked overall utility in that their irritant qualities would exceed their usefulness as local anesthetics. It appeared to the CCPA that a chemist of skill in the art, reading the cited reference for *all* that it disclosed, would *not* have been motivated to make additional close analogues based on the likelihood that they too would be irritants and thus have no practical utility.[30]

What if the prior art would motivate one of ordinary skill in the art to make the applicant's claimed invention but for a different reason than that relied on by the applicant? Should it matter? The following example, taken from a CAFC opinion decided this exact question.[31] The named petitioner, Dillon, was an assignor of the invention to Union Oil Company, whose application for patent was rejected in the USPTO. After unsuccessfully appealing to the USPTO Board of Appeals, the petitioner appealed to the CAFC and the rejection of the patentability of the claims was affirmed. The subject matter of the relevant claims was directed to compositions containing *tetra*-orthoesters in hydrocarbon fuels and methods of reducing particulate emissions using those *tetra*-orthoesters in hydrocarbon fuel. To clarify the subject

---

[29] *In re Steminski*, 444 F.2d 581, 586 (CCPA 1971). In the case of *In re Albrecht*, an additional issue was woven into the fabric of the case. An affidavit submitted by the applicants demonstrated additional activities of the claimed compounds not shared by the prior art. The claimed compounds of the Albrecht application possessed antiviral activity that the prior art compounds were allegedly lacking. It appears that the CCPA found this to be persuasive secondary evidence of nonobviousness, further supporting the nonobviousness of the claimed compounds.

[30] Albrecht et al. thus received their patent, U.S. 3,932,456.

[31] *In re Dillon*, 919 F.2d 688 (CAFC 1990).

2. A composition comprising: a hydrocarbon fuel; and a sufficient amount of at least one orthoester so as to reduce the particulate emissions from the combustion of the hydrocarbon fuel, wherein the orthoester is of the formula:

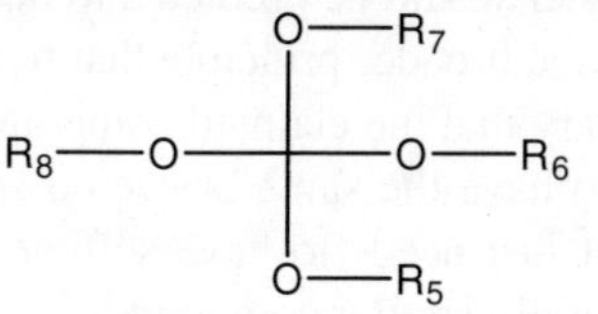

$$
\begin{array}{c}
O\text{---}R_7 \\
| \\
R_8\text{---}O\text{---}\!\!\!\underset{|}{\overset{|}{}}\!\!\!\text{---}O\text{---}R_6 \\
| \\
O\text{---}R_5
\end{array}
$$

where: R_5, R_6, R_7, and R_8, are the same or different monovalent organic radical comprising 1 to about 20 carbon atoms.

FIGURE 8.3   Claimed *tetra*-orthoester from Dillon.

matter of the dispute, the broadest claim from the patent application is shown in Figure 8.3.[32]

As a preliminary matter, the CAFC noted that the *tetra*-orthoesters were a known class of compounds but that their combination with a hydrocarbon fuel was undisputedly novel and therefore, that their use to reduce particulate emissions from combustion of hydrocarbon fuel was not shown *or suggested* in the prior art. Nevertheless, the CAFC found the claims obvious based upon the combined teachings of several U.S. patents. One patent (U.S. 4,395,267) described hydrocarbon fuels containing various additives for dewatering, including *tri*-orthoesters, and another patent (U.S. 4,395,267) described *tri*-orthoesters for use in hydrocarbon fuel to prevent phase separation between a hydrocarbon fuel and alcohol.[33] The CAFC also listed some U.S. patents as secondary references that disclosed both *tri*- and *tetra*-orthoesters for use as water scavengers in hydraulic (nonhydrocarbon) fluids. The CAFC held that one of ordinary skill in the art would have been motivated to add *tetra*-orthoesters to a hydrocarbon fuel to remove water and, moreover, that there was reasonable expectation that such a compound would reduce water in the fuel since *tri*- and *tetra*-orthoesters had been shown to work in other fluids (brake fluids), thus establishing equivalent activity between *tri*- and *tetra*-orthoesters for purposes of water removal. Furthermore, there was a sufficiently close structural relationship between the two compounds to expect at least similar results for reduction of particulate emissions.

Although the CAFC had already acknowledged that there was nothing in the art to specifically suggest that a *tetra*-orthoester in fuel would act to reduce particulate

---

[32]Claims 2–14, 16–22; and 24–37 were the only remaining claims in the application by the time it was taken up on appeal. During the course of prosecuting a patent in the USPTO, it is very common for a patent applicant or their representative to cancel, add and/or amend claims. If and when the claims eventually issue, they will be listed in numerical order starting with claim number 1.

[33]A *tri*-orthoester differs from the *tetra*-orthoesters of Dillon's application by containing only three alkoxy groups bonded to the same carbon atom, with the fourth group consisting of either a hydrogen or alkyl group.

composition, the court found that there was sufficient motivation in the prior art to make a hydrocarbon fuel composition that contained a *tetra*-orthoester since the prior art already disclosed a *tri*-orthoester in fuel as a water-reducing agent and the prior art also disclosed that a *tetra*-orthoester would be predicted to behave the same way. In so finding, the CAFC introduced the broader principle that to make a prima facie case of obviousness, it is not necessary that the claimed compound will be expected by one of ordinary skill in the art to have the same or even a similar utility as the utility disclosed by the applicant; it just needs to have sufficient overall utility to motivate one to make and try structurally similar compounds.

One way of thinking about this finding is to appreciate that the *Dillon* application claims a composition, not a reason for making that composition. If all of the limitations of that claim are suggested or taught in the prior art, then that claim is nonpatentable. According to the court's reasoning, it does not matter why one of skill in the art would have been motivated to make Dillon's composition, only that they would have been motivated.[34]

In the context of ANDA litigation on pharmaceutical compound claims, a company challenging the innovator's patent will often allege that the claimed compound is obvious in view of one or more similar structures that have been disclosed in the prior art. The case of *Yamanouchi Pharmaceutical Co., Ltd. v. Danbury Pharmacal, Inc.* is representative of the federal courts' handling of compound composition claims covering novel drug substances that are at least somewhat similar to prior art structures.[35] H_2-receptor antagonists work to prevent acid indigestion by inhibiting histamine action on H_2-receptors on the basolateral membrane of parietal cells in the stomach. Yamanouchi was the assignee of a patent (U.S. 4,283,408) that claimed, inter alia, the structure of the H_2-receptor antagonist famotidine (sold in the United States as Pepcid®). In January 1997, the generic drug manufacturers Danbury (a division of Schein) and Marsam (herein after collectively referred to as Danbury) filed an ANDA with the FDA, seeking approval to bring generic famotidine to market. In their certification to the FDA, the generic defendants alleged that claim 4 (specifically covering famotidine) of the patent was invalid (Figure 8.4).

Yamanouchi filed suit against Danbury for infringement, and the case was heard in the Federal District Court of the Southern District of New York, where Yamanouchi prevailed. Danbury appealed to the CAFC, still alleging that claim 4 of U.S. 4,283,408 was obvious in view of a combination of references. In its assertions, Danbury cited

---

[34] You might perceptively have noted that claim 2 also requires that the *tetra*-orthoester be present in an amount sufficient to reduce particulate emissions. This is also a limitation and cannot be ignored, and the CAFC did not ignore it. In rejecting patentability made on the additional amount-limiting term in the claim, the CAFC explained that the amounts of *tetra*-orthoester that Dillon actually claimed in later dependent claims were very broad in range and overlapped significantly with the amounts of *tri*-orthoester disclosed in the prior art patents and thus did not provide any additional opportunity for differentiation (more will be said about obviousness of ranges and amounts later). The Court also rejected petitioner's contention that the hydraulic fluid example was nonanalogous art and thus should not have been considered. In brushing aside the petitioner's urging, the Court asserted that the use of *tri*-orthoesters as dewatering agents in hydrocarbon fuels or *tetra*-orthoesters used in brake fluids were both in the field of the inventor's endeavor.

[35] *Yamanouchi Pharmaceutical Co., Ltd. v. Danbury Pharmacal, Inc.*, 231 F.3d 1339 (CAFC 2000).

1. Guanidinothiazole compounds of the formula

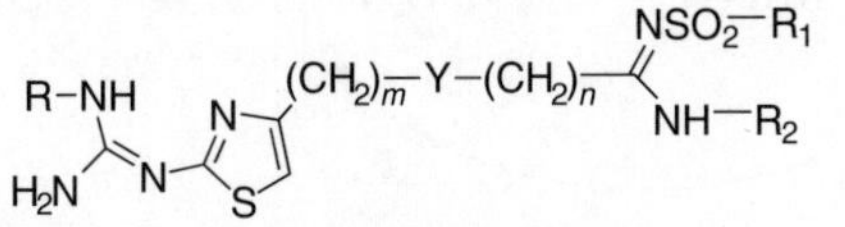

where: R represents a hydrogen atom or a lower alkyl group, R_1 represents an amino group, a lower alkyl group, a halogeno lower alkyl group, a phenyl, or naphthyl group that is unsubstituted or substituted by halogen, hydroxyl, amino, or alkoxy, a mono- or di-lower alkylamino group, an arylamino group, or an aralkylamino group; R_2 represents a hydrogen atom, a lower alkyl group, a lower alkenyl group, or a lower alkynyl group; Y represents a sulfur atom or a methylene group; m and n each represent an integer of 1–3, and the pharmacologically acceptable acid addition salts thereof.

4. The compound as claimed in claim 1 which is N-sulfamoyl-3-[(2-guanidinothiazol-4-

yl)methylthio]-propionamidine.

FIGURE 8.4   Claim to genus and species covering famotidine.

a compound from a prior art patent awarded to Yamanouchi (example 44 in U.S. 4,252,819) and the prior art H_2-antagonist tiotidine (both structures are compared to famotidine in Figure 8.5).

Danbury argued that it would have been obvious to combine the polar tail piece from example 44 of U.S. 4,252,819 with the substituted heterocycle of tiotidine

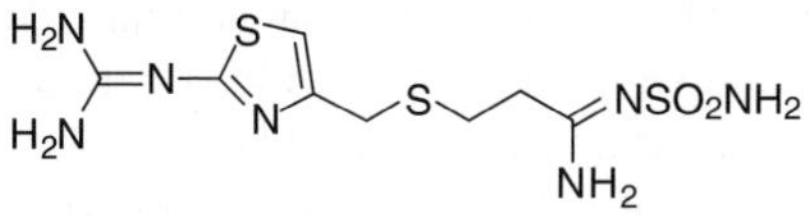

Famotidine

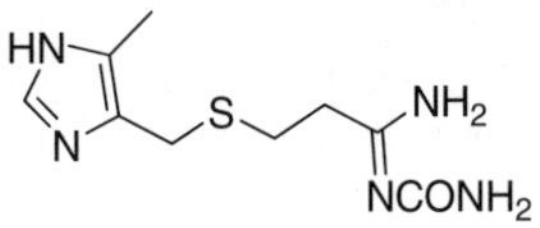

Example 44 (prior art) 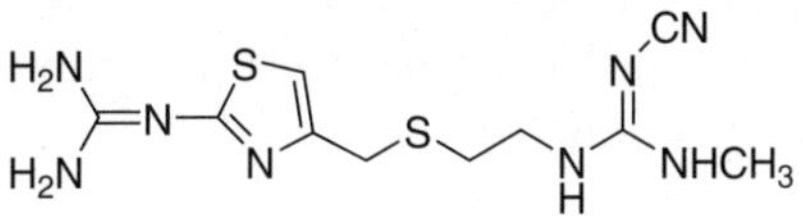   Tiotidine (prior art)

FIGURE 8.5   Structures of famotidine and prior art.

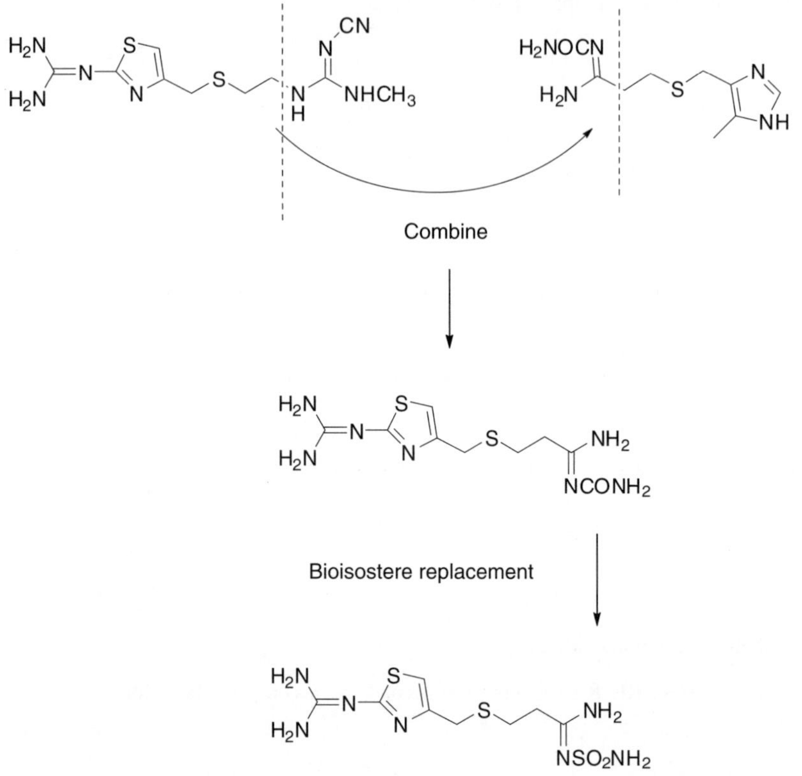

FIGURE 8.6   Danbury's obviousness argument.

and then make a bioisosteric substitution of the carbamoyl ($CONH_2$) group in the intermediate compound with a sulfamoyl group (SO_2NH_2) (see Figure 8.6).[36]

In its affirmation of the District Court's holding that rejected Danbury's obviousness argument, the CAFC asserted that Danbury failed to provide motivation for selecting compound example 44 to pursue as a lead. The court pointed out that there were more potent compounds than example 44 known in the art, and thus contrary to Danbury's assertion, the simple potency of example 44 was not motivation enough to select it for further modification. Beyond this, Danbury failed to explain what would have motivated one to use the polar tail from compound example 44 in combination with the heterocycle of tiotidine and then substitute the carbamoyl group with the sulfamoyl group. If Yamanouchi had not performed all of these steps, the end result would not have been nearly so successful. For example, if the sulfamoyl group had been

[36]In the parlance of medicinal chemistry, bioisosteres are groups of atoms that might be expected to impart similar physical and chemical properties when substituted from one biologically active molecule to another. They are frequently employed to try to generate additional active molecules from a known active lead—occasionally they even work!

substituted for the carbamoyl group on example 44 without substituting the hetero-cycle ring of tiotidine, then the resulting compound would have had approximately 1/4000 the potency of famotidine.

The CAFC also commented on the absence of a reasonable expectation of success. In this particular instant, the CAFC rejected Danbury's reasoning that an expectation of success merely required an expectation of minimal activity. Rather, the CAFC held that the success of discovering famotidine was not the expectation of discovering one of the "tens of thousands of compounds that exhibit baseline H_2 antagonist activity" but "was the finding of a compound that had high activity, few side effects, and lacked toxicity."

8.6   PRIOR ART AS A WHOLE MUST BE CONSIDERED FOR TSM TESTS

Recall from our introductory obviousness discussion that the obviousness analysis must be *based on the subject matter as a whole*. Accordingly, it should not be surprising that the components of our obviousness inquiry, such as motivation, must also be examined on the basis of the subject matter as a whole. Thus it would be improper to consider only references or portions of references that support motivation for making a modification and/or combination while ignoring additional portions or references that might conflict with those references or portions of references. One category where this can occur is where a reference actually teaches away from making the particular modification or combination. In a recent case, *Ecolochem, Inc. v. Southern California Edison Company* the CAFC supported a finding of nonobviousnss on just this point of teaching away in the prior art.[37]

The patentee in this case was Ecolochem, Inc., the holder of two patents (U.S. 4,556,492 and U.S. 4,818,411) which were allegedly infringed by Southern California Edison Company (defendant) through the defendant's deoxygenation of water in one of its nuclear power stations. The claimed invention was related to pressurized water reactors that used a primary system and a secondary system. Since a small amount of water from the secondary system is lost in each cycle, the water must be replaced, and that water must be of very high purity, purged of both minerals and oxygen. The patentee and defendant had a prior agreement where the patentee provided purified water to the defendant, but the patentee was eventually relieved of its service due to defendant's introduction of its own water purification and deoxygenation apparatus, which the patentee, in turn, alleged infringed their two patents. The patentee's process comprised a hydrazine/carbon process for removal of oxygen, followed by passage of the water through ion exchange beds to remove ionic impurities leached from the carbon bed on reaction of the hydrazine with oxygenated water. The cited prior art described several methods for deoxygenating water, including the use of hydrazine in the presence of carbon. However, in its recitation of the various methods, the prior art reference mentioned that the hydrazine/carbon process will "release salts

---

[37]*Ecolochem, Inc. v. Southern California Edison Co.,* 227 F.3d 1361 (CAFC 2000).

into the demineralized water" and that an alternative process using a hydrogen-based reductive process is energy saving and significantly less expensive. Other prior art references either failed to refer to the carbon/hydrazine process as practical or touted other processes in its stead. The patentee's solution, which consisted of using the hydrazine/carbon process followed by passage through negative and positive ion exchange beds, solved the problems in the art and provided a practical solution to the problem at hand. In reversing a lower court's findings, the CAFC upheld the patents and criticized the way that the lower court appeared to focus on the ways that the prior art *could be* combined to obtain patentee's claimed invention, while ignoring the considerable amount of evidence in the literature that would have discouraged one in the art to proceed in the first place.

Another manner of considering the prior art for all that it might teach asks that we consider how the motivation of a PHOSITA would be affected if the proposed modification or combination would destroy or hinder the operability of what was intended in the reference. For example, consider the following hypothetical example.

During the course of her research, a chemist at the small biotech company, Imaginit Pharmaceuticals, succeeds in synthesizing a series of compounds containing a core 1*H*-10-thia-1,3-diaza-cyclopenta[*a*]fluorene. Some of the compounds in the series demonstrate antibacterial activity and some of the compounds even demonstrate significant activity against vancomycin-resistant bacterial strains. Consistent with their structure activity relationship, the chemist draws up the Markush structure represented by formula I in Figure 8.7 and in consultation with her representative files a patent application claiming the compounds as well as their methods of use. In the course of the prosecution of the patent application, the USPTO examiner issues a rejection citing an issued patent from the country of Imaginarylanda. The cited patent reference includes compounds of the Markush formula II (also in Figure 8.7).

The cited prior art reference discloses that the compounds of formula II have significant cytotoxic and anticancer activity, which is attributed to the nitrogen mustard moiety since the reference also discloses that hydroxyl groups in the place of the

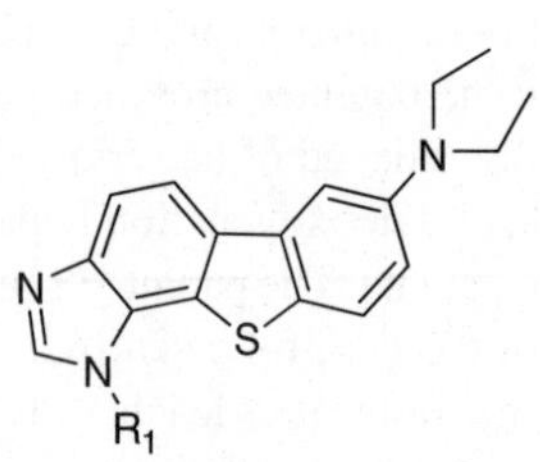

I

R_1 is hydrogen or C_1-C_6 alkyl

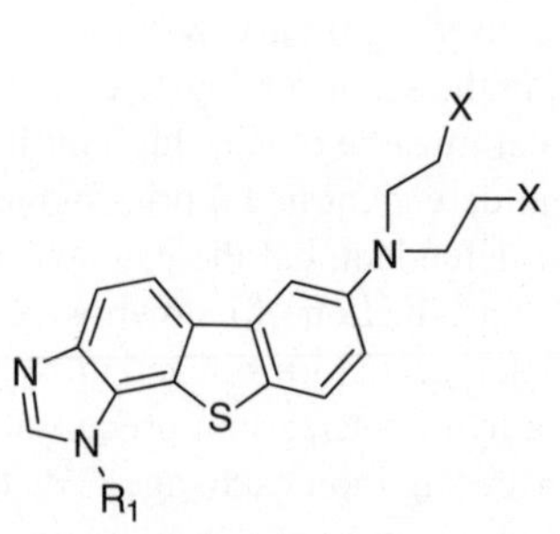

II

R_1 is hydrogen or C_1-C_3 alkyl
X is independently selected from Cl, Br, or I.

FIGURE 8.7   Claimed genus I and prior art genus II.

halogens at position X result in relatively noncytotoxic compounds.[38] The relevant parts of the USPTO first office action rejection are reproduced in Figure 8.8.

After reading the rejection, the chemist applicant grows despondent and calls her patent attorney to discuss how they can possibly patent the compounds of claim 1 absent some additional, expensive, and time-consuming tests to demonstrate the unexpected or superior results the examiner has requested. After a long, emotional conversation with the patent representative, the Imaginit chemist feels much better. The chemist feels even better when she sees her attorney's response to the previous office action rejection (Figure 8.9). As you can see, the attorney's response focused on the teaching away from removing the halogens that was present in the literature and the Imaginarylanda patent specification. If a PHOSITA were making cytotoxic compounds, he would not be motivated to remove the halogens. The examiner is convinced, for the next correspondence issuing from the patent office is a final notice of allowance of all five claims.[39]

8.7 OBVIOUSNESS AND UNPREDICTABILITY IN THE ART

Implicit in a rejection for obviousness based on similarity of structure to prior art structures is the assumption that close chemical structures should have similar chemical properties such that a change in a chemical structure should yield a somewhat predictable result, the smaller the change, the more similar the expected molecular properties. In an absolutely predictable art, one might be able to calculate the effect that a given change will have on the function or performance of the object or material being modified, for example, an increase in serial resistance in an electrical circuit yields a proportionate decrease in current. Therefore, a prior art disclosure of an electrical circuit might render a later claimed circuit obvious where the only thing that was changed in the claimed circuit was the resistance, where that change would be expected to yield a highly predictable result. In contrast, an art that is highly unpredictable would make the predicted effect of any given change less obvious. We have already been introduced to the notion in chemical patent law that compounds that are similar in structure are predicted to have similar properties; this is the core basis for a prima facie obviousness challenge on similar structures. Assuming the prior art disclosed a meaningful utility for the disclosed structures, one of ordinary skill in the art is

---

[38]The term *mustard* in this context refers to the halogen (Cl, Br, or I) located on the aliphatic second carbon from the nitrogen. Mustards typically are highly reactive moieties due to the presence of the nitrogen heteroatom, which can assist in the displacement of one of the halogen atoms. Such agents have found use as alkylating agents and often react with various biological molecules, including DNA-bases, resulting in cytotoxicity. Such cytotoxicity is sometimes useful in chemical warfare and the treatment of cancer. Mustards may also use heteroatoms other than nitrogen, for example, sulfur was in the original Mustard gas and is quite toxic as well.

[39]In this example, the examiner *attempted* to build a prima facie case of obviousness based on the structural similarity of the prior art genus combined with additional references that suggested that, at least in some instances, a hydrogen can substitute for a halogen.

DETAILED ACTION

Claims 1–5 are pending in the application. Claim 1 is directed to compounds and compositions and is rejected as obvious over reference 1 (Imaginarylanda patent 336,4069) in view of references 2–5 (various prior art documents). Claims 2–5 directed to methods of using compounds and compositions for the treatment of bacterial infections are allowed if rewritten with proper dependency to nonrejected base claims.

The following is a quotation of 35 U.S.C. 103 (a) which forms the basis for all obviousness rejections set forth in this office action:

(a) A patent may not be obtained though the invention is not identically disclosed or described as set forth in section 102 of this title, if the differences between the subject matter and the prior art are such that the subject matter as a whole would have been obvious at the time the invention was made to a person having ordinary skill in the art which said subject matter pertains. Patentability shall not be negatived by the manner in which the invention was made.

The subject of applicant's claim 1 is a $1H$-10-thia-1,3-diaza-cyclopenta[a]fluorene substituted at the 7-position with a diethylamine group. Reference 1 discloses a $1H$-10-thia-1,3-diaza-cyclopenta[a]fluorene substituted at the 7-position with a di-β haloethylamine (Cl, Br, or I) alleging cytotoxic activity useful as antitumor agents in mammals (including humans). The substitutions at the 7-position differ only by the presence of a Cl, Br, or I on the β-carbons of the diethylamino group. References 2–5 (all prior art) disclose various biologically active molecules where a direct comparison can be made between compounds substituted with one or more halogens (Cl, Br, or I) and hydrogen. Although the structures in references 2–5 are unrelated to the claimed invention, they demonstrate that halogen can be substituted for hydrogen and biological activity maintained or improved. Moreover, references 2, 4, and 5 all teach antibacterial activity for their compounds, and reference 3 discusses compounds with cytotoxic activity (although not containing a nitrogen mustard functionality). Having knowledge of references 2–5, one of ordinary skill in the art would have been motivated to substitute the halogens of the compounds of formula II of reference 1 with hydrogens to obtain additional biologically active compounds. Furthermore, one of ordinary skill in the art would have predicted that substituting the halogens of formula II with hydrogens would have been successful given the teaching in the literature (as illustrated by references 2–5) that such substitutions are often successful in producing additional compounds with biological activity.

It should be noted that it is of no matter that the motivation would not have been to make antibacterial compounds, as the motivation used to modify the prior art compounds does not need to be the same motivation used by the applicant to make their compounds. See *In re Dillon*, 919 F.2d 688 (CAFC 1990).

The applicant can still submit comparative data demonstrating unexpected activity or superior results associated with compounds of formula **I** and the examiner will reconsider the rejection in light of all of the evidence of record.

Conclusion

Claim 1 is not allowed, claims 2–5 are allowable if rewritten as properly independent and/or dependent claims.

FIGURE 8.8 First USPTO office action.

motivated to make close structural analogs to make additional active compounds. However, chemistry-related arts are often unpredictable due to the fact that small modifications in a molecule's structure or components of a composition can very often lead to large and unpredictable changes in the molecule or composition's function. Notice a contradiction here? If not a contradiction, clearly these two principles are in tension to each other.

APPLICANT'S RESPONSE TO OFFICE ACTION REJECTION

In the instant action, the examiner has alleged that claim 1 directed to a compound of formula I is non-patentable for obviousness over reference 1 in view of references 2–5. Applicants have carefully considered the examiner's reasons for nonpatentability and respectfully traverse the rejections for the following reason.

The examiner alleges that one of skill in the art would have been motivated to modify the compound formula of genus II as described in reference I. In particular, the examiner appears to believe that one of skill in the art would have been motivated to make additional cytotoxic compounds by substituting hydrogens for the required halogens. We start by respectfully drawing the examiner's attention to the nature of the compounds cited in reference 1. In particular, we note that the β-haloamino group of genus II is a nitrogen mustard, and it is well appreciated by those of ordinary skill in the art that nitrogen mustards are highly reactive moieties, capable of alkylating biomolecules both ex vivo and in vivo. We further note that this functionality, through its inherent reactivity, leads to compound cytotoxicity. Reading even the teaching of the specification itself, we see that substitution of halogens by hydroxyl groups led to compounds without demonstrable cytotoxicity. This is not surprising since one of skill in the art, upon examination of the disclosed molecules of formula II, would expect that the presence of the nitrogen mustard would have been responsible for most if not all of the cytotoxicity demonstrated by said compounds. In this regard, we believe the examiner's reliance on *In re Dillon* is mistaken. The court in *Dillon Id.* did not conclude that Dillon's compound was obvious despite a lack of motivation to modify the prior art structures but rather found that the motivation need not be the same as that relied on by the applicant. However, there still must be *motivation* to make the change that the applicant has made. One of ordinary skill in the art, upon reading the cited references would have no reason to exchange the halogen for hydrogen because they would understand that substitution to destroy the purpose of the prior art compounds. Insufficient motivation exists to modify a reference if that modification would render that art inoperable for its intended purpose. See, for example, *In re Gordon*, 733 F.2d 900 (CAFC 1984).[*]

Conclusion

For all of the forgoing, applicant respectfully requests reconsideration of the present rejection and earnestly solicits a timely notice of allowance.

[*] In the case of *In re Gordon*, the CAFC overturned the USPTO's rejection of applicant's invention. The subject matter of the claimed invention was a blood filter that consisted of ports for inflow and outflow of the blood sample through the filter, with a vent for releasing air at the top of the apparatus. In rejecting the claimed invention, the examiner cited a prior art strainer used for gasoline. The prior art device resembled applicant's device except that the outlet and inlet were located at the top and a petcock for draining fluid was located at the bottom. The examiner asserted that it would have been obvious to flip the device over and use it as applicant's had. On appeal, the CAFC overruled the USPTO, pointing out that if the prior art device were so flipped over and used, it would not serve its intended purpose as the petcock at top would serve no function and the water, which was meant to be drained by the petcock would now not be able to be drained off and dirt would clog up the inlet and outlet.

FIGURE 8.9   Applicant's response to USPTO rejection.

How can this tension be reconciled? First, it's clear that a more nuanced approach is necessary. In this regard, it is always preferable to refer more specifically to the predictability (or lack thereof) where enough is known about the particular chemical subject matter being discussed. For example, it is often the case that a class of drugs are known to relate to their target receptor or enzyme in a particular way. It may be understood in the art that certain areas of the compound can be varied considerably without greatly affecting compound activity, whereas other areas of the compound may be highly sensitive to small structural changes. A PHOSITA would expect the variation in the less critical region to yield compounds possessing the prior art disclosed activity whereas the changes in the critical or sensitive area would be less predictable since the art taught that even small variations could lead to inactive compounds. In real-world patent prosecution, the more nuanced analysis will often

17. An azatetracyclic compound of the formula

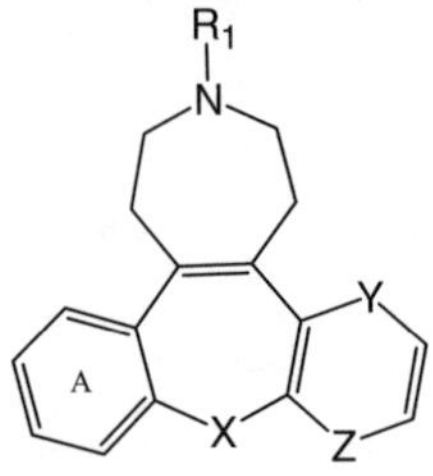

where: R1 represents various substituents; ring A is unsubstituted or substituted with various substituents; X is O, S, methylene, a direct bond, or N-R$_3$; (where R$_3$ represents hydrogen or lower alkyl); and one of Y and Z represents vinylene or S and the other is a direct bond.
(Much of the description of the variables has been left out.)

FIGURE 8.10   Claimed azatetracycle.

be offered and explained by the patent applicants themselves since they will be most familiar with the prior art and also because they will be advocating their position. For example, a patent office examiner might allege that a claimed structure is prima facie obvious and will point to the structural similarities between the claimed compound or genus and the structure(s) disclosed in the prior art. The applicants may then argue that the specific art the examiner is relying on is actually highly unpredictable and may point to additional teaching in that or other references that support their assertion.

To see how the specific unpredictability of the art can argue in favor of nonobviousness, consider a case heard before the USPTO Board of Appeals that dealt with this issue. *Ex parte Blattner* concerns a set of claims that were rejected by an examiner at the USPTO.[40] The rejected claims were subsequently appealed to the Board. The claimed subject matter in dispute concerned compound claims and methods directed to a set of azatetracyclic compounds for which the abbreviated claim 17, shown in Figure 8.10, was considered representative. Additional claims were directed to various structural subembodiments and pharmaceutical compositions containing the compounds of the invention as well as particular methods of use of the compounds including the treatment of agitation in a mammal.

The examiner rejected applicant's claimed invention for obviousness, citing U.S. 4,002,632 (the '632 patent), which was prior art to applicant's claimed invention and disclosed the genus in Figure 8.11 as having antidepressant activity. [41]

[40] *Ex parte Blattner*, 2 USPQ2d 2047 (Bd. Pat. App. & Inter. 1987).
[41] Notice that when *Y* is vinylene and *Z* is a bond, the right-hand aryl group of applicant's claimed genus is a substituted phenyl like that cited in the prior art. Although these genera are close, they do not overlap

1. A compound of the formula:

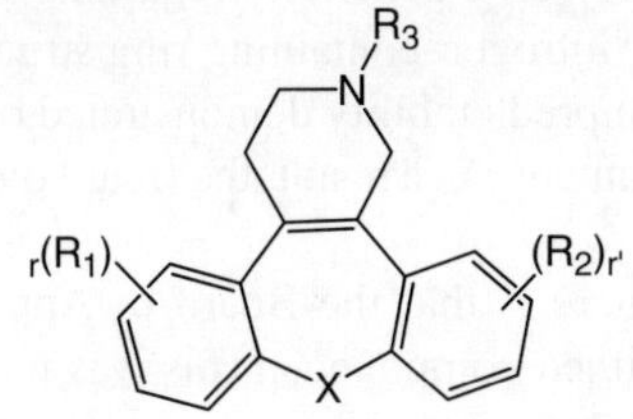

Or a pharmaceutically acceptable salt or nitrogen oxide thereof, in which
  R_1 and R_2 represent hydroxy, halogen, alkyl (1-6 C), alkoxy (1-6 C), alkylthio (1-6 C), or trifluoromethyl,

  R_3 stands for hydrogen, alkyl (1-6 C) or aralkyl (7-10 C),

  X represents oxygen, sulphur, the group >NR4 or hethylene (CH_2),

  R_4 stands for hydrogen or alkyl (1-4 C) and

  r and r′ represent the number 0, 1 or 2.

FIGURE 8.11   Prior art genus from '632 patent reference having described antidepressant activity.

The prior art genus differs primarily from applicant's claimed genus by virtue of the fused six-membered piperidine ring instead of the fused seven-membered homopiperidine ring claimed by the applicant. The '632 reference further discussed the surprising finding that their piperidine-fused compounds demonstrated antidepressant activity in pharmacological assays whereas five-membered pyrrolidine-fused compounds had depressive effects in those same assays. Thus the prior art taught that going from the five-member pyrrolidine-fused ring to the six-member piperidine-fused ring caused the activity to switch from depressive to antidepressive activity. The examiner, by asserting that the claimed genus disclosed in U.S. 4,002,632 rendered the applicant's later claimed genus prima facie obvious was relying on the axiom that similar structures should result in similar activities. While this might be persuasive in the absence of additional information, the prior art cited by the examiner actually indicated that a small structural change in the exact same area of the compound that was the source of the difference between the applicant's claimed invention and the genus cited against it resulted in a complete reversal in the compound's functional activity. If a five-membered ring and a six-membered ring resulted in compounds

because the nitrogen ring of the applicant's claimed invention is a seven-member nitrogen-containing ring and the prior art patent genus is the six-member nitrogen-containing ring. It is important that none of the compounds disclosed within U.S. 4,002,632 falls within the applicant's claimed genus. If that had occurred, the claimed genus would have been anticipated and thus not patentable (recall Section 7.3).

with very different activities, what would one predict for a seven-membered ring? While the examiner wished to rely on a generality regarding the structural similarity of six- and seven-membered, nitrogen-containing ring structures, the Board was more persuaded by the specific unpredictability demonstrated by the very art that the examiner used to support his argument. As a result, the Board overruled the examiner, and the claims were allowed.[42]

What is worth appreciating here is that the Board of Appeals was not looking at the activity of applicant's claimed compounds. This was not a case in which the properties of the claimed compounds were being used to demonstrate a surprising or unexpected result. Rather, this decision by the Board was specifically focused on the motivation to modify prior art compounds. The actual activity of the claimed compounds was not relevant to the Board's decision, as they believed that the motivation to make the applicant's claimed genus was absent from the prior art given the cited unpredictability in the art, specifically with respect to a change in the modified region of the compound's structure. As the Board put it when rejecting the examiner's finding: "What motivation would a person of ordinary skill in the art have expected for the azepine or 7-membered ring compound? A tranquilizer? An anti-depressant? A compound which is neither because the depressant and anti-depressant effects counteract each other?" The Board's pointed questions on this point are very important as they help sharpen our own understanding of what *motivation* means. When we use the term *motivation* in the obviousness/patentability sense, we are referring to a specific motivation to make that compound or that genus with the reasonable expectation that such a change will be successful. In the instant case, one could see a general motivation to make additional homologues to explore what effect the change might have on the compound's activity. However, since the art already teaches that a small change could lead to a very different activity, one would not have the specific motivation necessary to pursue any of the utilities alleged in the literature as any number of activities (or none at all) could be the result.

Finally we should not confuse the motivation issue in this case with the previous discussion of *In re Dillon*. Recall that for *In re Dillon*, the applicant filed a claim for a composition comprising a hydrocarbon fuel with a *tetra*-orthoester, with the motivation being to reduce particulate emissions. Prior art taught compositions of hydrocarbon fuel with *tri*-orthoesters for dewatering fuels, and *also taught the equivalence* of *tri*-orthoesters and *tetra*-orthoesters for dewatering hydraulic fluids. In *Dillon*, the Board held that there was a reasonable expectation that the *tri*-orthoesters and *tetra*-orthoesters would behave similarly in hydrocarbon fuels as well. So the prior art "provided the motivation to make the claimed compositions in the expectation that they would have similar properties." In *Ex parte Blattner*, no such equivalence was taught, in fact, the prior art taught that a small change in ring size could greatly change the activity of the compound, thus making the results of additional modifications to the ring size unpredictable.

---

[42]The entire issued claim can be found in U.S. 4,707,476 as claim 1.

8.8 UNEXPECTED RESULTS AS SECONDARY INDICES OF NONOBVIOUSNESS

One of the critical differences between a finding of prima facie obviousness versus a finding of anticipation is that a prima facie finding of obviousness may be rebutted by a sufficient showing of secondary indicia of nonobviousness (Graham step 4). Keep in mind, that unlike a finding of a lack of novelty, which cannot be rebutted once established, obviousness is a more flexible standard that must take into account all of the prior art as a whole as well as how one of ordinary skill in the art would interpret this art in view of the claimed invention as a whole. Given the difficult and somewhat subjective nature of this inquiry, obviousness determinations are open to more information as to how to reach the ultimate conclusion. Thus, as we will see in this section, there is a large variety of information and evidence that can be entered into the obviousness determination. As we discuss each one of the secondary considerations on its own, you will better be able to see how these criteria are useful in assisting the fact-finder in making the ultimate conclusion regarding nonobviousness.

In the course of prosecution of a patent application before the USPTO, the examiner will accept or reject each claim in the application, and for claims rejected on the basis of alleged obviousness, the applicant has a few options regarding how to proceed. The applicants and/or their representative may decide that the examiner is correct in the allegation of obviousness and either cancel the claim or amend the claim in such a way that they think that the allegation of obviousness will no longer be valid. Alternatively, the applicant may choose to argue that the examiner's argument is not valid and may highlight flaws in the examiner's argument, preferably citing legal precedent that supports the applicant's position. However, in some cases, the applicant may choose not to argue against the examiner's position, either believing that it is correct or that perhaps reliance solely on legal arguments is unlikely to prevail. Assuming he wishes to further pursue the claim, the applicant may decide to submit evidence of unexpected or superior results to rebut the prima facie case of obviousness. Such submissions are typically provided to the USPTO by way of a signed declaration, wherein the declarant provides a statement containing factual information regarding the claimed invention. The evidence so submitted must be sufficient, together with whatever other arguments and evidence of record that exists, to demonstrate that the invention would not have been obvious to one of ordinary skill in the art.

One could reasonably ask why a demonstration of results matters if the invention is obvious on its face. If it would have been obvious to make the invention that the applicant is attempting to claim, why does it matter if that invention turns out to have some unexpected or superior results regarding its performance? To better understand this, it is helpful to analyze this inquiry as directed to the patenting of molecules or chemical compositions. Whether it would have been obvious to claim a particular compound or genus of compounds is typically made by an examination of other molecules or genera of molecules that have been taught in the art. As a general matter, the closer the prior art structures are to the claimed structures, the more likely it is that a prima facie case of obviousness will be successfully asserted against the claimed structures.[43] However, the obviousness of a compound or composition cannot be considered independent of its properties, for as the CCPA (CAFC predecessor court) explained:

> A compound and all of its properties are inseparable; they are one and the same thing.... But a formula is not a compound and while it may serve in a claim to identity what is being patented, as the metes and bounds of a deed identify a plot of land, the thing that is patented is not the formula but the compound identified by it. And the patentability of the thing does not depend on the similarity of its formula to that of another compound but of the similarity of the former compound to the latter. There is no basis in law for ignoring any property in making such a comparison.[44]

Since the obviousness or nonobviousness of a compound should be determined on the basis of whether one of ordinary skill in the art would have appreciated the "invention as a whole," the obviousness determination of a claimed molecule or composition cannot exclude its properties in view of the prior art. Thus if an allegation is made that a prior art disclosure or combination of disclosures renders a later claimed compound obvious, then that allegation of obviousness must hold true for the compound as a whole, including its properties. A showing that the claimed compound or composition provides unexpected results or superior benefits (which necessarily relate to the properties of the claimed compound in question) when compared to the prior art compound or composition rebuts the basis of the obviousness allegation.[45] For example, a chemist may wish to patent a compound or genus of compounds where the compounds are structurally similar to a compound or genus of compounds with some utility that have been disclosed in the prior art. The USPTO examiner may assert that the structures are obvious in view of the prior art due to the structural similarity and disclosed activity and may issue a rejection on that basis. If the applicants believe that they can demonstrate superior or unexpected results for their claimed compound or genus, they may wish to provide that additional evidence to the examiner; often in the form of an affidavit or declaration.[46]

---

[43] "Structural similarity, alone, may be sufficient to give rise to an expectation that compounds similar in structure will have similar properties" (*In re Payne*, 606 F.2d 303 (CCPA 1979)). Later in this section, we will examine in some detail through actual examples what sort of structural differences might result in a prima facie case of obviousness.

[44] *In re Papesch*, 315 F.2d 381 (CCPA 1963).

[45] Unexpected results could also be the *absence* of a property that a claimed invention would have been expected to possess based on the prior art. *Ex parte Mead Johnson & Co.* 227 USPQ 78 (Bd. Pat. App. & Inter. 1985).

[46] Factual declarations submitted to the USPTO usually come in the form of what is referred to as a §1.132 declaration. The declarants are typically people who assist the legal representative in establishing the additional evidence necessary to help overcome the examiner's rejection, and their declarations often contain data comparing the claimed compound or composition to the prior art compound or composition that is being cited against it for obviousness. Or if the necessary data are already in the patent application, the declarant may affirm that the data presented are unexpected results for the claimed compound or composition compared to what was known or expected of the prior art. Declarations can be fertile ground for later findings of inequitable conduct so the applicants need to carefully consider the data they present,

This discussion raises two key questions. First, what type of unexpected or superior results are necessary to overcome an obviousness rejection in chemical cases? Second, what part of the claimed invention must demonstrate unexpected results and what part of the art are the unexpected results determined in comparison to? To answer these two questions, we'll break the inquiry down into the following four aspects:

1. Unexpected results must be taught by, or flow from, the patent application.
2. Unexpected results in one area may be sufficient.
3. Unexpected results may be a difference in degree or kind.
4. The claimed invention must be tested against the closest prior art and not all of the prior art.

8.8.a   Unexpected Results Must Be Taught by, or Flow from the Patent Application

At the outset, it is important that we appreciate that any unexpected results that the applicant hopes to rely on in rebutting a prima facie case of obviousness need to be explicitly disclosed or at least inherently flow from a described feature of the invention. This means that one cannot expect to overcome a rejection based on alleged obviousness by demonstrating unexpected properties that were not disclosed in the patent application as filed or do not flow naturally (inherently) from the disclosure in the application.

As an example, the case of *In re Davies* dealt with claimed subject matter related to "toughened styrene polymers" that had been prepared using a butadiene/styrene toughening agent.[47] In appealing claims that were rejected by the USPTO, the applicant did not argue the allegation of prima facie obviousness in view of the several references cited, but rather argued that the USPTO should have considered certain indices of unexpected or superior results. In particular, the applicants wished to enter affidavits containing evidence of not only improved mechanical strength but also improved gloss, transparency, and processability. The CCPA upheld the USPTO's rejection of the proffered evidence, stating that the "the basic property or utility must

---

including how and what they present, mindful that full disclosure of all the details is better than selective disclosure. To allow the examiner to fully assess the credibility of the declarant, the applicant's legal representative should be sure to describe any relationships between the declarant and the issue she is commenting on. For example, if the declarant is an employee of the applicant's or has received any money or support of any type, past or present, from the applicant, then that relationship needs to be disclosed. Likewise, if the declarant has any personal involvement or relationship to the subject matter then that will need to be disclosed as well. The failure to do so can jeopardize the validity of any claims to issue from that patent (or closely related ones whose prosecution might have been colored by the failure to disclose the conflicting interests of the declarant). In *Ferring v. Barr*, 437 F.3d 1181 (CAFC 2006), the applicants failure to disclose that some of the declarants had at one time or another received financial remuneration from the company (though not related to the patent in suit—e.g., one declarant's institution had at one time received grant money from Ferring) resulted in a finding of inequitable conduct and unenforceability of the subject patent.

[47] *In re Davies*, 475 F.2d 667 (CCPA 1973).

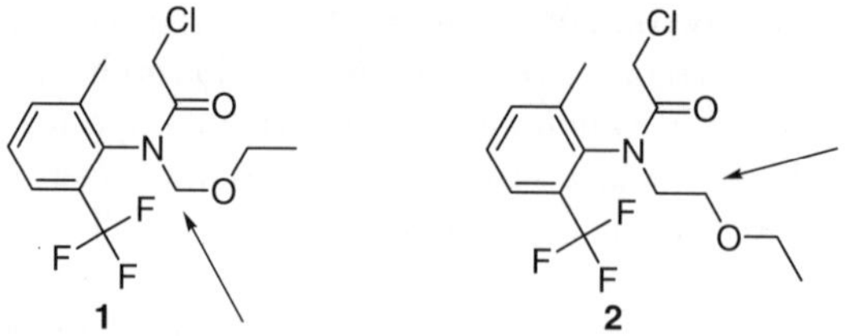

FIGURE 8.12   Claimed compound 1 and prior art compound 2.

be disclosed in order for affidavit evidence of unexpected properties to be offered."[48] The properties in question, the allegedly unexpected combination of improvement of strength together with the improved gloss, transparency, and processability, were not disclosed in the application, and the CCPA did not believe that such properties would inherently flow from the disclosure either. The CCPA noted that it was difficult to understand how unexpected properties of improved gloss, transparency, and processability find a basis in or inherently flow from the disclosure that appeared to be limited to the discussion of the use of toughening agents as well as the fact that the application appeared to be limited in discussion of improved mechanical properties, such as impact resistance.[49]

8.8.b   Unexpected or Superior Results Can Be Demonstrated Through a Single Property

Before considering the type of unexpected or superior results that are sufficient for overcoming a legal conclusion of obviousness, it is important to appreciate that unexpected or superior properties do not have to be demonstrated for every tested aspect of the invention. In other words, the claimed invention may prove superior in only one property but still be found to demonstrate unexpected or superior results enough to merit patenting. In the case of *In re Chupp*, the applicant for patent was attempting to claim compound 1 shown in Figure 8.12, where the prior art disclosed compound 2.[50]

The specification of the application disclosed that compound 1 and related compounds were useful as herbicides, and the examiner rejected the claimed compound alleging obviousness over a prior art reference (a Swiss patent) that included a specific example with the structure 2. As rebuttal, the applicant submitted a declaration that discussed results obtained when compound 1 was compared to compound 2.

---

[48] *Id.* at 670.

[49] One can readily conjure up additional, more extreme hypothetical examples to further drive home the point. For example, a patent applicant disclosing compounds with pharmaceutical utility would not be able to make a showing whereby compounds in the application show unexpected results as polymerization catalysts. He might, however, make a showing that the compounds have unexpectedly lower toxicity than the prior art compounds, which could should in turn translate to better pharmaceutical agents.

[50] *In re Chupp*, 816 F.2d 643 (CAFC 1987).

In particular, the compounds were tested for their ability to control two weeds—quackgrass and yellow nutsedge—in fields of two crops—corn and soybean. The tests demonstrated that the claimed compound gave superior results compared to the prior art compound, exhibiting selectivity factors (crop safety combined with weed-killing activity) at least five times better than the prior art compound. The examiner persisted in rejecting the claimed compound on the grounds that the applicant's own specification indicated that the claimed compound would not show superior activity in crops other than corn and soybean. The applicant appealed first to the USPTO Board of Appeals, where the examiner's rejection was upheld, and subsequently to the CAFC, where the obviousness rejection was overturned and the claim was ultimately allowed to issue (U.S. 4,731,109). In its written opinion, the CAFC held that it was not essential that a claimed compound demonstrate superior activity over a prior art compound in all common activities but rather that it might be sufficient if the compound is unexpectedly superior in one of the spectrums of a common property, as was the case here.

8.8.c Unexpected Results: Different in Degree or Different in Kind?

As we have seen for so much of our obviousness analyses so far, the obviousness determination is made on a weighing of the evidence as a whole. Thus when a prima facie case of obviousness is made and a rebuttal attempted with a demonstration of unexpected results, it is incumbent on the examiner (or fact-finder) to make the determination in view of *all of the evidence.* Having said this, it is possible to make some qualitative assessments regarding the types of showing that have or have not been found capable of overcoming a prima facie case of obviousness by examining several cases, bearing in mind that the rebuttal evidence of unexpected results cannot be isolated out as the dispositive variable since all of the evidence and arguments must ultimately be considered together.

Before we get started in our analysis of actual court cases dealing with this issue, let's first consider some general propositions. In the cases in which unexpected results have been submitted to attempt to rebut a prima facie case of obviousness, the courts have generally characterized the evidence as either a "difference in kind" or a "difference in degree." To differentiate between a difference of kind and a difference of degree, first imagine the situation in which a compound is compared to a prior art compound for which a prima facie case of obviousness has been alleged. In supporting the prima facie case of obviousness, the USPTO examiner references a prior art compound that is structurally very similar to the claimed compound and has disclosed utility as an agonist at a particular receptor. The examiner further notes additional prior art where similar structural changes have retained compound activity and thus concludes that one of ordinary skill in the art would have been motivated to make the small structural change with the reasonable expectation that such a change will result in additional compounds with similar activity. However, the applicant submits a declaration with evidence comparing the two compounds, demonstrating that the prior art compound was an agonist while the claimed compound was an antagonist at the same receptor.[51] In this instance, the unexpected results are not simply one of degree but rather are one of kind. Such a complete change in the molecule's function should carry considerable weight as it strongly rebuts the presumption that molecules with similar structures are presumed to have similar activity.

Next, assume the claimed compound is also an agonist at the same receptor but that the applicant's compound has 30% greater potency for the receptor than the prior art compound.[52] Again the examiner makes the same case for obviousness as before, and the applicant prepares a rebuttal declaration documenting the 30% greater affinity of the claimed compound for the receptor. Here, it is much less likely that the applicant will prevail with such a showing of results. Rather, the applicant is likely to be met with yet another rejection, this one stating that the results are only a difference of degree and not a difference of kind. The examiner will likely point out that such differences in activity are expected as one of ordinary skill in the art would expect that small changes in structure are likely to result in small changes in potency.

So you may well ask, at what point does a difference of degree become a difference of kind? While a 30% increase in receptor affinity might not be a difference in kind, what about a 100%, 200%, 400%, or 800% increase? What if those numbers are or are not statistically significant? How about a 50% difference, but with differences across multiple receptors? How about other properties of the compound such as solubility, lipophilicity, or toxicity? As you may have guessed, there is no easy answer to any of these questions. Further complicating our analysis is that we are considering these questions in the abstract. In the actual situations for which we are concerned, the examiner or the court will be considering the weight of *all* the evidence. In so doing, the determination of obviousness is not based solely on the basis of the unexpected results submitted but on the strength of the prima facie case as well. To the extent that the characterization of unexpected results as a matter of degree or a matter of kind is a conclusion, then that conclusion must be consistent with the ultimate conclusion in regard to obviousness. Accordingly, it is often difficult to parse the actual weight afforded the unexpected results relative to the entirety of the evidence. In other words, evidence that might be deemed a matter of degree for one case in which the overall conclusion is going to be that the claimed invention is obvious could very well be deemed as a matter of kind for a different case in which the overall conclusion is going to be that the claimed invention is not obvious. Please keep these thoughts in mind as we analyze two different cases where evidence of unexpected results were submitted.

In the first case, *In re Soni*, the CAFC heard an appeal from the applicants for a patent where several claims to a high molecular weight conducting polymer were rejected by the USPTO as being obvious over a number of selected references, citing

---

[51] An agonist is a molecule that binds to and alters the activity of a receptor; an antagonist is a molecule that binds to a receptor and inhibits an agonist's effects.

[52] *Potency* refers to the concentration of compound required to achieve a particular effect at a given receptor. With all other things being equal, a compound with greater potency at a given receptor will require less compound to achieve a given effect. In biological systems, actual activity depends on many variables, of which receptor affinity is only one.

lower molecular weight polymer compositions.[53] Claim 1 of the patent application was considered to be illustrative of the invention by the court and is shown below.

1. A melt-processed composition which comprises

    (i) an organic polymer which is not cross-linked and has a molecular weight which is greater than 150,000 when measured by high temperature gel permeation chromatography, and

    (ii) a particulate conductive filler which is dispersed in the polymer and which is present in an amount sufficient to render the composition electrically conductive.

In the specification, the applicants for patent had included a number of tests comparing the properties of a polyethylene polymer having a molecular weight of 203,000 with a polyethylene polymer having a molecular weight of 148,000. The data provided in the specification for the two polymers indicated that the higher molecular weight polymer had at least a 50-fold increase in tensile strength as well as at least a 5-fold increase in peel strength in addition to improved resistivity and recovery behavior. The applicant conceded that the prior art cited by the USPTO was sufficient to support a prima facie case of obviousness but argued that the data they provided in the specification were sufficient to demonstrate superior results beyond what one of ordinary skill in the art would have predicted. In overruling a final rejection from the USPTO and finding for the applicants, the CAFC stated: "Mere improvement in properties does not always suffice to show unexpected results. In our view, however, when an applicant demonstrates substantially improved results, as Soni did here, and states that the results were unexpected, this should suffice to establish unexpected results in the absence of evidence to the contrary."[54]

In contrast, in *In re Merck & Co.* the court found that the applicants for a reissue patent did not demonstrate unexpected results sufficient to overcome the prima facie case of obviousness established by the USPTO.[55] In this example, claim 1 of the patent in question (U.S. 3,428,735) will be considered illustrative of the matters at issue:

1. A method of treating human mental disorders involving depression which comprises orally administering to a human affected by depression 5-(3-dimethylaminopropylidene) dibenzo[a,d][1,4]cycloheptadiene (amitriptyline) or its non-toxic salts in daily dosages of 25 to 250 mg of said compound.

The chemical formula referred to in claim 1 is also known as amitriptyline, and its chemical structure is given in Figure 8.13, along with the chemical structure of imipramine used as an antidepressant and additional prior art compounds having central nervous system activity from which the method of claim 1 was deemed obvious.

The CAFC upheld the USPTO's finding of a prima facie case of obviousness. In so doing, the court first noted the structural similarity between amitriptyline and

[53] *In re Soni*, 54 F.3d 746 (CAFC 1995).
[54] *Id.* at 751.
[55] *In re Merck & Co.*, 800 F.2d 1091 (CAFC 1986).

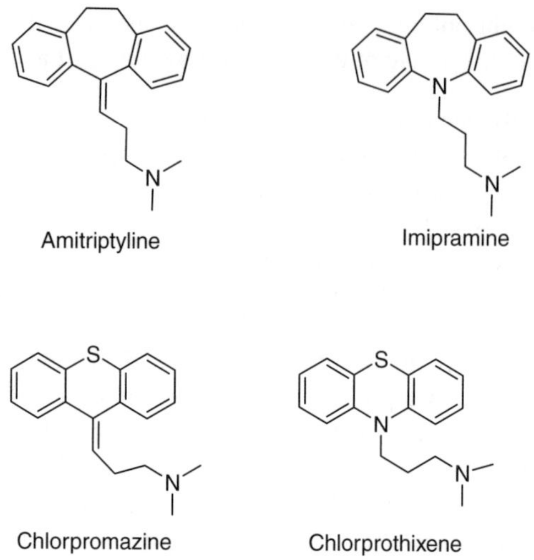

FIGURE 8.13 Structure of claimed compound (amitriptyline) and prior art compounds.

imipramine (a compound with known antidepressant activity) but also that the literature taught bioisosterism between the bridgehead nitrogen in imipramine and the alkene carbon of amitriptyline. The court further highlighted the interchangeability of the alkene and the aminomethyl moiety through the prior art structures of chlorpromazine and chlorprothixene (structures also shown in Figure 8.13), both of which were reported to possess similar pharmacological properties to each other, including a strong central nervous system (CNS) effect. The CAFC also discussed a set of prior art documents that revealed data from several tests of amitriptyline and imipramine indicating the two compounds were very similar in a variety of properties, including their use as tranquilizers having narcosis-potentiating effects. The prior art document went on to further suggest that amitriptyline should be tested for depression alleviation, a property that imipramine was known for, and that side effects were likely to be similar to those seen with imipramine.

In response to the USPTO prima facie case for obviousness, Merck attempted a number of approaches. First, they argued that the USPTO had relied on an "obvious to try" standard and that there was not motivation in the prior art to create their invention. However, the court noted that obviousness does not require absolute predictability but rather only a reasonable expectation that the desired result will be achieved. Merck further argued evidence of unexpected results, including data that amitriptyline has an "unexpectedly more potent sedative and a stronger anti-cholinergic effect." As part of this approach, Merck introduced an affidavit by an expert witness that some patients responded better to amitriptyline than imipramine. The CAFC was not convinced. They highlighted the fact that the prior art of record "clearly" taught that amitriptyline was a known sedative and, further, that all tricyclic antidepressants, in general, possess the secondary properties of sedative and anticholinergic effects. The fact that some patients were helped more by one drug than another was not held to be significant. In citing a reference indicating that amitriptyline was "highly sedative" while imipramine was "somewhat less sedative than amitriptyline," the court stated that the difference was a "matter of degree rather than kind," and affirmed the finding of obviousness.

To better understand the different holdings in *In re Soni* and *In re Merck & Co.,* it is helpful to examine a few points that are not explicitly of the record but merit discussion. It appears from its holding that the *Merck* court's overall decision was at least in part due to a very strong prima facie case. Besides the structural similarity, the court was able to point to a number of references that appeared to provide a direct suggestion that amitriptyline be tested as an antidepressant. Such a strong and direct suggestion was due to the fact that this was a special circumstance where the applicant was attempting to patent a method for a compound that was already well known in the art as a CNS-active agent. Many of the primary activities of the compound were already understood, and some discussion had already occurred in the prior art regarding the possibility (or even desirability) of using it as an antidepressant. Obviousness as a whole means that the elements of structural similarities, prior art teaching of equivalence, and unexpected results must be considered together. Where a prima facie case of obviousness is very strong, a greater degree of unexpectedness of the demonstrated results will be required in rebuttal.

8.8.d The Claimed Invention Must Be Tested Against the Closest Prior Art

When the applicant for patent is attempting to make a showing of unexpected results or benefits to rebut a prima facie case of obviousness, the question often comes up as to what prior art does the claimed invention need to be compared against. The short answer is "the closest prior art."[56] However, the closest prior art does not need to be the same prior art that the patent examiner relied on but rather can be different prior art if that different prior art is in fact closer to the claimed invention. For example, in the appeal of *Ex parte Humber*, claimed 13-chloro substituted compounds were found prima facie obvious in view of prior art nonchlorinated compounds. Instead of making a showing over the prior art nonchlorinated compounds, the applicant showed unexpected results over more closely related 9-, 12- and 14- chloro substituted compounds; this was deemed sufficient to rebut the prima facie case of obviousness.[57]

In situations in which there is more than one prior art reference and both are equally close to the claimed invention, the applicant will be required to make a showing in regard to both of the prior art disclosures, unless the teachings of the prior art references are each sufficiently similar to each other that making a showing against one will suffice for the other. For example, let's assume an applicant is attempting to claim the structure in Figure 8.14 and the two prior art structures have each been cited against the claimed structure.

---

[56]MPEP 716.02 citing *In re Burckel* 592 F.2d 1175 (CCPA 1979), "A comparison of the claimed invention with the disclosure of each cited reference to determine the number of claim limitations in common with each reference, bearing in mind the relevant importance of particular limitations, will usually yield the closest prior art reference."

[57]MPEP 716.02 citing *Ex parte Humber*, 217 USPQ 265 (Bd. App. 1961).

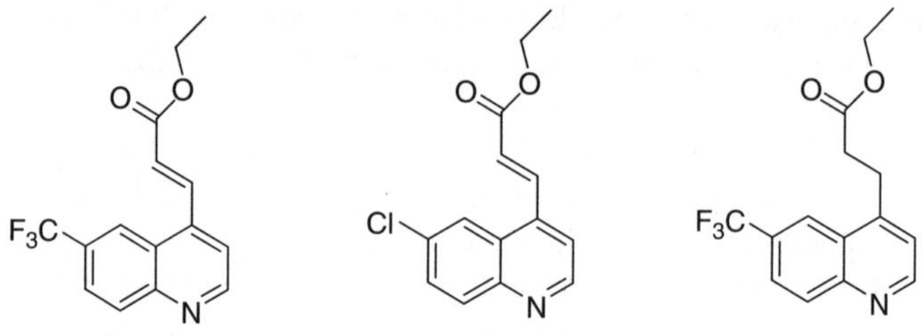

| Claimed compound | Prior art 1 | Prior art 2 |

FIGURE 8.14   Comparison of claimed compound to more than one prior art compound.

Assuming the both prior art compounds have a disclosed utility, the examiner alleges that one of ordinary skill in the art would have been motivated to make close structural analogs of the disclosed prior art compounds in order to find additional compounds having the disclosed activity.[58] Further, the examiner has cited additional references from the literature in which saturated propionic esters have been replaced with unsaturated propionic esters and where chlorides have been successfully replaced with trifluoromethyl groups, and the molecules so derived have had similar if not better activity. Thus the examiner argues that it would have been obvious to make those modifications to the prior art compounds. Sounds like a strong prima facie case of obviousness unless the applicant can show her claimed compounds demonstrate unexpected properties or superior results, but compared to which structure? The answer in this example would most likely be both. The reason is that if the applicant compares the claimed compound to prior art compound 1, she will be demonstrating an unexpected result for the trifluoromethyl group relative to the chloro group. However, this still does not answer the question of whether the unsaturated ester can demonstrate unexpected results compared to the saturated ester of the prior art. Both prior art compounds will need to be prepared or otherwise acquired and compared to the claimed compound.

8.9   PRIMA FACIE OBVIOUSNESS BASED PRIMARILY ON SIMILARITY OF CHEMICAL STRUCTURE

Obviousness of chemical structures is premised on the principle that compounds that are similar in structure are predicted to have similar properties. As a result, when a compound is disclosed in the prior art and an utility for that compound is also

---

[58]Remember from *In re Dillon* that the disclosed utility need not be the same utility that the applicant was interested in. What if no utility had been disclosed for the prior art compounds? In that situation, one of ordinary skill in the art would not have been motivated to make close analogs because there would have been no reason to do so. Close analogs would not be obvious when the prior art provides no motivation to make analogs.

disclosed or predicted, the motivation for modifying that structure is usually found in the desire to obtain additional compounds having the previously disclosed activity. Thus, for example, if the prior art disclosed a compound as its *methyl* ester and explained that compound was useful as an analgesic, than one of ordinary skill in the art might surmise that an *ethyl* ester homologue would also be active. An applicant who later attempts to claim that ethyl ester in a patent application would have to deal with the fact that the prior art already disclosed the homologue as well as an activity for that homologue. As a result, the patent examiner could make a persuasive argument that the close structural similarity of the prior art compound together with the disclosure of its activity would have made the homologue an obvious analogue to make; one of ordinary skill in the art would have been so motivated because he could make additional compounds with analgesic activity.

Until now, our primary organizing principle has been understanding a prima facie case of obviousness, the TSM test, and rebutting the prima facie case with unexpected or superior results. To enhance our understanding further, we will see how this obviousness framework has been implemented in regard to different types of chemical subject matter. By presenting the remaining material in this fashion, we can expand on the analytical concepts we have already introduced as well as provide additional specific examples of how that those concepts have been interpreted and applied in the chemical arts.

8.9.a   Isomers and Homologues

Isomers in the broadest sense are compounds with the same molecular formula but different structures. Isomers are most broadly categorized as either structural isomers or stereoisomers. Compounds that are structural isomers have a different arrangement of atom connectivity relative to each other, whereas compounds that are stereoisomers have the same atom connectivity but the atoms still arrange differently in a spatial sense. We'll begin this section with structural isomers, for which one example is shown in Figure 8.15.

Despite the fact that the simple compounds in the figure have the same formula, their properties differ very significantly one from another. Dimethyl ether is a highly flammable gas at room temperature, whereas ethanol is a semiconsumable liquid at room temperature. Even though these two compounds have the same formula, the connectivity of the atoms results in very different properties. For an indication of how some structural isomers might be treated, let's briefly review the case of *Ex parte Mowry*, which concerned an appeal to the Board of Patent Examiners at the

$$H_3C^{\diagup O \diagdown} CH_3 \qquad\qquad CH_3CH_2OH$$

$$C_2H_6O \qquad\qquad\qquad C_2H_6O$$

Dimethyl ether   Ethanol

FIGURE 8.15   Two formula isomers.

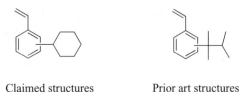

Claimed structures   Prior art structures

FIGURE 8.16 Claimed styrene and prior art styrenes.

USPTO over rejected claims directed to, inter alia, monocyclohexylstyrene isomers (-*ortho*, -*meta*, and -*para*) of the type shown in Figure 8.16.[59]

The claimed structures were disclosed as having utility in polymerization reactions (with other monomers) for the preparation of various plastic copolymers. The applicant's claims were rejected by the examiner under certain grounds, including alleged obviousness. The obviousness argument made by the examiner cited the disclosure of generic structures having an allegedly *isomeric* relationship with applicant's claimed structures (cited structure also shown in Figure 8.16).[60] In other words, the patent examiner's view was that the cyclohexylstyrenes were prima facie obvious due to the existence of the prior art isohexylstyrene genus.

Despite having the same *alleged* molecular formulae, the Board found that the claimed structures were patentable and that "that isomers in the broad sense, that is, those having the same empirical formula but with different radicals, would *not* necessarily be equivalent, or that one would be suggestive of the other." In its opinion, the Board appears to suggest that a different treatment might be afforded to the compounds in suit where the compounds are "positional isomers" rather than isomers of formula only. If we glance back to the basic formula isomer compounds shown in Figure 8.15, we can appreciate that one of ordinary skill in the art would not per se assume formula isomers to be obvious in view of each other, given the poor relationship between a chemical formula as written and its actual structure and especially its function. The chemical formula is much too basic a level to apply any broad rule regarding obviousness.

In contrast to chemical formula isomers, positional isomers are a narrower subclass where the compounds contain the same appending groups, but those groups may be connected in a different manner one from another. Positional isomers must have the same substituents, but those substituents occupy a different position on the molecule.

---

[59] *Ex Parte Mowry* 91 USPQ 219 (Bd. App. 1950).

[60] The Board cited these compounds in their discussion of isomers, which they characterize (correctly) as compounds "having the same empirical formula," which they apparently thought characterized the relationship between applicant's claimed compounds and the prior art compounds. It appears that the Board (as well as the *Manual of Patent Examining Procedure*) failed to take into account that two hydrogens are lost on formation of a carbocyclic ring, so that even though the carbon count is the same, the hydrogen count differs by two. So in fact, not only are these compounds *not* positional isomers as the Board noted but they are not structural isomers at all. This simple oversight, however, would not affect the Board's ultimate conclusion and, in fact, would serve only to reinforce it as the compounds are even further removed than they assumed in their analysis.

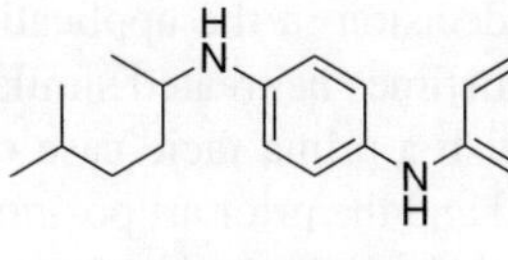

Claimed compound

N-(1,4-dimethylamyl)-N-phenyl-p-phenylenediamine

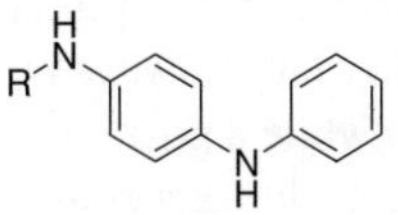

Genus (R = secondary alkyl 3-11 C) 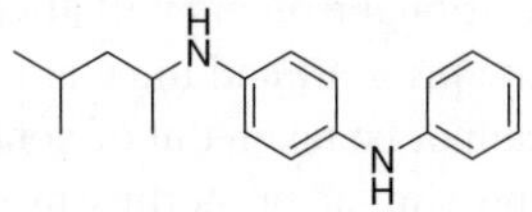 Species 1 (homologue) 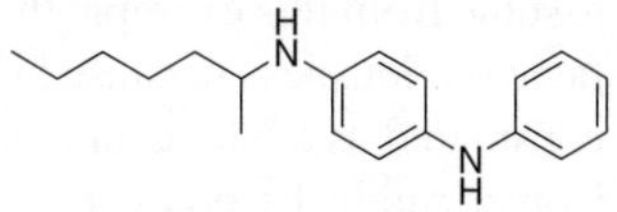 Species 2 (positional isomer)

FIGURE 8.17 Claimed compound and prior art compounds from '808 reference.

Let us turn to *In re Wilder* which focused on the issue of positional isomers.[61] This decision arose from a patent application with a claim to the structure shown in Figure 8.17.

The applicant stated in the specification that the compound is useful as an antidegradant in rubber and had minimal toxicity to human skin. The USPTO rejected the application asserting that the prior art patent U.S. 2,734,808 (the '808 reference) rendered the applicant's claimed compound obvious. The '808 reference disclosed a genus of compounds and two specific compounds of note, one a positional isomer and one a homologue, all of which are represented in Figure 8.17. The compounds of the '808 reference were described as having utility as gasoline additives that inhibit gum formation and retard the degradation of tetraalkyl lead in leaded gasoline. The applicant appealed the USPTO rejection to the CCPA.

On the basis of the close structural similarity, the Court found that both the homologue and the structural isomer rendered the applicant's claimed compound prima facie obvious. The applicant, in an attempt to rebut the alleged prima facie case, submitted evidence that the claimed compound was lacking in skin toxicity, which was a property not demonstrated by the prior art positional isomer. While the Court recognized this difference as being sufficient to rebut the prima facie case with respect to the positional isomer, they still found the claimed compound obvious in view of the *homologue*. Unlike the positional isomer in this case, the homologue demonstrated the same property as the claimed compound (no skin toxicity) and thus the applicant could not rebut the prima facie case raised by the homologous compound. The claim to the compound was thus denied.[62]

---

[61] *In re Wilder*, 563 F.2d 457 (CCPA 1977).

[62] The applicant raised additional arguments for nonobviousness, which were not persuasive to the Court. One was that that the two prior art compounds had very different skin toxicities despite their similarity in

As can be gleaned from the decision in the application of *Wilder*, homologues and positional isomers might sometimes be treated similarly—in this example, both were deemed sufficient to establish a prima facie case of obviousness. In the case of *Wilder*, the claimed compound and the prior art positional isomer and homologue were very similar in structure. The homologue is missing one carbon atom and the positional isomer has the methyl group at the five-position of the hexyl chain moved over one carbon so that the chain is no longer branched. However, one should not assume from this example that positional isomers will always lead to a case of prima facie obviousness because both the prior art and the claimed isomer must always be considered as a whole and thus cannot be taken out of context. So while a positional isomer might be predicted to lend similar properties in general, such a generality can sometimes be overridden by a more specific understanding of how such changes might relate to the specific area being analyzed.

The facts of *In re Wilder* also dealt with a prior art homologue, where a homologue is defined as a compound differing from another by the successive addition(s) of the same chemical group (e.g., CH_2). This means that a homologue could be 1 methylene removed from its comparator or 20 (or even more) methylene groups removed from its comparator. Because a homologue can vary so much in its degree of added repetition, the degree of predictability for homologues further removed in structure would be expected to decrease with the increasing difference in size between the claimed structure and the prior art structure. Although they are still homologues no matter the number of additional repeating units, at some point the tail begins to "wag the dog."[63] So while homologues or positional isomers are often close enough structurally to create a structural presumption of obviousness, nothing is absolute. As pointed out in the case of *Ex parte Blattner* discussed earlier, this assumption must still be balanced with all other factors, including the predictability of the art in view of what the art as a whole would teach as well as the particular aspects of the compound being claimed. Thus homologues and positional isomers in the prior art can be significant evidence for prima facie obviousness, but are not always adequate to guarantee that conclusion.[64]

8.9.b Enantiomers

Chirality is a property inherent to certain molecules, both organic and inorganic, in which two molecules, identical in absolute structural terms, are nevertheless non-superimposable. The word *chirality* itself is derived from Greek stems that relate to the term "handedness," and this is very appropriate as your own two hands are probably the most easy to relate to example of chirality. If you look at your two

---

structure. This, the applicants asserted, was evidence that there was no predictability in the related art and thus rebutted the presumption that structurally similar compounds are expected to have similar activity, a reasonable argument. The Court did not agree, essentially restating its belief that the inability to rebut the presumption with respect to the homologue was sufficient for a finding of obviousness based on the disclosure of the homologue.

[63] At what point the tail actually wags the dog probably depends on the size and type of dog as well as the particular nature of the tail itself.

[64] MPEP 2144.09 contains the USPTO instruction to examiners on this and related topics.

hands, you can appreciate that they are essentially identical (assuming your manicuring is highly symmetric) but yet, your left hand is not superimposable on your right hand. Instead, they relate as one object relates to its own mirror image. While an enantiomeric pair of molecules (the molecule and its mirror image) have the same gross physical properties, they interact differently with other molecules possessing chirality—think of trying to shake somebody's left hand with your right hand; it does not work the same way as when shaking their right hand with your right hand. As a result, two enantiomers may act differently toward important classes of chiral biomolecules such as receptors, enzymes, and membrane components.

Over time, the appreciation of the importance of chirality in drugs has grown.[65] In many circumstances, a drug containing a single enantiomer has a different and, very often, a preferred biological profile over a drug containing both enantiomers in equal amounts (referred to as a racemic mixture). In fact, as our medical sophistication has grown, we increasingly appreciate that an enantiomer and its mirror image rarely have the same pharmacological properties and that administration of a racemic mixture is often tantamount to administering two different drugs or perhaps one drug and an equal amount of an impurity.[66] A technological evolution has enabled the industrial scale production or separation of many single enantiomers, resulting in the revisiting of many older, racemic mixture drugs, this time with an eye to evaluating and possibly redeveloping the mixtures as single enantiomer drug products.[67] By way of example, albuterol has been marketed for some time as a racemic mixture having β_2-agonist activity useful for inducing bronchodilation in order to alleviate asthma attacks. The R-enantiomer of albuterol provides the bronchial relaxant activity while the S-enantiomer appears to contribute to side effects only.

The determination of nonobviousness with respect to a single enantiomer in view of a prior art racemic mixture requires one to look at the claimed subject matter (in this case, a single enantiomer) in view of the closest prior art (the racemic mixture) and ask the question of whether one of ordinary skill in the art would have been motivated to prepare the enantiomer in view of the prior disclosure(s) of the racemic

---

[65]The trend in the last couple of decades has been toward the use of single enantiomer drugs over drugs for which it is possible to have a racemic mixture of both enantiomers.

[66]The exceptions to this generality are cases where a single enantiomer rapidly and completely racemizes under physiological conditions rendering the administration of a single enantiomer less effective or even pointless. For example, it has been hypothesized that the administration of the R-(+)-enantiomer of thalidomide might have avoided the tragedy of many birth defects caused by administration of the drug as a racemic mixture. However, this may not be true as oral administration of the drug as a single enantiomer has been demonstrated to lead to very rapid acid-catalyzed racemization of the compound's only chiral center. The drug is currently marketed as a racemic mixture used in the treatment of leprosy and multiple myeloma (although not in pregnant women or those planning on becoming pregnant). In other cases, each enantiomer might possess a unique action for which the combination of the two is viewed as a desirable outcome. For example, the commercially successful antidepressant venlafaxine (Effexor®) is marketed as its racemic mixture. It has been reported that one enantiomer is responsible for most of the compound's serotonin-reuptake inhibition, while the other enantiomer is responsible for most of the drug's norepinephrine-reuptake inhibition. Together, the two enantiomers provide a drug with combined serotonin and norepinephrine-reuptake inhibition, both of which are believed to contribute to the compound's desired pharmacology.

[67]C.P. Miller and J.W. Ullrich. *Chirality* 20(6) (2008), 772–780 and references therein.

mixture. Since it has been appreciated for some time that a single enantiomer can have beneficial advantages when compared to a racemic mixture, it is no great leap to predict that the enabled disclosure of a racemic mixture for use in pharmaceuticals will likely provide the motivation for the preparation of each of the single enantiomers for the same use. The motivation would be the close structural similarity coupled with a reasonable expectation that at least one of the enantiomers will provide a similar or superior activity over the racemic mixture. Thus the attempt to patent a single enantiomer when the racemic mixture is known in the art will *likely* result in a case of prima facie obviousness.[68] However, a prima facie case of obviousness can still be rebutted by a secondary considerations of nonobviousness (Graham step 4), including the demonstration of unexpected or superior results.

Issues related to prior art enablement, prima facie obviousness and secondary considerations of nonobviousness often intertwine in obvious determinations during patent prosecution and litigation since their definitional boundaries are often not well demarcated. In the case *Sterling Drug Inc. v. Watson,*[69] the USPTO rejected claims to various single enantiomer compounds,[70] and this decision was appealed to the U.S. District Court for the District of Colombia. The claims in question—10, 12, and 14—are presented in Figure 8.18 together with the claimed structures (L-arterenol is the neurotransmitter norepinephrine).[71]

---

[68] Keep in mind that a prima facie case of obviousness is never automatic and the prior art as a whole must be considered when determining whether the burden of establishing a prima facie cases has been met. For example, the prior art as a whole might teach away from using a specific single enantiomer of a racemic mixture, suggesting the racemic mixture for that instant might be preferred over a single enantiomer. Or, if prior art suggested a single enantiomer was prone to recemization under useful conditions (e.g., in vivo), there might not be sufficient motivation to prepare a single enantiomer. While these examples are only hypothetical, they reinforce the importance that per se rules of obviousness are not permissible and that a successful assertion of prima facie obviousness must consider the particular facts regarding the application at hand. See, for example, *In re Ochiai*, 71 F.3d 1565, 1572 (CAFC 1995). Still though, an enantiomer is much more likely to be the subject of a successful assertion of prima facie obviousness than where the particular compounds shared less structural similarity.

It would not be unreasonable to ask why a prior art disclosure of the racemic mixture does not anticipate (i.e., result in a §102 rejection) a later claim to one of the enantiomers. After all, a prior art racemic mixture does contain each of the two enantiomers. The difference is subtle but important. The prior art disclosed a mixture whereas a later claim to the single enantiomer is claiming one or the other enantiomer, essentially by itself. The prior art must meet each and every element and limitation of the later claimed invention to anticipate it. In the case of a single enantiomer, the prior art does not provide the limitation that one or the other enantiomer be present primarily by itself. See e.g., *In re May*, 574 F.2d 1082 (CCPA 1978), "The novelty of an optical isomer is not negated by the prior art disclosure of its racemate."

[69] *Sterling Drug Inc. v. Watson*, 135 F. Supp. 173 (D.D.C. 1955).

[70] Pure enantiomers and mixtures enhanced in one enantiomer have the property that when plane polarized light is passed through a solution containing the enantiomer or mixture enhanced in one enantiomer, the plane polarized light is rotated in one direction or the other. The L- and D-notations refer to the direction that the solution will rotate the plane polarized light (levorotary (L), left; dextrorotary (D), right).

[71] Norepinephrine is used pharmacologically for its ability to cause peripheral vasoconstriction and increase blood pressure, which is useful to stem the effects of large drops in blood pressure as can occur in trauma or shock. Drugs affecting norepinephrine release and/or reuptake also have significant CNS utility, including the treatment of depression and attention deficit disorder.

10. A substance selected from the group consisting of D-arterenol, L-arterenol, and their acid addition salts, said substance being in crystalline and substantially pure form and being substantially free from its optical antipode.

12. An acid addition salt of L-arterenol, and their acid addition salts, said substance being in crystalline and substantially pure form and being substantially free from its optical antipode.

14. L-arterenol acid D-tartrate in crystalline and substantially pure form and substantially free from D-arterenol acid D-tartrate.

L-arterenol (norepinephrine)   D-arterenol

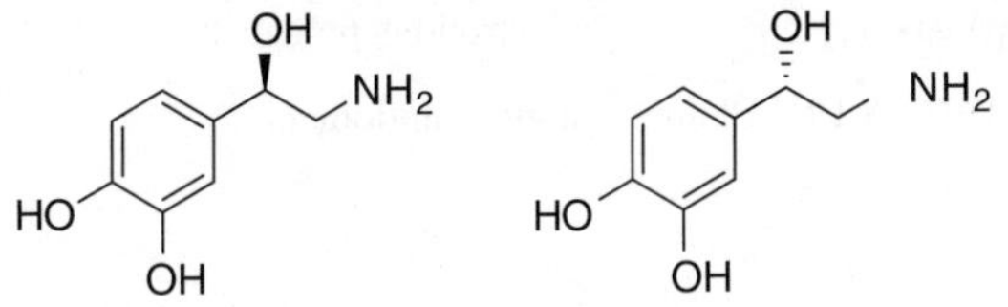

FIGURE 8.18   Claims to single enantiomer of L-arterenol (norepinephrine) and D-arterenol.

In the beginning of its analysis in regard to the three claims, the court summarized the USPTO's argument that the claims were not patentable in view of (1) a prior art reference that disclosed certain salts of racemic arterenol, (2) L-arterenol (norepinephrine) is naturally present in the human body, and (3) another prior art reference that described the next known homologue of L-arterenol. The court, in finding claims 12 and 14 valid, quickly disposed of the USPTO arguments by first noting that L-arterenol had "phenomenal therapeutic properties" with respect to the treatment of irreversible shock and occluded arteries.[72] The court further noted that it is the L-form and not the D/L (racemic)-form that has the therapeutic value. In dispensing with the closest homologues of the prior art, which apparently had some effect in raising blood pressure, the court noted that "there is a vast amount of difference between the therapeutic value of something that will so control the blood pressure that successful surgery can be accomplished and something else that falls so short of doing so that the patient dies." In contrast, the court found that claim 10 was not patentable as written since the claim was also directed to D-arterenol. The court remarked that there was no showing of any "unexpected and unobvious properties" of the claimed D-arterenol, and thus apparently that claim could not be considered nonobvious in view of the prior art, including the known racemic mixture.

Finally, the court noted that the existence of L-arterenol in the human body "ignores the fact that it has no therapeutic value unless isolated and available in its pure

---

[72] It is not clear at what point the line is crossed between expected and unexpected results, but "phenomenal therapeutic properties" is apparently sufficient.

1. A levo-isomer of a compound selected from the class consisting of 1-cyclohexyl-1-phenyl-3-piperidinopropan-1-ol and 1-cyclohexyl-1-phenyl-3-pyrrolidinopropan-1-ol and their acid addition salts and quaternary ammonium salts substantially separated from their dextro-rotary isomers.

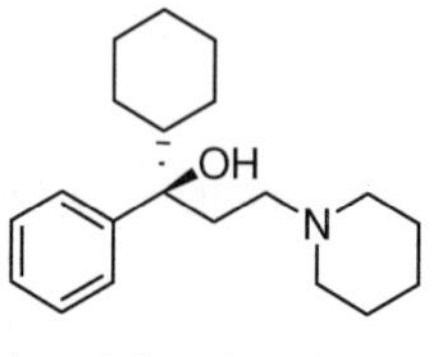

1-cyclohexyl-1-phenyl-3-piperidinopropan-1-ol

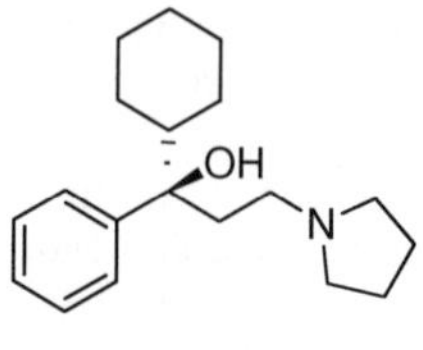

1-cyclohexyl-1-phenyl-3-pyrrolidinopropan-1-ol

FIGURE 8.19 Claimed single enantiomers.

form ..."[73] and that no compound of the type had ever been derived from its natural source unless seeded with pure crystals of the type claimed in applicant's invention. It appears implicit from the court's last finding of nonobviousness on the fact that the preparation of pure L-arterenol compound in the prior art was not enabled.[74]

Let's quickly juxtapose this result with an enantiomer case having a different outcome. *In re Adamson* resulted from an appeal to the CCPA of denial of patentability by the USPTO Board of Appeals, which rejected claims 1–11 of the applicant's application directed to compounds having spasmolytic activity and processes for making same.[75] The stated reason for the rejection is that the compounds were obvious in view of the prior art disclosure of the racemic mixture. Claim 1, which was deemed by the court to be a typical claim of the invention, is reproduced in Figure 8.19 together with the structures of the claimed compounds.

---

[73]Can you understand why the acknowledged presence of L-arterenol (norepinephrine) in the human body did not *anticipate* (render not novel) any of the three listed claims? If not, remember that anticipation requires that each and every limitation of the claim must be met. Note especially the requirement of each claim that requires the substances be in "crystalline and substantially pure form." Products of nature are not patentable subject matter per se, but here the inventors provided the compounds in a form that they did not naturally exist in (crystalline and substantially pure).

[74]Or as was stated in the later case of *In re Hoeksema*, 399 F.2d 269, 274 (CCPA 1968), "if the prior art of record fails to disclose or render obvious a method for making a claimed compound, at the time the invention was made, it may not be legally concluded that the compound itself is in possession of the public." An interesting question in the *Sterling Drug* decision remains as to claim 10, which included the D-enantiomer, which the court rejected patentability thereof in view of the prior art. However, it appears that the preparation of the "substantially pure" D-arterenol would have been no more enabled than that of the L-arterenol. If that is indeed the case, it is not evident why the claim to the substantially pure D-arterenol would not have been allowed as well (provided that the D-arterenol of claim 10 still meet the utility requirement). In any event, the claims at question ultimately produced U.S. 2,774,789 claiming L-arternol but not D-arternol. The entire patent is only three pages long, perhaps the value of an invention inversely proportional to the patent length.

[75]*In re Adamson*, 275 F.2d 952 (CCPA 1960). A spasmolytic drug is used to prevent muscle spasms, particularly spasms in smooth muscle.

The applicants argued that their single enantiomer compounds were nonobvious for two reasons. First, they discovered that the racemic mixture of the prior art could be separated. Second, they submitted an affidavit alleging that the claimed levorotary compounds had substantially higher spasmolytic activity than either the dextrorotary isomer or the racemic mixture while having only slightly higher toxicity. It was the applicant's contention that the prior art suggested that the levorotary and dextrorotary forms of the compounds produced should have had similar therapeutic activity. However, the Court was not persuaded by the applicant's arguments regarding the unexpected and superior activity of the claimed single enantiomers compared to the racemic mixture. The Court noted that the claimed levo-isomers were only twice as potent as the racemic mixture and that the dextro-isomer was inactive. In regard to enablement, the Court roundly rejected the applicant's contention, asserting that the prior art of record showed there were numerous techniques for separating a racemic mixture into its constituent enantiomers. Although the reference did not describe compounds of similar structures as those claimed by applicants, the Court held that such specificity was not required. The obviousness rejection was thus upheld.

The difference in facts between cases (stated and especially unstated) always makes generalizations difficult, but it still is worthwhile to consider why the Court in *Sterling Drug* found one of the enantiomers nonobvious in view of the racemic mixture whereas the court in *Adamson* did not. First, it is useful to appreciate the similarities. In both cases, the Courts indicated that the enantiomers in question would be prima facie obvious absent a showing of useful or unexpected properties. The motivation to prepare a pure enantiomer is the expectation that the enantiomer will have similar or even somewhat improved properties then the mixture. In *Sterling Drug*, the Court was very impressed with the "phenomenal therapeutic properties" of L-artenerol with respect to the treatment of irreversible shock and occluded arteries, and it is important that the court noted that it is the L-form and not the D/L (racemic)-form that has the therapeutic value. In the case of *Sterling Drug*, it appeared that the difference was not one of degree but rather a difference in kind (the difference between working and not working). In contrast, the Court in *Adamson* appeared to take the twofold increase in activity of the claimed enantiomer over the racemic mixture as a predictable and not unexpected result. Likewise, the only moderate increase in toxicity relative to the racemic mixture despite having twofold greater activity was also shrugged off. The Court noted that the racemic mixture had toxicity between the two isomers which is "to be particularly expected."[76]

Beyond the differences in activity between the claimed enantiomers of each case relative to the prior art racemic mixture, there was also the different treatment afforded

---

[76] If the toxicity of the compounds in *Adamson* were linked to their activity, then it would not be surprising that the enantiomer possessing the spasmolytic activity was the more toxic enantiomer, which appeared to be the case. Moreover, the fact that the receptor or enzyme targets of most drugs are chiral lends credence to the notion that one might expect that one enantiomer might be twice as active as the racemic mixture since the other enantiomer is often inactive or close to it. Since the racemic mixture contains equal amounts of the active enantiomer as the inactive enantiomer, one might reasonably expect that the active enantiomer could be twice as potent as the racemic mixture.

by the Courts regarding the question of whether the prior art would have enabled the production of the claimed enantiomers. In *Sterling Drug*, the Court appeared to suggest that the prior art would not have enabled the production of the single enantiomers of the claimed invention whereas in *Adamson*, the actual production of the individual enantiomers was deemed to be able to be produced by techniques within the purview of those of ordinary skill in the art. Without the actual experimental details involved in both of these cases, it is not possible to make any hard conclusions. Since enablement requires that one take into account whether one of ordinary skill in the art could have made the claimed invention (without undue experimentation) *at the time of the invention*, the technological time frame must also be considered. From the latter half of the 20th century to the present time, much progress in synthesizing and/or separating racemic mixtures into single enantiomers has taken place. As a corollary, a court or examiner in the patent office is more likely today than several decades ago to find the prior art enabled for a claimed enantiomer in view of the prior art disclosure of a racemic mixture. The decision in *Adamson* was 5 years after that of *Sterling Drug*, but from the facts, it is not clear when the inventions were made. If they were also separated by a similar frame of time, than that fact may also help explain the different findings in enablement between the two cases.[77]

With earlier cases like *Sterling Drug* and *Adamson* paving the way for modern determinations of nonobviousness in view of prior art disclosures of racemic mixtures, we can discern that the courts are less likely to find the prior art nonenabled as they were in the days of *Sterling Drug*. However, finding that a single enantiomer is nonobvious due to unexpected or superior results is still possible, as we will see in a recent case addressing this issue, *Ortho-McNeil*.[78] This case arose when the generic drug producer, Mylan Laboratories, Inc. (Mylan) filed an ANDA with a paragraph IV certification that asserted, inter alia, that U.S. 5,053,407 (the '407 patent), which covered the antibiotic levofloxacin, was invalid for alleged obviousness. The plaintiffs in the case, Daiichi Pharmaceutical Co., Ltd., Johnson & Johnson, the parent company of Ortho-McNeil Pharmaceuticals, Inc., and Johnson & Johnson Pharmaceutical Research and Development (Ortho) sued Mylan for constructive patent infringement and the present action ensued. Levofloxacin is the subject of claim 2 of the '407 patent, which is dependent from claim 1, both of which are reproduced in Figure 8.20.

In its analysis of Mylan's assertion that levofloxacin was obvious in view of the prior art disclosure of the racemic mixture (ofloxocin), the Court agreed that levofloxacin was *enabled* as of 1985 (when levofloxacin was prepared). The Court

---

[77] Although the temporal component is an important consideration generally, specific difficulties encountered in synthesizing a single enantiomer or resolving a racemic mixture are still possible. Some racemic mixtures might be readily separable when applying a given technique while others might not. Likewise, the synthetic preparation of some compounds as single enantiomers may be much more easily accomplished by techniques available to one of ordinary skill while others may not.

[78] *Ortho-McNeil Pharmaceuticals, et al. v. Mylan Laboratories, Inc.* 348 F.Supp.2d 713 (N.D. W. Va. 2004). Affm'd on appeal (CAFC 2005). While you might be thinking we've flogged the enantiomer issue quite enough already, the *Ortho-McNeil Pharmaceuticals* case is especially instructive because of the wide array of secondary evidence that was considered and discussed by the court.

1. An S(-)-pyridobenzoxazine compound represented by the formula (VI)

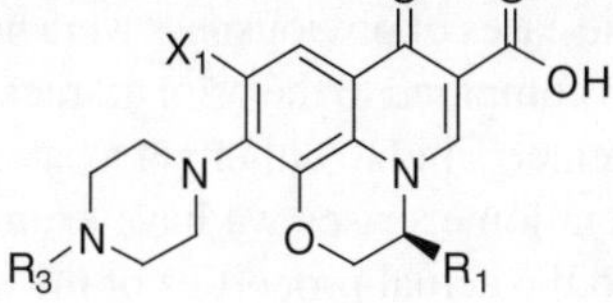

(VI)

Where: X_1 represents a halogen atom, R_1 represents an alkyl group having 1 to 4 carbon atoms, and R_3 represents an alkyl group having 1 to 3 carbon atoms.

2.   S(-)-9-Fluoro-3-methyl-10-(4-methyl-1-piperazinyl)-7-oxo-2,3-dihydro-7H-pyrido[1,2,3-de][1,4]benzoxazine-6-caroxylic acid according to claim 1.

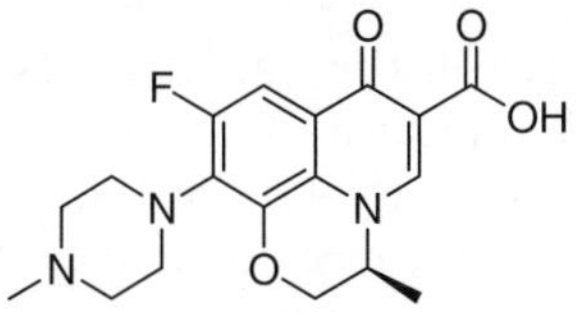

(Levofloxacin – structure for compound ofclaim 2)

FIGURE 8.20   Claims to levofloxacin.

noted that one of the compound's own inventors, in a chapter written for a book, stated:

[W]ith the development of synthesis methods via stereoselection and improvement in the analytical methods of optical isomers in the recent years, many came to believe that only one of the enantiomers is the important substance and that the other one is, if bluntly said, an almost impure substance. Influenced by ideas like these, we decided to focus on the antibacterial activity of the two [ofloxacin] enantiomers, resulting in the application of optical resolution.

To the Court this was an important admission as it rebutted an earlier assertion by the inventor's representatives that structure–activity relationship principles would have completely discouraged skilled artisans from preparing the separate isomers of the racemic mixture (ofloxacin) or from reasonably expecting that one enantiomer would exhibit greater activity than the racemic mixture. The publication also indicated that the prior art adequately enabled methods of preparing the compound, so one of ordinary skill in the art would therefore have been motivated to make levofloxacin.

The Court next turned to the properties of levofloxacin itself. From our previous enantiomer cases where prima facie cases of obviousness were held to exist, the actual properties of the enantiomer, when compared to the prior art racemic mixture, held the key to success or failure in the patentee's bid to either get a patent issued or to defend its validity. Like the other two enantiomer cases we have examined, the decision in the instant case also came down to the actual properties of the enantiomer compared to the racemic mixture of the prior art.[79] In this regard, the Court was persuaded that the properties of levofloxacin were sufficiently unexpected to find the invention nonobvious. In particular, the solubility of levofloxacin was highlighted as being especially surprising given that the patent assignees provided evidence demonstrating levofloxacin to be approximately 10 times more soluble than the racemic mixture. In addition to expert witness testimony documenting the significance of this large difference, the court noted that prior to the discovery of levofloxacin, the largest reported difference in solubility between an enantiomer and its racemate was five-fold.[80] This difference alone, the Court indicated, was an unexpected result and presumably sufficient by itself to rebut the prima facie evidence presented against its patentability.[81]

The Court also documented the fact that levofloxacin was approximately two times more potent than the racemate ofloxacin. While this fact alone might not have been persuasive, it was persuasive to the court when coupled with the fact that levofloxacin is less toxic than ofloxacin. This appeared to contradict the conventional scientific wisdom in the art that the toxicity of the tricyclic quinolone antibiotics moved in parallel with the therapeutic activity.[82] The Court also noted levofloxacin's unexpectedly higher effectiveness against *S. pneumoniae* when taking into account its exposure levels compared to oflaxocin as well as levofloxacin's unexpected potency against quinolone-resistant *S. pneumoniae*.

[79] In the instant case, the Court apparently did not find a prima facie case of obviousness because the Court believed that Mylan had failed to provide evidence that the prior art would have provided a reasonable expectation of success for levofloxacin. In this regard, the Court recited the exemplary properties of the compound and argued that those properties could not have been reasonably expected. The Court further explained that even if a prima facie case of obviousness did exist, the case would be effectively rebutted by the many secondary indicia of obviousness, including unexpected results, which are the subject of the rest of the case summary presented here. This interpretation of reasonable expectation of success by the District Court may have set the bar too high for establishing prima facie obviousness and has been criticized. See J.J. Sparrow, *Stan. Tech. L. Rev.* 2 (2007).

[80] The difference in solubility between a pure enantiomer versus a racemic mixture of the same compound relates to the stability of the compound in its crystalline state. It is not surprising that a crystalline enantiomer can pack differently from a crystalline racemate, and thus differences in solubility are not per se unexpected. In this case, it is the magnitude of the difference that the Court found to be unexpected.

[81] In this regard, the Court quoted from a case we discussed previously, *In re Chupp*, 816 F.2d 643, 647 (Fed. Cir. 1987) ("Evidence that a compound is unexpectedly superior in one of a spectrum of properties . . . can be enough to rebut a prima facie case of obviousness.")

[82] We saw in the earlier case *Adamson*, *id*., that the Court was not surprised by the fact that the single enantiomer was more toxic than the racemic mixture as it is often the case (although certainly not always), that a toxic effect of a drug correlates to its activity.

While we have focused extensively on unexpected results as one of the secondary considerations of nonobviousness, there are several additional considerations that can be argued. In this opinion, the Court addressed these several additional secondary considerations of nonobviousness as well. These additional important factors include commercial success, simultaneous invention, fulfillment of long-felt need, prior failure, others copying of the invention, third-party praise and recognition, and skepticism of persons skilled in the art.

Commercial success of an invention can be useful secondary evidence of nonobviousness. The logic is that a putatively obvious invention would be less likely to experience commercial success since the preexisting, obvious progenitors would be too competitive in the market place. Absent some advantage in marketing or other factors, it is sometimes possible to assign the success of the product to its advantageous and nonobvious properties (letting the market speak). To establish the commercial success of a product for purposes of providing evidence to help rebut a prima facie case of obviousness, the applicant or patentee must establish a nexus between the merits of the invention in view of the prior art and that product's commercial success. As a procedural matter, the party wishing to establish the commercial success firsts needs to demonstrate that the "product enjoys significant sales in a relevant market, and that the successful product is the invention disclosed and claimed in the patent."[83]

Once the applicant or patentee has established the commercial success, the burden then switches to the party asserting that the patent is obvious to prove that the commercial success is actually due to other factors such as, for example, effective marketing. In *Ortho-McNeil*, the Court highlighted the fact that levofloxacin (sold as Levaquin®) was a very good selling antibiotic with tremendous growth in sales from $170 million in its launch year 1997 to more than $1 billion in 2003. Further noting that the successful product was the compound claimed in the patent in suit, the Court then allocated the burden to Mylan to prove that the commercial success was in fact due to "extraneous factors" separate from the virtues of the compound itself. Although Mylan presented "voluminous data" attempting to demonstrate that levofloxacin's commercial success was due to other factors, the Court discounted their evidence as being insufficient to "sever the nexus between the merits of levofloxacin and its commercial success."[84] In particular, the Court noted numerous statistics demonstrating that levofloxacin was much more successful than the racemate (marketed as Floxin), a difference that could not be explained by marketing muscle alone.

---

[83] *Ortho-McNeil Pharmaceuticals, et al., id.,* citing *J.T. Eaton & Co. v. Atlantic Paste & Glue Co.,* 106 F.3d 1563 (CAFC 1997), a case that really was about building a better mousetrap.

[84] According to the Court, Mylan presented evidence of "exogenous forces including delays in FDA regulatory approval, changes in FDA clinical studies requirements, unanticipated emergence of a new competing class of anti-infectives, subsequent development of resistance in that class, Ortho's lack of prior anti-infectives experience, FDA-approved indications, ongoing clinical development, publication strategy, pricing, rebates and discounts, samples, product positioning, sales force size and quality, investment in brand marketing expenses, thought leader development, microbiological support, formulary access, and managed care strategy." The Court further noted that "most of these forces involve deliberate marketing strategy by Ortho; the remaining forces relate to the medical and regulatory context in which levofloxacin has been sold."

In its conclusion to the portion of its decision recognizing levofloxacin's commercial success, the Court stated:

> the Court is unpersuaded by Mylan's exogenous forces arguments. In the competitive anti-infective market, the overwhelming response by doctors and hospitals to levofloxacin cannot be attributed exclusively to clever marketing strategies and a "perfect storm" of unforeseen FDA delays and regulatory changes. Levofloxacin was a commercial success on its own merits, a factor which weighs in favor of unobviousness.

The next secondary consideration that the Court took up was near simultaneous invention. If two or more separate inventors almost simultaneously discover an invention, this can be regarded as secondary evidence that the invention is obvious. Mylan presented evidence that multiple other groups had synthesized levofloxacin within one year after the inventors of the challenged patent. The Court only assigned moderate emphasis to this secondary evidence, however, because Mylan provided no corroborating evidence as to the exact date of four out of five of the near simultaneous discoveries. This was an issue because one of the inventors of the patents in suit published on some of levofloxacin's properties and general resolution methods soon after they made their discovery of levofloxacin. Thus it was not clear whether the near-simultaneous invention from the other four groups was due to the obviousness of the invention at that time or because of the inventor's own disclosure.[85]

Secondary evidence of nonobviousness can also be found where an invention fulfills a long-felt need. Where a long-felt need has existed, presumably any obvious solutions would have already been tried and from a strict policy perspective, fulfilling the long-felt needs of society should rank pretty high on the things that should be encouraged. There was significant evidence of the limitations in fluoroquinolones (the class of molecules that includes levofloxacin) before the invention date of levofloxacin. For example, it had been noted that the fluoroquinolones to date had efficacies against gram-positive bacteria that were "less than optimal." Levofloxacin fulfilled that need "in light of the drug's demonstrated effectiveness against *S. pneumoniae* (a gram positive bacteria)."

Evidence that others had tried to make the invention at issue and failed is also probative as secondary evidence of nonobviousness. The fact that others had failed where the patentee or applicant has succeeded can persuasively argue that the invention was not so obvious, and in particular, its enablement or reduction to practice required skill beyond what is ordinary in the art. It is interesting, in the present case, that Ortho presented evidence that the inventors themselves had failed in numerous attempts to produce the claimed invention. The Court rejected this evidence, pointing out that previous courts had generally focused on the failure of others and not the applicants or patentees themselves. As a result, the Court did not consider this evidence.[86]

---

[85] An untypical example in which the inventor's own publication compromised its *opponent's* evidence.

[86] If the question of enablement of the prior art is the question, then it is not entirely clear why the inventor's own testimony could not be given at least some weight. Any credibility issues related to his own interest in the outcome could be factored in accordingly; perhaps the courts do not want to motivate inventors to initially fail so as to support nonobviousness arguments later.

It has been said that imitation is the sincerest form of flattery, and apparently this is true for inventions as well. If others attempt to copy an invention after it is made public, this can be taken as a sign that the invention bears copying due to some meritorious aspect of that invention that was unavailable for copying prior to the disclosure of the invention itself. Here the Court focused on the fact that the racemate oflaxocin had been known as a drug for some time, yet Mylan chose levofloxacin as the drug it wished to make as a generic. Thus the Court asserted that, notwith-standing this fact, Mylan wished to copy levofloxacin despite Mylan's contention that levofloxacin and oflaxocin were "virtually indistinguishable." Mylan countered that it chose levofloxacin because it was sold in tablet form, was manufacturable at Mylan, and presented a "good business opportunity, was available from a quality raw material provider, and provided a first-to-file Paragraph IV opportunity for exclu-sivity."[87] In reply the Court deflected Mylan's arguments by asserting that the fact that levofloxacin was a good business opportunity was evidence that the compound had special properties that accounted for its business success, thus supporting the nonobviousness of the compound.[88] The Court further noted that it would be hard to believe that Mylan would not copy a compound without "heavily weighing its respec-tive properties."[89] The Court further noted that Mylan had previously attempted to market its own proprietary antibiotic sparfloxacin (Zagam), which apparently did not achieve much commercial success due to phototoxicity and the fact that the drug was potentially cardiotoxic. As a result of these factors, the Court decided that Mylan's choice to copy levofloxacin was significant secondary evidence of nonobviousness.

Another category of secondary evidence of nonobviousness considered by the Court was the extent of third-party praise and recognition for levofloxacin and its discovery. For support, the Court quoted from a recent CAFC case that discussed this factor, stating "Appreciation by contemporaries skilled in the field of the invention is a useful indicator" of nonobviousness.[90] Finding that this factor weighed in the favor of nonobviousness, the Court recited some awards the inventor had received, including the Molecular Chirality Award and recognition of the inventor by the Pharmaceutical Society of Japan, an Education Science Minister Award, and the "prestigious purple ribbon medal" from the Emperor of Japan for his outstanding contributions. According to the Court, Mylan's rebuttal contention that all of the awards were for the development of the racemate ofloxacin (which was invented by the same person) suggested but did not prove the whole of their contention, and

---

[87]The first-to-file a successful paragraph IV abbreviated new drug application is awarded 180 days exclusivity, during which it will be the sole generic entrant to the market.

[88]In reality, a generic drug company's business model is to copy innovator drug company's products. If this form of secondary evidence were to hold significant weight, it would be a significant detriment to any generic company's attempt to establish the innovator's product as obvious. Given the fact that the Hatch-Waxman provision exists at least partially to encourage generic companies to challenge invalid innovator patents, it might be appropriate for the generic company to argue that their motivation to copy comes from the Hatch-Waxman legislation itself, and not the nonobvious nature of the drug.

[89]Do you think Mylan was more interested in the superior properties of levofloxacin or its commercial success?

[90]*Vulcan Eng'g Co. v. Fata Aluminum, Inc.*, 278 F.3d 1366, 1373 (CAFC 2002).

accordingly, the evidence of recognition of the inventor and invention were also indicative of nonobviousness of the invention.

Expressions of skepticism toward an invention by those of skill in the art can be used as secondary evidence of nonobviousness.[91] However, evidence of skepticism in the form of an article presented by the inventor's representatives was deemed by the Court to be *not relevant*. Despite the fact that the presented article was pessimistic about the possibility that an optically pure tricyclic fluoroquinolone antibacterial would obtain certain desired properties, the Court found that it could not have discouraged an inventor at the time of levofloxacin's invention since the article was published 4 years after the date of the invention. As a result, the article could not accurately reflect or influence the state of mind of one of ordinary skill at the time of the invention and thus skepticism of those of skill in the art was not established.

Finally, the Court opened its findings for the last indicia of secondary evidence that would be contemplated in the instant case by quoting from precedent: "Licenses taken under the patent in suit may constitute evidence of nonobviousness; however, only little weight can be attributed to such evidence if the patentee does not demonstrate 'a nexus between the merits of the invention and the licenses of record.'"[92] The levofloxacin patent was licensed to Ortho-McNeil/Johnson & Johnson, Glaxo and Hoechst. However, the Court largely discounted this evidence by noting the fact that all three companies also owned licenses to ofloxacin, and Mylan offered evidence that the three companies bought licenses to levofloxacin for its extended patent life as well as other reasons. The Court held that the inventor's representative had failed to show that the licenses were purchased because of the superior properties of levofloxacin itself and thus placed little or no weight on the license evidence.

As a result of its analysis, the Court concluded that the relevant claims to levofloxacin were nonobvious over the prior art (including the racemate). The Court, in reaching its decision, systematically reviewed and applied the obviousness/nonobviousness factors and weighed these in view of the claims to levofloxacin. It is not surprising that the Court spent a considerable amount of time reviewing secondary evidence of nonobviousness, since, as we saw before, the likelihood of finding a prima facie case that an enantiomer is obvious is high so the consideration of the compound's properties as well as other secondary indicia becomes a central concern.

8.10 OBVIOUSNESS OF A SPECIES OR GENUS IN LIGHT OF A PRIOR ART GENUS

In Chapter 7, we discussed the effect that disclosure of prior art genera and species would have on the novelty of a later claimed genera and species containing very similar subject matter. For example, we found that in certain cases, a prior art species could anticipate a later claimed genus. We also learned that an attempt to claim a

---

[91] See, for example, *Ruiz v. A.B. Chance Co.*, 234 F.3d 654, 668 (CAFC 2000) citing *In re Hedges*, 783 F.2d 1038, 1041 (CAFC 1986), "[P]roceeding contrary to the accepted wisdom . . . is 'strong evidence of unobviousness".
[92] *In re GPAC*, 57 F.3d 1573, 1580 (CAFC 1995).

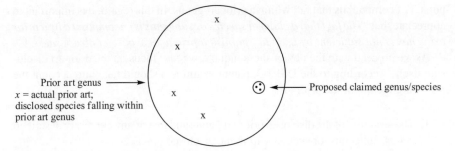

FIGURE 8.21   Venn diagram of relationship between prior art genus that encompasses a later claimed genus and species.

species that fell within an earlier disclosed genus might be anticipated, if the earlier claimed genus was small enough so that each and every member of the earlier claimed genus could be at once envisaged. However important these two situations are to patentability, they do not account for many possible situations that can occur when the prior art discloses a genus and an applicant is later attempting to claim a species or genus that somehow relates to that prior art genus (meaning the claim is either enclosed, overlaps or abuts the earlier claimed genus). Given the importance of genera in the chemical arts, evidenced especially by the common adoption of Markush descriptions, it will be worthwhile to delve further into understanding these relationships in view of how they affect obviousness determinations.

To start, a situation that very often occurs in the chemical arts can be depicted symbolically (Figure 8.21) in which a prior art genus is represented by the large circle. Most often, the prior art genus will be part of a patent or patent application publication but does not necessarily have to be; it might also be part of a disclosure in the general chemical literature. The prior art genus may or may not contain actual examples within its scope that have been prepared or otherwise singly identified in the prior art disclosure. The presence or absence of actual examples within the genus will often be an important aspect of the obviousness determination that we undertake, as is the size of the genus together, of course, with their structural relationship and related teachings with the genus and compounds we wish to claim. To "put some meat on the bones," let's take a look at the ways that a prior art genus may relate to our attempt to later claim a species or genus that is somehow related to the prior art genus.

In this first scenario, a chemist has found that either a single compound or set of compounds that he has made falls entirely within the scope of a genus disclosed in the prior art. The applicant wishes to claim his own genus of compounds that he has also populated with a few working examples. In his patent application, he would most likely try to claim both the genus and the individual examples separately.[93] The instinctive reaction of the practicing chemist in such a situation is to write off the

---

[93] Recall that applicants often claim their inventions in multiple ways. If an applicant wishes to claim a genus of compounds he will do so and then will often separately claim any individual compounds examples as narrower claims. That way, if the claim to the broader genus is not allowed (or held nonenforceable for some reason), he might still be able to assert the claims to the individual compounds.

pursuit of compounds that fall within a prior art genus. In this regard, it is important to appreciate that *"The fact that a claimed species or subgenus is encompassed by a prior art genus is not sufficient by itself to establish a prima facie case of obviousness."*[94]

As we proceed with the obviousness inquiry, we *first* focus on the prior art disclosure itself. According to the USPTO, patent examiners should take into account the following factors of the prior art:

1. The structure of the disclosed prior art genus and that of any expressly described genus, subgenus, or species within the disclosed genus.
2. Any physical or chemical properties and utilities disclosed for the genus as well as any suggested limitations on the usefulness of the genus and any problems alleged to be addressed by the genus.
3. The predictability of the technology.
4. The number of species encompassed by the genus taking into consideration all of the variables possible.[95]

First, we will want to note the structure of the disclosed prior art genus, including any expressly described genus, subgenus, or species within the disclosed genus. If the genus is like the broader map, then any subgenus and/or species contained therein are like streets and house locations. As we determine the prior art effect that this genus map might have on any later attempt to claim compounds or subgenera falling within its purview, we need to closely consider the extent to which the map is already filled in and, accordingly, whether the details of the genus and its contents are sufficient to render the later claimed compounds obvious in view of the collective teachings of the prior art.

Next, we will need to consider any physical or chemical properties and utilities disclosed for the prior art genus, as well as any suggested limitations on the usefulness of the genus, and any problems alleged to be addressed by the genus. This is important, for example, because when we consider what motivates one of ordinary skill in the art, we cannot ignore the presence or absence of a disclosed utility. From *In re Dillon* we learned that the particular utility disclosed in the prior art was not important, only that a utility was disclosed such that one of ordinary skill in the art would have been motivated to prepare the later claimed compounds that are in question. However, there still must be a disclosed utility sufficient to motivate one to modify the prior art compounds or genus in such a way as to render a later claimed compound or genus

---

[94]MPEP 2144.08 II, citing *In re Baird*, 16 F.3d 380, 382 (CAFC 1994). Of course we are referring only to the disclosure for its effect on the *patentability* of our compounds. This is a separate question from a freedom-to-operate analysis, which asks whether we could actually *practice* our invention (without an agreement or license). In particular, if the prior art disclosure is an issued patent, and our compounds of interest fall within the scope of a genus that the other party holds an issued valid claim for, then we not only have to ask if our compounds are patentable in view of the prior art claimed genus but also whether we might be infringing their patent if we decide to practice a use of our compounds; patentability and freedom to operate are two separate inquiries. See Section 1.4.

[95]MPEP 2144.08 (II)(A)(1).

obvious. Thus, for example, if a genus of compounds was disclosed together with information suggesting that genus of compounds was inactive or not useful, then it is highly unlikely that one of ordinary skill in the art would be motivated to make additional analogs or compounds within such a genus given its alleged lack of utility.[96]

The third point requires that we consider the predictability of the art or technology in question. As we have previously discussed, where one is ascertaining the relative obviousness of a compound or genus of compounds relative to a prior art disclosure of a genus of compounds and related compounds, the obviousness or not of making changes to the prior art genus to arrive at the later claimed invention must take into account how predictable the result is likely to be—less predictability equals less likelihood of finding the motivation to make specific changes. Different areas of chemistry might have different relative degrees of predictability, depending on the narrower question of what is being looked into.

The last point concerns the size of the prior art genus. All things being equal, a large prior art genus is less instructive then a small prior art genus. An allegation that a very large prior art genus makes later claimed species or subgenus falling within that disclosure obvious is tantamount to saying the proverbial haystack makes every hidden needle obvious.

Once the prior art disclosure has been properly characterized according to the four factors just discussed, the next step is to compare the species or subgenus that the applicant wishes to any subgenus or species disclosed in the prior art according to the following process:

1. Compare the claimed species or subgenus to the prior art genus.
2. Identify the closest prior art species or subgenus to the claimed species or subgenus.
3. Make explicit findings of the similarities and differences in terms of structure, properties, and utility between the prior art genus, subgenus, species, and the claimed species or genus.[97]

Then, looking at all the evidence before us, we will need to decide whether the prior art disclosure would have motivated one to make the later claimed invention and whether she would have a "reasonable chance of success" in doing so.

Of course, should this analysis result in a sufficient finding to make the later claimed species or genus obvious, then the applicant will have the opportunity to establish secondary evidence of nonobviousness, such as unexpected properties, as we have already discussed.

To reduce these theoretical considerations to an example, let's take a look at a hypothetical situation in which the prior art discloses a genus together with some compounds falling within the genus. In view of the prior art disclosure, we'll do an analysis using the principles we have just discussed.

---

[96] *In re Steminski*, 444 F.2d 581 (CCPA 1971).
[97] MPEP 2144.08.

A chemist working for a small pharmaceutical company has discovered that a class of compounds she created upregulates insulin secretion in animal models of hypoinsulinemia and may be useful for the treatment of diabetes. The compounds appear to work by a different mechanism then currently used therapies and appear to be safe, having little or no apparent toxicity in various preclinical toxicological models. The chemist creates a Markush representation of molecules that she believes will work for the desired utility and for which she believes she has provided sufficient enablement for their preparation and use. She has also singled out one particularly good compound that may become a development candidate, a compound that she would like to claim separately from the genus. However, in her prior art search, she has become aware of a document that she is worried will impact the patentability of her invention. The prior art disclosure is a published patent application with a claim to a genus of compounds as well as two compounds that fall within the genus. It discloses that the compounds have a number of different utilities, including use as potential treatments for various maladies, but diabetes is not specifically mentioned. The relevant structures are shown in Figure 8.22.

The beginning of our analysis requires that we first appraise the prior art disclosure ("scope and content of the prior art"). The prior art describes a genus of compounds

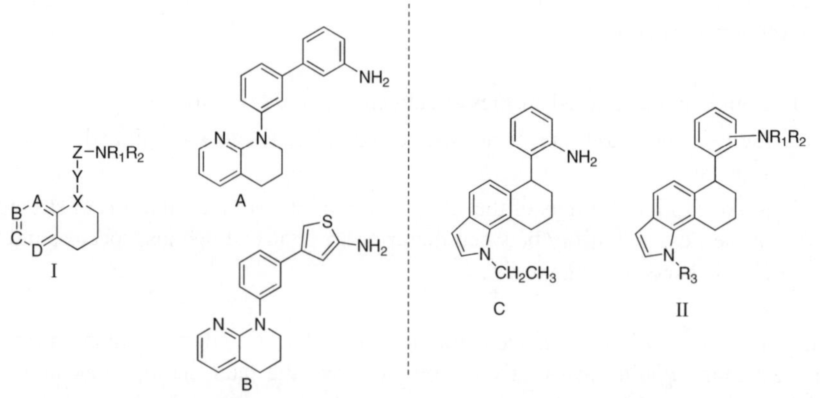

Prior Art Disclosure

Where A, B, C, D are each independently N or CH; or A and B; B and C; or C and D together, with a linker chain together form an additional fused ring containing from 5 to 7 atoms, where said ring contains a total of up to 3 heteroatoms and said ring may be optionally substituted with from 1 to 3 additional substituents selected from C_{1-3} alkyl, halogen, OH or NH_2 and said ring maybe aromatic or nonaromatic; and X is N or C; and Y is an aryl group containing from 6 to 10 ring atoms or a bond; Z is an aryl group containing from 6 to 10 ring atoms or a heteroaryl group containing from 5 to 7 ring atoms where at least one but no more than 2 of said ring atoms are selected from N, O, or S; and R_1 and R_2 are each independently hydrogen or C1-3 alkyl.

Proposed Genus and Species

Where R_3 is hydrogen or C_{1-3} alkyl; and R_1 and R_2 are each independently hydrogen or C_{1-3} alkyl.

FIGURE 8.22 Prior art disclosure and proposed genus and species.

with a bicyclic or tricyclic core, wherein when that core is tricyclic, the optional third ring is further optionally substituted with numerous possible groups including C_{1-3} alkyl. Appended to the core, attached via a carbon or nitrogen atom (the X group) of the core structure is a linker comprising an optional aromatic group (it is optional since Y can also be a bond) and a second aromatic or heteroaromatic group, where that second aromatic or heteroaromatic group is substituted with a further optionally substituted amino group.

Next, we move to step 2 of the evaluation of the prior art. There is utility for the prior art genus and compounds; they are reported to have activity as medicaments although a specific teaching regarding diabetes is not present. From *In re Dillon*, we know that the disclosure of any utility is enough to motivate one to modify the prior art structures. Had there been no disclosed utility for the compounds or if the disclosure had revealed an undesired activity or aspect of the compounds, then it is possible that a motivation to modify the prior art compounds would not have been present, but this is not the case here.

In regard to the predictability of the technology (step 3 of prior art analysis) we can probably make some generalizations but are limited by the facts as to how far we might go. In particular, we know that the chemical sciences are generally not very predictable; medicinal chemistry is particularly problematic on that score. One is trying to understand not only the attributes of one's own compounds or genus but also the interaction that those compounds will have in complex biological systems where the fundamental nature of the drug target(s) interaction is generally not well understood. It has been shown time and time again that even very small changes in a drug's structure can have profound effects on a molecule's biological profile.[98] Since an unpredictable technology is less likely to lead to obvious solutions based on prior art structures, we recognize that this factor cuts against a finding of obviousness based on those prior art structures.

The last step in the appraisal of the prior art genus is to examine the size of the genus, meaning the number of hypothetical members contained within the genus. A large genus dilutes the effect of whatever structures it is teaching because there are so many possibilities to consider.[99] Since the number of possible structural embodiments of a genus is the product of the number of possible variations at each position, the present prior art genus is tremendously large, containing a very large number (at least many millions) of hypothetical compounds.

Having completed the prior art analysis, we next must appraise the compounds and genus that our chemist wishes to claim. Let's first verify that our chemist's proposed genus (and a fortiori the species falling within it) fall within the scope of the prior art genus. Starting with the left-hand side of the prior art genus first, we will set variables A and B to CH. Next, we note that variables C and D may be connected via a linker such that the result can generate a five-member heteroaryl group, including one containing a single nitrogen in the backbone. Further, the five-member heteroaryl

---

[98] However, this is still a generalization since, depending on the particular target, the particular class of drugs, etc., the predictability might vary (yes, even unpredictability is unpredictable).

[99] That which discloses everything discloses nothing.

group is optionally substituted with from one to three groups, including a C_{1-3} alkyl. The variables on the left-hand side of the prior art genus allow for a selection that would cover the left-hand side of the claimed species and subgenus.[100] Moving to the right-hand side, the variable X can be set to a carbon atom, thus matching the proposed genus and species atom at that position. Y can be an aromatic ring or a bond. When Y is a bond, the meaning is that Z and X will be directly bonded to each other. By selecting Y as a bond, the Z group can then be selected as a six-member aromatic (benzene ring), and R_1 and R_2 can be hydrogen or C_{1-3} alkyl. Through the process of picking and choosing the appropriate groups from the prior art Markush genus, we are able to confirm that it does circumscribe the subgenus and species that our chemist wishes to patent, and so we understand the relationship between the compounds and the genus the chemist wishes to claim and the prior art genus.

Now that we've confirmed the relationship between the prior art disclosure and the chemist's genus and species (matching that depicted in Figure 8.22) we can begin to make the critical point-by-point analysis to see if the prior art would render the claimed species and/or genus obvious.

Our first observation is that the prior art genus is very large, containing a very large number of hypothetical members. As we have already discussed, the larger the prior art genus (all things being equal), the less likely it is to suggest any particular embodiment falling within it. Second, the genus contains certain embodiments that are only optionally present and contains no additional instruction that would direct one to actually include such an embodiment. For example, the core structure that is specifically delineated in the prior art Markush is a bicyclic ring system that can be one of many possibilities, depending on whether the variable backbone members are selected as carbon or nitrogen. The ability to fuse a third ring on is only an optional embodiment, where that fused ring can be in one of three positions and where that fused ring can be chosen from potentially hundreds of different heteroaromatic ring systems. The Markush genus that is sought to be claimed requires a specific heteroaromatic ring (pyrrole) to be fused in a specific position, which together make up the benzo[g]indole ring system. In addition, we can see that the bicyclic core of the prior art Markush genus contains numerous optional nitrogen atoms at various positions. Thus what may at first look to be a single bicyclic core contains the possibility of dozens of different bicyclic ring systems, depending on the selection of A, B, C, and X. Further, there are additional optional substituents available to the prior art Markush that are not present in the proposed genus or species. For example, the prior art Markush allows for an additional aromatic linker Y to be present, whereas the proposed genus and species do not. Notice further that the aromatic and heteroaromatic choices in the linker (-Y-Z) piece of the prior art are defined only

---

[100]The term *optionally substituted* means that the atom or group in question may or may not be substituted by one or more of the groups listed (e.g. C_{1-3} alkyl, halogen, OH or NH_2). Where a Markush of an organic molecule genus does not specify what atom should be attached to a carbon or nitrogen atom, one of skill in the art would understand that to mean that the valence is filled by hydrogen. Ordinarily, carbon is a fixed four-valence atom and nitrogen is three (unless carrying a charge), and it is customary to not draw in every hydrogen but rather to understand that if nothing is specified, then hydrogen is the default fill for the valency on that atom.

generically, and in fact, each one of Y and Z must in turn be selected from a very large number of ring systems.[101] Taken in sum, there are a very large number of possible ways that one of ordinary skill in the art could proceed from the prior art disclosure without ever finding their way to the later claimed genus. This means that the prior art genus, *by itself*, fails to render the later claimed genus obvious. How could it? Look at the prior art genus, now look at the proposed genus to be claimed. Would the prior art genus motivate one to make the proposed genus or species? Of course not! What would have motivated, taught, or suggested one of ordinary skill in the art to make the proposed genus (and species) rather than any of the millions of other possibilities that the prior art genus might suggest? A synthetically skilled organic chemist could work many lifetimes randomly making analogs falling within the prior art genus and never even come close to making a compound falling within the later claimed genus.

We are not done yet however. We need to remember that the prior art is good for all that it teaches and thus we need to look at what else the prior art document provides in the way of further direction. Of special interest, notice that not only does the prior art provide a genus of compounds having various asserted utilities but it also provides a couple of actual examples that fall within the genus scope. The question is whether these specifically disclosed prior art compounds direct us closer to our proposed genus and compounds? First, we can see that both of the exemplified species are similar in structure to each other, containing tetrahydro naphthyridine core ring systems. Thus neither represents a benzo[g]indole as is required by the proposed genus and species. In fact, neither prior art species even possesses a tricyclic core. Furthermore, the two prior art examples contain the optional linker where Y is phenyl and such a group is not present on the proposed genus or species. So, would the specifically exemplified structures lead one closer to the later claimed Markush and species? The answer is almost certainly no, since if in fact these compounds are signposts, they are actually pointing away from the direction taken by the applicant's claimed species and genus. So, given the very large size of the genus, and the unpredictability of the art coupled with a lack of additional teaching that would motivate one to make the proposed genus or compound, the prior art disclosure in this hypothetical will not render the later claimed genus and species obvious.

We have just analyzed the situation in which a prior art genus, due to great breadth and insufficiency of any additional teaching, would not render obvious a later claimed narrow genus and compound falling within its scope. To clarify through comparison and distinction, let's change the facts of the previous hypothetical scenario in a way that will help highlight the type of scenario that more readily avails itself to an allegation of obviousness. As we explain this alternate situation, we'll keep closely in mind the factors that we discussed in our previous scenario, remaining vigilant

---

[101] You'll notice in the prior art definition, the aromatic and heteroaromatic ring systems are defined by only the number of atoms they contain. Typically, patents and patent applications will further clarify such definitions, including the possibility for additional optional substitutions on those ring systems, either in the claim itself or in the patent specification. Such further subdefinitions were left out to simplify the Markush example.

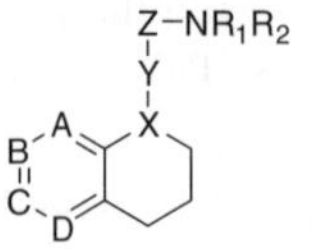

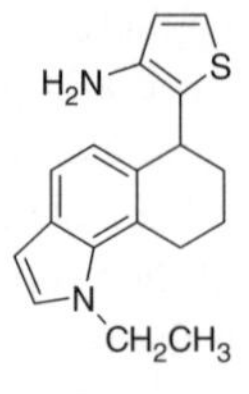

Prior art reference 1: A preferred embodiment where A and B are each CH; C and D are carbon and together, with a three atom linker makes a third ring where said linker is -CH=CH-NR$_3$- or -CH=CH-S-; X is C or N; Y is a bond; Z is phenyl, naphthyl, thiophenyl or benzofuranyl; and R$_1$ and R$_2$ are each independently selected from hydrogen and C$_{1-3}$ alkyl.

Prior art reference 2: A thiophene group is a common bioisostere for a phenyl group and may often be used interchangeably in the search for alternative analogues with similar activity."

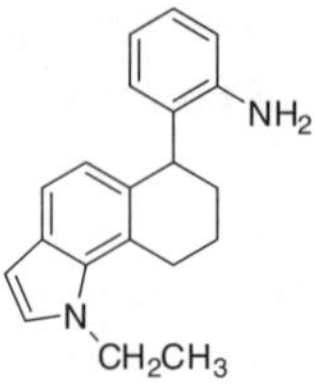

Proposed genus and species

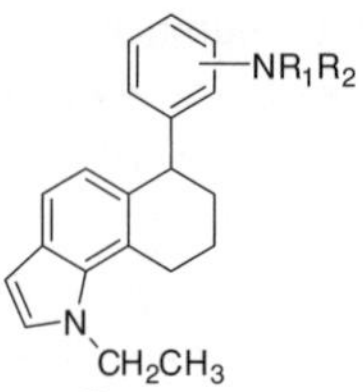

FIGURE 8.23   Narrower prior art genus and species.

to how they differ from the current scenario and why that makes a difference. In addition, to more approximately mimic what might happen in a real-world scenario, we will combine additional teachings from another prior art reference.

Let's assume that reading the prior art further, our chemist notes an additional Markush genus is one of several preferred embodiments in the prior art reference and that a single compound D falling within the preferred embodiment is also disclosed (Figure 8.23).[102]. Also, a quote from a second prior art reference has been included, which suggests the interchangeability of a thiophene for a phenyl group.

This preferred prior art Markush genus disclosure is much narrower than the previous genus disclosure of Figure 8.22. Many of the variable choices have been significantly narrowed or eliminated, resulting in a genus that creates a much tighter cage for the compounds and genus our chemist wishes to claim (thinking back to our original hypothetical Figure 8.22, the size of the outer circle representing the

---

[102]The preferred language is commonly used in patent application specifications. A patent drafter will often place such language into a document to lay out fallback positions should it be desired to claim narrower embodiments during the prosecution of the patent application. While the use of the preferred language is generally not necessary to stake out such fallback positions, it is still commonly used, and one should appreciate the purpose of such clauses and the importance of them in patent drafting. If one drafted a broad genus claim without any alternative embodiments, an inability to get a valid claim to the broadest embodiment for any reason could potentially lead the patentee with nothing. By crafting one or more narrower embodiments of the broadest Markush genus, the invalidation of the broader claim through, for example, anticipation or obviousness in view of a prior art reference, could still leave one or more additional fallback embodiments valid and enforceable. These alternative, narrower embodiments also serve as prior art teachings to others. If a prior art document progressively narrows toward a genus or species that you wish to patent yourself, you have a progressively more difficult time distinguishing the compounds you wish to patent.

prior art genus has now been made much smaller). However, there is more. The prior art also teaches a specific example that contains a thiophene moiety directly bonded to a tetrahydrobenzo[g]indole (the same core as the proposed genus and species). From this, we can see that the claimed compound C in Figure 8.23 differs from prior art example D only in that the pendant amino-substituted thiophene group has been replaced by a pendant amino-substituted phenyl. Assuming utility was alleged or disclosed for compound D, a case of obviousness still requires that one have some reason (i.e., be motivated) to make the claimed compound or genus. The essential question in the instant case then boils down to whether the prior art would motivate, teach, or suggest one to substitute the thiophenyl group of the prior art with the phenyl group of the claimed compound and genus; in other words, would it be obvious to substitute a phenyl group for the thiophene in the prior art? First, we can look to the express teaching of the genus itself. From the description of Z, it is apparent that a phenyl group is one of the four possible groups. So in addition to the teaching of a structurally very similar compound, the same reference also teaches a preferred subgenus where the thiophene is interchangeable with a phenyl group. Further yet, there is a second prior art reference that teaches the common interchangeability of a thiophene group for a phenyl group.[103] Since prior art reference 2 teaches that one of ordinary skill in the art would appreciate that a phenyl group and a thiophene share a bioisosteric relationship and the prior art Markush provides a phenyl group as one of a limited number of choices for that position, one could make a very strong prima facie case that it would be obvious for one of ordinary skill in the art to prepare a phenyl analog of the compound D, which is the same structure as that desired to be claimed. The motivation is clear: the preparation of additional, biologically active compounds for use as medicines. Of course, the type of obviousness "proved" by the analysis of the prior art is prima facie only and still subject to the consideration of secondary evidence of nonobviousness (Graham step 4). So it is likely that our chemist needs to find some secondary consideration, such as unusual results to differentiate her compound with a phenyl group from the prior art compound with a thiophene.

8.11 OBVIOUSNESS OF RANGES

To this point, we have focused our attention primarily on understanding how to define obviousness and how it relates to chemical structure similarity in the prior art. Chemical claims often though go far beyond the claiming of a particular compound. In some cases, we may wish to include as one of our claimed elements a range variable. For example, homogeneous or heterogeneous mixtures often claim their component ingredients in a range. Chemical processes often include claimed ranges directed to temperatures, components, or times. Even single component compound claims may include a purity limitation that can be represented by a range. Thus, for example, a claim to a compound having a purity of >98% must have its obviousness analyzed for its structure together with the purity limitation. If a prior art reference disclosed

---

[103] See *In re Merck & Co.*, 800 F.2d 1091 (CAFC 1986), *supra*, where a reference teaching bioisosterism was used to support an obviousness finding.

Prior art disclosure anticipates

Prior art discloses point falling within
later claimed range.

Prior art discloses range falling within
later claimed range.

Prior art disclosure is considered for obviousenss

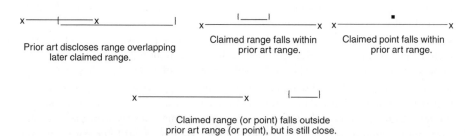

Prior art discloses range overlapping
later claimed range.

Claimed range falls within
prior art range.

Claimed point falls within
prior art range.

Claimed range (or point) falls outside
prior art range (or point), but is still close.

FIGURE 8.24   Effect of prior art ranges on later claimed ranges.

the same compound but only as a minor component of a complex mixture, that reference might not suggest the same compound in its purified form and thus might not render the claimed compound obvious. Since ranges can be such an important part of chemical claims, it is useful to see how prior art disclosures of ranges might affect the obviousness of a chemical claim containing one or more range limitations.

In Chapter 7, we saw that the prior art disclosure of a point or set of points falling within a later claimed range (or set of ranges) will anticipate the later claimed range.[104] Likewise, a prior art disclosure of a range that falls entirely within a later claimed range will also anticipate the later claimed range. However, there are other possibilities occurring with ranges (Figure 8.24). A later claimed range may overlap a prior art range, fall outside of a prior art range, or fall entirely within a prior art range. These last three possibilities do not anticipate the later claimed range or point but might affect the obviousness if other criteria of obviousness are met.

From Figure 8.24, we can discern that if a later claimed range overlaps or falls close to an earlier disclosed range or point, then we need to consider the prior art disclosure for obviousness but not anticipation. Likewise, if the claimed range (or point) falls entirely within the prior art range, then the prior art disclosure will also have to be

---

[104] *Titanium Metals, id.*

considered in the context of obviousness. One thing to keep in mind throughout this initial discussion is that we will initially be focusing on a single range as if the only thing that were being claimed was a single element falling within a single range. While this might be the case in some situations, in many others the claimed invention will consist of one or more ranges along with additional elements and limitations. For a claim to be rendered obvious, the claimed invention as a whole must be obvious, meaning that *each and every* element of the claim must be present, taught, or suggested in the prior art. Thus we still must analyze all of the elements of the claim in view of all of the art. So, for example, where a prior art data point is shown to fall within a later claimed range, that point will anticipate that particular range; however, that does not automatically mean that it will anticipate the claim since the claim may contain additional limitations, each of which must also be considered. Moreover, the individual limitations in a claim may be either rendered nonnovel or obvious or a combination of these. If all of the claim limitations and elements are rendered not novel by a single reference, then the claim can be considered for anticipation and secondary showings of nonobviousness are not available.[105] However, if some of the limitations are anticipated and others are allegedly obvious, then the claim "as a whole" will need to be considered for obviousness, including the consideration of secondary evidence of nonobviousness.

If a claimed range overlaps or falls within a prior art range, strong evidence for a prima facie case of obviousness exists, as demonstrated in the case of *In re Woodruff*.[106] This case came to the CAFC by way of the USPTO, where claims 27–34 of the application were rejected by the patent examiner and the rejection was upheld by the USPTO Board. The patent application was directed toward methods of storing vegetables to retard fungal growth. It was reported that storing vegetables in an atmosphere containing reduced oxygen and increased carbon monoxide can greatly reduce the amount of fungi growth on the vegetables. The contested claims consisted of the following independent claims.[107]

27. A process for inhibiting the growth of fungi on fresh leafy and head vegetables comprising maintaining said fresh leafy and head vegetables in modified gaseous atmosphere including carbon dioxide in an amount from 0 to about 20% by volume, molecular oxygen in an amount of about 1% to about 20% by volume, carbon monoxide in an amount of about 3% to about 25% by volume, with the remainder being substantially all molecular nitrogen, for a time sufficient to inhibit the visible growth of fungi on said fresh leafy and head vegetables, and at a temperature in the range of about 29°F to about 60°F.

(Claim 31 is the same as 27 except it provides that the carbon monoxide be present "in an amount of more than 5% to about 25% by volume.")

---

[105] This assumes that the anticipating disclosure also presents the limitations and elements in such a manner that one of ordinary skill can at once envision the claimed invention from the prior art disclosure. No mixing and matching elements and limitations to arrive at the claimed invention.

[106] *In re Woodruff*, 919 F.2d 1575 (CAFC 1990).

[107] Since the applicant did not argue the separable patentability of the dependent claims, it was not further evaluated in the court's opinion; those claims were not put into issue by the applicants.

TABLE 8.1 Comparison of the Claimed Ranges of the Patent
Application and the Prior Art Disclosure

Claim 27	Claim 31	'119 Patent
0–20% CO_2	0–20% CO_2	0–5% CO_2
1–20% O_2	1–20% O_2	1–10% O_2
3–25% CO	>5–25% CO	1–5% CO
Balance N	Balance N	Balance N
29–60°F	29–60°F	32–40°F

CO_2, carbon dioxide; O_2, molecular oxygen; CO, carbon monoxide.

The sole piece of prior art cited by the examiner was issued U.S. patent 3,453,119 ('119 patent), which was likewise directed to storing vegetables in a modified atmosphere at below room temperature to prevent various deteriorative changes but did not specifically teach or suggest the prevention of fungi as a utility. The '119 patent, like the instant application, disclosed various ranges for the elements present. Table 8.1 lists the elements of the instant claims alongside those of the '119 patent.

Comparing the prior art '119 patent to the two sets of claimed ranges in the contested application reveals that each of the prior art gas ranges falls entirely within the ranges of the instant application, except for the claimed CO range, which overlaps the prior art range in claim 27 and is slightly outside the range of claim 31.[108] As Figure 8.24 demonstrates, if the prior art discloses an embodiment range that falls within a later claimed range, then the later claimed range is anticipated. But only when each and every element is anticipated is the entire claim anticipated. In the present example, each and every element is anticipated except one. But that one means that the claim as a whole is not anticipated and therefore is novel. However, the claim was still deemed prima facie obvious because each and every range is contained or overlaps with the prior art disclosure.[109]

---

[108] In the court's opinion it was noted that although the *claimed* range in the prior art '119 patent was 1–5% CO, the specification used language of "about 1–5%" which the court pointed out overlaps with applicant's claim 31 of CO >5%. From the standpoint of evaluating patentability of the later claimed invention in view of the prior art, it is not important whether the range comes from the prior art claims or the specification. Although the prior art disclosure of about 1–5% was deemed to overlap with >5%, it is not clear whether the decision would have been any different if the CO ranges were close but not overlapping.

[109] In support of the contested claims, the applicants for patent submitted several declarations, including a declaration from the inventor of the prior art cited patent. The inventor of the prior art patent reference explained that he was not concerned with fungi but rather with bacteria and its effects on the vegetables, and he further stated that "processes that control bacteria and slime in leafy head vegetables do not necessarily control fungi and vice-versa." The court did not find the declarations persuasive, pointing out that the purposes were similar enough that they would not affect the functioning of the claimed invention. The prior art '119 patent dealt with preventing the deterioration of vegetables of which the prevention of fungal growth could be seen as a subset of the previously disclosed method; presumably allowing fungi to grow on vegetables is a form of deterioration of the vegetables. The court stated the general rule that "merely discovering a new benefit of an old process cannot render the new process again patentable." While the court acknowledged that the presently claimed process was not entirely old, apparently it was close enough in the court's determination to not give patentable distinction to the claimed limitation "for a time sufficient to inhibit the visible growth of fungi" as found in claims 27 and 31.

The court upheld the USPTO rejection of the claimed invention, noting that in a case such as this where the range overlaps, an applicant must be able to demonstrate that the particular claimed range is critical, "generally achieved by showing that the claimed range achieves unexpected results relative to the prior art range." The court noted that the applicants failed in their submitted comparison, since they only compared their invention to a commercial embodiment of the '119 patent. In the comparison, the carbon monoxide in the test group that represented the '119 patent was allowed to drop its CO concentration to 0% after 4 days, which does not match the disclosed range from the '119 prior art patent (1–5% CO). Since the '119 patent did not teach such a dropping out of range, the comparison failed in persuading the court.[110] Using *In re Woodruff* as our example, we can see that where the prior art contains a description of a later claimed invention except that one or more of the ranges overlaps, the later claimed invention will likely be found prima facie obvious but can be rebutted with a sufficient showing of unexpected results in the claimed range, something the applicants failed in doing.

In our next example, we will see that if a prior art reference discloses a range that a later claimed range falls within, the later claimed range is likely to be regarded as prima facie obvious. This rule applies even in some examples where a very large number of claimed ranges make up the claimed invention. In the case of *In re Peterson* the applicant for patent appealed a decision of the USPTO Board of Appeals, which found the claimed composition obvious.[111] The application in question was directed toward a nickel-based single crystal alloy used in the manufacture of industrial turbines, which are exposed to high temperatures. The Court focused on claim 5, reproduced here, which was agreed on by the applicant and the board to be representative of the application.

5. A nickel-base superalloy having special utility in the production of single crystal gas turbine engine blades consisting essentially of about 1 to 3 percent rhenium, about 14 percent chromium, about 9.5% cobalt, about 3.8% tungsten, about 2 percent tantalum, about 1.5% molybdenum, about 0.05% carbon, about 0.004% boron and respectively, from about 3 to 4.8% aluminum, from about 4.8% to about 3 percent titanium, and balance substantially nickel.

The USPTO examiner had rejected the claim for obviousness, citing 3 prior art references. The federal circuit, during the course of the appeal, focused on one of the references, a published European patent application, 240,451 (the '451 application). In its discussion of the '451 application, which also disclosed superalloys, the Court juxtaposed one of the disclosed compositions from the '451 application and the subject matter of claim 5 from the instant application (Table 8.2).

Every single one of the claimed elements fell within the range set out in the '451 prior art application. In its holding, the Court affirmed the obviousness finding by the USPTO and explained that "[S]electing a narrow range from *within* a somewhat

---

[110]Presumably, if the applicants had shown that CO in ranges of 1–2% were insufficient to protect the vegetables from fungi, than they would have had a better argument as they then would have shown that their chosen range was critical and that the prior art range failed at least in part of its disclosed range.

[111]*In re Peterson*, 315 F.3d 1325 (CAFC 2003).

TABLE 8.2   Comparison of the Compositions of Claim 5 and '451 Application

Element	Claim 5 (Approximate)	'451 Cited Art
Rhenium	1–3%	0–7%
Chromium	14%	3–18%
Cobalt	9.5%	0–20%
Tungsten	3.8%	0–18%
Tantalum	2%	0–15%
Molybdenum	1.5%	0–4%
Carbon	0.05%	0.002% (at least)
Boron	0.004%	0.002% (at least)
Aluminum	3–4.8%	3–8%
Titanium	3–4.8%	0–5%
Nickel	Balance	Balance

broader range disclosed in a prior art reference is no less obvious than identifying a range that simply *overlaps* a disclosed range. In fact, when, as here, the claimed ranges are completely encompassed by the prior art, the conclusion is even more compelling than in the case of mere overlap." Further discussing the principles of the case before it, the Court noted that it was a "normal desire" for scientists to improve on what is already known and that this desire provides the motivation to optimize within a given set of percentage ranges.[112]

Since by now we are well aware that a finding of prima facie obviousness can be rebutted by a demonstration of unexpected results, it should come as no surprise that the Court was willing to consider such a showing by the applicants with their claimed ranges. For a demonstration of unexpected or superior results in an invention that claims ranges, the unexpected or superior results must be demonstrated throughout the claimed range. In the instant case, examples from the application were used in an attempt to demonstrate that the claimed range of rhenium was crucial to the superior performance of the claimed alloy. The Court analyzed the numerous examples presented in the specification of the application at issue and compared them. In one example, no rhenium was present, and the alloy was not as good at avoiding rupture for extended periods as were the alloys that contained 1% rhenium and 2%

---

[112] You may have found some analogy between the prior art range situations described here and the previous discussion of Markush structures. If you did, you may wonder why a prior art disclosure of a large series of ranges is any different from a prior art disclosure of a very large Markush structure. More particularly, you may ask why it seems that the prior art disclosure of multiple ranges has apparently more clout in rendering later claimed ranges prima facie obvious than is apparently the case for Markush structures as prior art; this is a good question, and the answer is not all together clear. One possible expectation is the greater presumed unpredictability of varying actual compound structure within a Markush embodiment versus altering the relative amount of a substance within a list of ingredient ranges. For example, when we make subselections within a broader genus, we are actually selecting groups of completely different chemical structures. In contrast, when we subselect within a disclosed range of an ingredient or list of ingredients, we are still dealing with that same ingredient.

rhenium, both of which would appear to support unexpected results for at least part of the claimed range (1–3% is the claimed range for rhenium). However, the Court affirmed the USPTO board's finding that this result was not unexpected, since the extra protection afforded was only 57 hours free from rupture at 1% rhenium versus 34 hours without rhenium. The Court also noted that the most significant improvement appeared to take effect at 2% rhenium (114 hours without rupture). Finally, the Court noted that for the same set of examples, no example was presented describing an alloy with 3% rhenium. In fact, the Court drew the opposite conclusion regarding 3% rhenium from another set of examples where an alloy containing 3% rhenium ruptured in a shorter time than one containing no rhenium.[113]

As an important consideration, the Court mentioned the limit to the range within a range rule that it espoused in this case. To this end the Court noted that especially broad prior art ranges that can encompass a number of distinct compositions are not within the purview of its prima facie obviousness holding. If this were not the case, one can imagine the possibility of somewhere in the art there being a reference disclosing a number of compositions with ranges from 0.0000% to 99.9999% for each and every element of the periodic table. Generating such a table would take a few minutes. Such a disclosure could conceivably capture almost any alloy imaginable within its disclosed ranges, thus making them all prima facie obvious. Such a result would hardly seem appropriate, and the Court recognized this difficulty by acknowledging that the previously disclosed composition ranges to which the rule applies are "not especially broad." Of course, *not especially broad* is itself a generalization and not a strict guideline and is likely to be open to considerable interpretation.

To this point in our discussion on prior art ranges, we have dealt with examples where the prior art ranges overlapped with or entirely encompassed later claimed ranges. Does this mean we are in the clear when claiming a range that does not overlap or is not encompassed by a prior art reference? The answer is, alas, not necessarily. Specifically, it depends on the variable that is being claimed and whether the prior art was such that the variable being claimed was understood to be a result-effective variable. For example, chemical processes that vary temperature of reaction or concentration of reagents are likely to find those changes considered to be routine and their variation part of a routine optimization falling short of nonobviousness as

---

[113]The Court did note that this set of examples were not exactly analogous because the example with 3% rhenium also contained less titanium than the example with no rhenium. The Court also addressed the applicant's allegation that the cited prior art "taught away" from applicant's claimed invention. As discussed previously, prior art that teaches away from making the claimed invention can be strong evidence against prima facie obviousness. However, the Court fairly quickly disposed of applicant's arguments, finding that the fact that the prior art mentioned a preferred alloy that contained no rhenium was not the same as disparaging or otherwise discouraging the use of rhenium to make the alloys of the claimed invention. In addition, some of the other references suggested limiting the amount of chromium content to avoid adverse effects on the alloy. However, the Court noted that despite the admonitions, those same prior art references disclosed alloys containing 12% and 14% chromium, which are close to the actual claimed range of applicant's alloy. Given the Court's logic, one might surmise that had the applicants claimed a single point for rhenium of about 2%, they might have been able to demonstrate the necessary unexpected or surprising results for that value of rhenium. However, by claiming the range more broadly the applicants appeared to be attempting to claim amounts of rhenium that simply did not show unexpected results.

happened in the case of *In re Aller*.[114] The subject matter of *In re Aller* is a patent application directed to a process for preparing phenol. The application claims were rejected by the USPTO and were subsequently brought to the CCPA on appeal. Claim 8 was considered illustrative of the issue before the court:

> 8. Process for decomposing isopropyl benzene hydroperoxide and the production thereby of phenol and acetone which comprises bringing said peroxides together into intimate contact with aqueous sulphuric acid of a concentration between 25 and 70% at temperatures between 40 degrees and 80 degrees.

The rejection by the USPTO was for alleged obviousness of the claimed process in view of a single prior art reference.[115] The prior art reference described a process for decomposing isopropyl benzene hydroperoxide and an example in the paper described conducting the decomposition of the peroxide at 100°C with a 10% sulphuric acid solution. The court affirmed the USPTO's finding that claim 8 was prima facie obvious, noting that "it is to be expected that a change in temperature, or in concentration, or in both, would be a non-patentable modification." However, the court did look at applicant's data for a showing of unexpected results. To this end, the court heard evidence regarding the improved yield of applicant's claimed process versus the prior art process example (83.7–100% versus 75%) and recovered acetone yield (71–88% versus 60%).[116] In rejecting the significance of the differences, the court noted that the least significant differences between the prior art and the claimed process were "but a few percentage points" and that the greatest differences were "still within the range of variation one might expect to result from changes in a reaction condition." Finding a lack of criticality in the claimed ranges of the applicants, the Court rejected all additional rebuttal evidence the applicants presented and refused granting the pending claims.

From the last few cases, it should be clear that manipulation of variables deemed to be results affecting for chemical processes as well as compositions will likely result in a prima facie obvious rejection, which, in turn, will require that the applicant demonstrate unexpected or surprising results throughout the claimed range to overcome. This does not mean that variation from a prior art condition or range will always result in a prima facie assertion of obviousness, especially when the prior art does not recognize the significance of the range or condition being varied. In some cases, it is possible that the prior art provides a list of conditions or ranges to effect a certain process or that comprise a particular mixture, but it might not be expected that altering one or more of those conditions or ranges can positively affect the result. In such a case, varying that condition or range may no longer be routine since the

---

[114]MPEP 2144.05 (II)(A). *In re Aller*, 220 F.2d 454 (CCPA 1955).

[115]The prior art reference was an article from the German Chemical Society in 1944. Just to reiterate, §102(a) notes published prior art is good on the day it publishes and it doesn't matter what language it is written in or that it was published in a country at war with the United States at the time.

[116]Apparently time to complete reaction was similar between the prior art and applicant's claimed process. The prior art process was finished in 1.5 hours; and applicant's claimed process examples were done between 20 minutes and 3 hours.

recognition of the particular variable was itself not routine. This is logical given that the invention as a whole is the touchstone comparator for applying the obviousness analysis.

This principal that a person of ordinary skill in the art must recognize a parameter as a result-effective variable before establishing an optimum range for this variable is considered obvious and routine arose from a case, *In re Antonie,* covering a patent application for the efficient treatment of wastewater.[117] The applicants' claimed invention was rejected at the USPTO for alleged obviousness in view of a single prior art patent. The claimed invention was a wastewater treatment device that comprised a tank that wastewater continuously passes through. The tank relies on semi-immersed disks (contactors) that continuously rotate, thereby aerating the wastewater to improve the metabolic efficiency of microorganisms growing in the tank and on the contactors. The microorganisms break down the waste products that are pumped through the tanks. The prior art that the patent office relied on to reject the application described a very similar apparatus. In the claims at issue, the applicant's tanks required a ratio of tank volume to contactor surface of about 0.12 gallons per square foot. The applicant for invention contended that this particular ratio provided maximum treatment capacity for all of the setups of the claimed device, and any ratio less than the claimed ratio would give a lower treatment capacity while any ratio greater would provide no increase in treatment capacity but would instead only increase cost.[118]

The patent office alleged that the cited prior art rendered the claimed device obvious, although the prior art did not mention an optimal tank volume to contactor area ratio. The cited art did mention that efficiency (reduction in impurities) of the device could be increased to 95% by increasing the area of the contactor. The examiner reasoned that although the specific ratio of contact area to tank size was not taught, it would be obvious since if one kept the tank volume constant while varying the contactor area, one could end up with the ratio taught by applicant's claims (0.12 gal./ft^2). The Court rejected the patent examiner's arguments, instead pointing out that the recognition of the importance of treatment capacity itself was not in the prior art. To optimize for treatment capacity, the Court reasoned that one would first need to understand that the ratio of tank volume to contactor area is essential before conducting experiments to optimize that ratio. In other words, the applicant's claimed invention was looking at maximizing treatment capacity *for a given contactor area.* The prior art taught that increasing the contactor size would increase efficiency, but that efficiency relates only to the reduction of impurities in the wastewater and does not take into account the contactor area. Thus it is possible that very large contactors in a given tank volume could decrease impurities to a greater extent than smaller contactors; however, they might not decrease the amount of waste per area of contactor to the same amount as a smaller contactor (where the ratio is <0.12 gal./ft^2). Conversely, applying the applicant's formula to a tank with oversize

---

[117] *In re Antonie,* 559 F.2d 618 (CCPA 1977).

[118] Treatment capacity refers to the reduction of impurities for a given throughput and unit contactor area. Thus the claimed ratio allegedly provided a maximum reduction of impurities for a given contactor area and throughput at a minimum cost.

contactors would teach one to increase the tank volume until the ratio of 0.12 gal./ft.^{2} was obtained, at which point maximum capacity for that tank would be achieved.

While the prior art suggested that increasing contactor size could increase the amount of waste products removed from a given tank, it is entirely possible one would increase the contactor size well beyond what is necessary for the 0.12 gal./ft^2 ratio claimed in applicant's invention. The only way one could have found the maximum capacity for a given tank as described in applicant's claimed invention is if one took a prior art situation and adjusted the volume of the tank and looked for maximum efficiency while maintaining throughput and contactor area constant or adjusted the volume while optimizing for maximum throughput while maintaining efficiency and contactor area constant. The prior art suggested neither of these two scenarios. While one certainly could have read the prior art disclosure and simply begun *trying* to optimize the tank for all of the different variables, the hindsight application of such a random optimization without an idea of what to look for does not amount to making the later claimed invention obvious. According to the court, "obvious to try" was not the standard, and they apparently weren't willing to embrace it for this case.

8.12 CHANGING THE SEQUENCE OF INGREDIENT ADDITION

In cases where the prior art discloses a process for accomplishing something and that process consists of a certain number of steps, it might be possible to rearrange the process so that the steps are performed in a different order. Such a variation from the prior art will be *novel* if the applicant claims the steps in an order different from the order in the prior art. However, the outcome with respect to prima facie obviousness might conclude in a less auspicious manner for the patent applicant. While our next case, *In re Burhans*, is not exactly a chemistry case, it is rather close as it is often said that practicing chemistry is like baking bread.[119] Assuming the patentability of chemical processes and bread making are also analogous, let's extend the metaphor a bit further by considering *In re Burhans* for what we might learn about changing the sequence of steps in a process.

The decision of *In re Burhans* came as the result of an appeal from a final rejection at the USPTO over a number of rejected claims. Claim 4, shown below, will be considered illustrative for our purpose.

Claim 4. The method of making genuine whole wheat flour which consists in separating the germs from the wheat, manufacturing flour from the remaining constituents of the wheat, aging the flour, incorporating thereinto finely divided nonrancid wheat germ constituents, and thereafter impregnating the flour with carbon dioxide.

In consideration of the appealed claim(s), the Court cited several prior art references. In particular, one of the cited references described the treatment of freshly prepared flour with carbon dioxide as a method of preventing its becoming rancid.

[119] *In re Burhans*, 154 F.2d 690 (CCPA 1946).

According to the reference, the addition of CO_2 "renders fresh flour fit for immediate use and prevents flour at a far advanced age from decomposition." The Court added that it was conventional and known in the art to age the flour before using it. The same reference as well as others cited described the separation of the germs from wheat, milling the wheat to flour, and then adding back the germ. Putting these together one can see that the main innovation in the claimed invention is that the CO_2 was added after aging versus before aging, as described in the prior art reference. The patent applicant argued that the references taken singly or together did not suggest his characteristic steps for making bread. Without much discussion, the court found applicant's argument without merit and rejected the claimed invention for obviousness. In the Court's words: "There is no merit in the point here in the absence of any proof in the record that the order of performing the steps produces any new and unexpected results."

In chemical processes, it is not only the order of steps that might be considered but many other factors as well. For example, chemical ingredients might be mixed together at different rates such that the ingredient could be added at once as in batches or continuously over a given period of time. By now you likely can guess the outcome of any such attempt to overcome the prior art by making such a change, but let's review a case, *In re Dilnot*, where just such a situation was encountered.[120] *In re Dilnot* arose as the result of an appeal from a final rejection at the USPTO over a number of rejected claims. The case concerned methods and compositions related to the production of cement. Only one of the claims is relevant to the present discussion and that claim is shown below.

22. The method of producing a cellular cementitious structure consisting of the steps of, preparing a slurry of cementitious material, generating a stable air foam, *continuously* introducing said foam into said slurry, and intimately mixing said foam with said slurry as said foam is introduced to produce a substantially uniform dispersion of said foam in said slurry, introducing said mixed foam and slurry into a form of desired configuration, and allowing said mixture to set and harden (emphasis added).

The Court affirmed the rejection of this claim over a prior art description of a similar process, quoting from the USPTO's rejection:

Jahjah [the prior art reference cited by the USPTO] combines the preformed foam into various cement and concrete mixtures wherein known aggregates may optionally be incorporated. The foam is introduced from a reservoir into a concrete mixer containing the cement mix, and, in our opinion, it would be obvious to a technician having ordinary skill in the art to add the foam gradually to the cement mix while it was being mixed, since, as pointed out by the Examiner, conventional cement mixers operate in this fashion. Further, Jahjah obtains uniform dispersion of the foam and produces a multicellular cementitious product not differentiated in this record from that obtained by appellant.

---

[120] *In re Dilnot*, 319 F.2d 188 (CCPA 1963).

In a subsequent rejection of the same claim upon reconsideration, the Court further provided that the "Appellant apparently relies upon continuous operation to differentiate over the Jahjah batch process. It is, however, well within the expected skill of the technician to operate a process continuously." So from the case of *In re Dilnot*, we learn that changing from a batch addition of an ingredient to a continuous addition will probably not be sufficient to escape a prima facie obviousness rejection. Of course this does not mean such a process is not patentable, only that unexpected results relative to the prior art batch addition would need to be demonstrated.

8.13 OBVIOUSNESS OF COMBINING EQUIVALENTS TOGETHER FOR SAME KNOWN PURPOSE

As a general matter, it is within the realm of one of ordinary skill in the art to combine compounds, compositions, materials, etc., where those compounds are known for the same purpose. The motivation and expectation are that the combination will be similarly useful in the additive sense. This is clearly laid out in *In re Crockett*, resulting from an appeal of claims rejected by the USPTO.[121] The claims in question were directed to methods and materials related to the casting of steel. The composition claim 96 is reproduced below.

> 96. A treating material for injection into molten gray cast iron to produce upgrade or nodular gray cast iron products and which comprises essentially a major proportion of finely-divided calcium carbide in admixture with a minor proportion of a finely-divided nodulizing agent selected from the group consisting of magnesium oxide and rare earth oxides.

The USPTO rejected the claim and cited three references in the rejection. The three references were patents that the Board of Appeals in the USPTO referred to as Morrogh I, II, and III. While Morrogh I is not relevant to the discussion of claim 96, Morrogh II disclosed a process for casting steel by which magnesium oxide is added to the molten steel before casting. Morrogh III describes a similar process by which a stream containing calcium carbide is injected into molten iron before casting. The subject matter and teaching of the instant application was directed toward processes and a composition useful in making cast steel where magnesium oxide *and* calcium carbide (or a rare earth oxide) are added together to molten steel before casting. In Morrogh II, the use of magnesium oxide is described and taught as promoting a desired nodular structure to the cast iron. In Morrogh III, calcium carbide is also described for accomplishing the same purpose. What is different about the claimed composition is that it would require both materials and also, that those materials would be present together in the composition so as to be suitable for adding to the molten steel at the same time. The court upheld the obviousness rejection of this

---

[121]*In re Crockett*, 279 F.2d 274 (CCPA 1960).

claim, finding that it would have been "obvious to use these two materials together in the production of nodular cast iron. The combining of additives is an old expedient." In this case, the two additives were both known to be useful for the same purpose and so the motivation would be to simply provide a material useful for achieving the same result as either one of the individual ingredients by itself.

Of course, we are speaking of prima facie obviousness rebuttable with a showing of any of the secondary considerations of obviousness. Unexpected results could come, for example, if a synergistic rather than an additive effect occurred. Alternatively, the abolition or even diminution of one or more undesired characteristics of the separate components might also be persuasive evidence of nonobviousness. Furthermore, each and every claim limitation must also be considered, especially where the claim language itself requires combination performance that exceeds the prediction of merely additive. Unexpected results from combinations of ingredients useful for the same purposes has been addressed in a recent case, *Knoll Pharmaceutical Co., Inc. et al. v. Teva Pharmaceuticals USA, Inc.*[122] The defendant in this decision was Teva, a drug manufacturer that was trying to launch a generic version of Knoll's FDA-approved painkiller, Vicoprofen®. The form of drug in suit was a tablet containing two analgesics, hydrocodone bitartrate (7.5 mg) and ibuprofen (200 mg). In the District Court, summary judgment was issued against Knoll Pharmaceutical, finding that all six claims of its U.S. patent 4,587,252 were invalid for obviousness. To appreciate the subject matter of the invention, claim 1, which is a process of treatment (interchangeable in name with a method of treatment claim), and the composition claim 4 have been selected for illustration. As you read them, pay special attention to the claim limitations (particularly "a greater analgesic effect") in regard to considering secondary evidence of nonobviousness.

1. A process for treating pain in a mammal which comprises administering to the mammal an amount of a pharmaceutical composition effective to provide an analgesic effect, said pharmaceutical composition comprising hydrocodone or a pharmaceutically acceptable acid addition salt thereof, the ratio of hydrocodone to ibuprofen being within the range that the administration of a therapeutic amount of said composition to a mammal will provide a greater analgesic effect than the effect obtainable by use of either hydrocodone or a pharmaceutically acceptable acid addition salt thereof or ibuprofen or a pharmaceutically acceptable addition salt thereof.

4. A pharmaceutical composition comprising a pharmaceutically acceptable carrier and an analgesically effective amount of:

(a) one part by weight of an analgesic agent selected from the group consisting of hydrocodone and pharmaceutically acceptable salts thereof, and

(b) about 20 to 80 parts by weight of ibuprofen or a pharmaceutically acceptable salt thereof.

---

[122] *Knoll Pharmaceutical Co. v. Teva Pharmaceuticals USA*, 367 F.3d 1381 (CAFC 2004).

In finding the claims invalid for obviousness on summary judgment, the District Court had held:

> The prior art expressly teaches one of ordinary skill in the art to combine an opioid with an NSAID.[123] Furthermore, based on the prior art, a person of ordinary skill in the art of pain management would have had a reasonable expectation of success in combining hydrocodone, a narcotic analgesic, with ibuprofen, an NSAID.

On appeal, the CAFC acknowledged that the prior art did "appear(s) to suggest combining an opioid, such as hydrocodone, with various NSAIDs, such as ibuprofen'" but a genuine factual dispute still existed because there was no evidence that the prior art taught or suggested an enhanced biological effect for the combination. For claim 1, this is important because the enhanced effect is actually one of the required elements of the claim, and of course, the prior art together with ordinary skills must teach or suggest each of the claimed elements. For claim 4, which does not have that explicit limitation of an enhanced effect, there still exists the right to present secondary evidence that the combination demonstrated surprising or unexpected results. In this regard, Knoll submitted to the CAFC clinical evidence demonstrating the synergistic effect of ibuprofen when combined with hydrocodone for both pain control and enhanced muscle repair.[124] These secondary considerations were sufficient for the CAFC to reverse the summary judgment of the District Court.

8.14 SUBSTITUTING EQUIVALENTS KNOWN FOR THE SAME PURPOSE

As we just learned, combining compounds or materials known in the art to be useful singly for the same purpose can be prima facie obvious. It is also instructive to consider the situation that occurs where one equivalent material or compound is substituted for another in a composition or process. In this context, it is important to not focus solely on the actual equivalence of the material being substituted but also to focus on whether one of ordinary skill in the art would appreciate this equivalency. These points are addressed in the case of *Smith v. Hayashi*, arising from an interference

---

[123] Ibuprofen is a nonsteroidal anti-inflammatory drug (NSAID).

[124] The District Court initially refused to consider the unexpected results evidence because the evidence was discovered after the patent had issued. The patent application was filed in 1984 and the patent issued in 1986. The studies demonstrating unexpected synergy of the combination were not conducted until long after, and not published until 2002. The CAFC rejected the District Court's refusal to entertain the later generated data, stating that evidence can be generated after the patent grant and should not be excluded from consideration since the full range of an invention does not have to be fully understood at that time the application is filed. The requirement is that the secondary evidence for nonobviousness must at least flow from the specification of the patent application and for this, the CAFC found "the specification expressly acknowledges that the efficacy of the combination is 'surprising' in that it provides an analgesic effect greater than that obtained by increasing the dose of either constituent administered alone." Presumably, the later determination of the magnitude and type of the 'surprising' results are seen as flowing from the teaching of the specification itself.

matter between the listed parties.[125] In the course of the interference proceeding, a "phantom count" was constructed from the subject matter corresponding to claim 1 of the Hayashi application and claim 25 of the Smith application.[126] The phantom count is as follows:

> Count: An electrophotographic plate for producing an electrostatic latent image on the top layer thereof which comprises from the bottom up:
>
> a. a conductive substrate;
> b. a layer of vitreous selenium, a vitreous selenium/tellurium alloy, a vitreous selenium/arsenic alloy or phthalocyanine having a thickness of from 0.05 to 20 microns;
> c. a top layer comprising a member selected from the group consisting of poly-N-vinyl-carbazole, poly-acenaphthylene, poly-9(4-pentenyl)carbazole, poly-9-(5-hexyl)-carbazole . . . (numerous other chemicals listed) . . . and naphthalene, said top layer being substantially visible ray transmissive and substantially non-light sensitive in the visible range in the absence of a sensitizer; the sensitizer for said top layer consisting essentially of vitreous selenium, selenium/tellurium, selenium/arsenic or phthalocyanine layer b.

During the course of the interference, Hayashi moved to limit the scope of the count. In particular, the inclusion of phthalocyanine (part b of the count) was objected to as being a distinct, separately patentable embodiment of the invention. Hayashi wished to remove phthalocyanine from the count since it originated from only his application. By doing so, Hayashi would not have to subject that embodiment to the outcome of the interference but rather could own an invention that separately claimed phthalocyanine. In other words, Hayashi wanted a patent to essentially the same claim as was defined in the count, but where only phthalocyanine was claimed in the layer described by part b of the count. By arguing that phthalocyanine did not belong with the other materials listed in section b of the count, Hayashi was effectively arguing that a claim containing only phthalocyanine in part b would be patentably distinct (novel and nonobvious) over the rest of the claim where vitreous selenium, a vitreous selenium/tellurium alloy, and a vitreous selenium/arsenic alloy would then make up the remaining choices for part b.[127]

---

[125] *Smith v. Hayashi*, 209 USPQ 754 (Bd. Pat. App. & Inter. 1980).

[126] A count is a constructive claim used during the interference to define the contested subject matter between the two parties. A phantom count refers to a constructive claim that does not necessarily correspond to a single claim from either subject application but is assembled to capture the subject matter from two substantially similar claims from the two applications. For purposes of the present discussion, we can think of the count as an ordinary patent claim.

[127] Here the issue of obviousness has come up in a different way then we usually see it. Normally, we have looked at a court's review of a USPTO rejection of a claim or set of claims. Alternatively, we have looked at a court's analysis of claim validity in view of litigation that seeks to invalidate the patent. The principles of patentability do not change in the present context; only the context has changed. Thus the Board is reviewing the issue of whether the claim proposed by Hayashi would be obvious in view of the count as presented. Think of the count (minus phthalocyanine from part b) as prior art, and the proposed claim (the count with only phthalocyanine as part b) as the subject matter sought to be patented. The Board is determining whether such a claim as proposed would be nonobvious in view of the proposed modified count.

In support of the separate patentability of a claim directed to having a part b containing phthalocyanine only, Hayashi argued that the material was "not equivalent" to the other selenium-containing materials listed in the Markush (vitreous selenium, a vitreous selenium/tellurium alloy, or a vitreous selenium/arsenic alloy). Hayashi argued that phthalocyanine, a polycyclic hydrocarbon, had a multitude of properties that are different from the element selenium. The Board acknowledged as much, but they pointed out that the patentable distinction between the two materials had to be analyzed within the environment as defined by the patent claim. In other words, one has to focus on the actual function of the material when being used in the material or device that is being claimed. The Board emphasized through numerous examples, including those taught in the actual patent application specifications themselves, that the materials shared the same use in prior art photosemiconductors as comprising their "b" layers. Thus while selenium alloys and phthalocyanine are not viewed as equivalents for *every* possible purpose, they are equivalents for the purposes of the claimed invention. However, the Board noted that more was needed to deny separate patentability to a phthalocyanine part b layer in view of a selenium-containing part b layer, stating "the mere fact that they may be "equivalent" for this particular purpose does not establish in and of itself that one is obvious in view of the other." For example, if a material was shown to be equivalent after the invention was made (e.g., in a later judicial proceeding), this does not mean that the equivalence was obvious at the time that the invention was made.[128] The question always reduces down to what one of ordinary skill in the art would perceive at the time of the invention, not what might be demonstrated later. However, in finally rejecting Hayashi's attempt to separately claim phthalocyanine in the part b layer, the Board noted that the evidence of equivalence was noted in the prior art, and accordingly, the replacement of one equivalent with another is strong evidence of obviousness.

8.15 PURIFIED FORMS OF COMPOUNDS OR MATERIALS

From our earlier discussions on novelty, you will recall that a claim could be antici-pated if and only if each and every element was present in a prior art reference. Some of the earlier claim examples we analyzed included claims to compounds where the

---

[128]It is interesting that unexpected results can be demonstrated to show nonobviousness after the invention (but provided they flow from the invention disclosure). In reverse, we determine prima facie obviousness at the time of the invention. Despite what might seem like a deviation from expected symmetry, it is nevertheless reasonable because one can only determine the prima facie obviousness of an invention judge from the standpoint as to what knowledge and skills were available at the time the invention was made. The inventor does not have the benefit of a crystal ball or time machine when she makes the invention, she must be judged by what was available to her at the time of the invention. However, it would be unfair to expect an inventor to understand every unexpected property of the invention at the time the invention was made. Very often, all of the advantages or unexpected results take quite some time to demonstrate. Moreover, the inventor cannot always know in advance what prior art she will be challenged with and accordingly, what sort of prior art she will need to compare her invention to.

claimed compound needed to be present with a particular purity. *If the prior art disclosed that same compound but not in the purified form, the claimed form of that compound is novel.* Although of high potential value, most natural products have yet to be isolated or purified, so one area of chemical patent law where purity or related claims are especially important is in the claiming of natural products, including small molecules (e.g., taxol).[129] However, as we learned earlier in our discussion on patentable subject matter, products of nature are one of the categories that cannot be patented—Mother Nature was first to invent. However, a claim to a natural product other than how it is present in nature should be novel. Of course, as we now so well know, novelty alone doesn't confer patentability, and so it is important to examine USPTO and federal court decisions for opinions relating to the obviousness/nonobviousness of purified forms of previously disclosed compounds. As we consider the question of whether a previously known, impure (present in mixture) compound in the prior art makes a later claimed, purer or isolated form of the compound obvious, one needs to consider whether a person or ordinary skill in the art would have a reason to want to purify the prior art mixture; there has to be some indication that one of ordinary skill in the art would be aware of the compound and, further, that one would expect some use or utility for the purified or isolated form.

The second consideration asks if one of ordinary skill in the art would have been able to carry out the purification even if they were motivated to do so. Remember that obviousness also requires that the prior art provide not only the motivation to make the later claimed invention but also a likelihood that one of ordinary skill in the art would have been successful in doing so. Isolation of complex and/or sensitive materials can be very difficult and the outcome not at all certain. Thus very often a wish to make something happen is a far cry from being able to make it happen, and the art must be able to provide both the wish and the means to make that wish come true.

On the surface, this might all appear somewhat contrary to the immediately preceding section related to the prima facie obviousness of optimization of ranges. One might surmise that claiming a compound in simply a purer form is akin to claiming a different range or amount that was either known in the art or in nature. However, recall that the ranges being optimized in the previously discussed cases were prima facie obvious because they were *known* to be results effective variables.[130] Thus where one of ordinary skill would know that the variable is results effective and be in the possession of the means to do so, one would possess the motivation and the means to do so. The motivation would be the mechanic's desire to optimize what is known together with a workmanlike ability to manipulate the range. This is in contrast to the generic situations we have described for claiming a more purified form where the

---

[129] In this context, small molecule compound simply refers to a nonprotein or nonbiopolymeric material, though the subject matter also can be seen as applying to naturally occurring proteins, gene sequences, etc.

[130] Contrast these with *In re Antonie*, 559 F.2d 618 (CCPA 1977), where the claimed variable (0.12 gal./ft.^{2}) was not art appreciated and could only have been arrived on somewhat inadvertently.

natural form was simply not previously known or identified. Or if it were identified, a use for it was not appreciated. Or even if it was identified and a use appreciated, the means to obtain the material might be beyond one of skill in the art. This does not mean that claiming a purified form cannot be prima facie obvious; it means we need to keep in mind the considerations just mentioned when making that assertion.

Our next case, *In re Kratz*, provides an opportunity to examine the obviousness inquiry as it relates to an applicant's claimed invention directed to compositions containing a compound already found in the natural state.[131] This case focuses on the purification of compounds responsible for the flavor of strawberries and arose from a reissue application (reissue of U.S. 3,499,769, ultimately reissued as Re. 30,363). The appealed claims of interest are shown below.

13. A process for imparting a strawberry flavor or aroma to a foodstuff which comprises adding thereto from about 0.5 up to about 25 parts per million, based on the weight of said foodstuff of a composition consisting essentially of synthetically produced substantially pure 2-methyl-2-pentenoic acid.

14. A flavor modifying composition useful in imparting a strawberry flavor to a foodstuff consisting essentially of (i) from 1 to about 20% by weight of said flavoring composition of synthetically produced substantially pure 2-methyl-2-pentenoic acid and (ii) the remainder of said composition being at least one adjuvant for said 2-methyl-2-pentenoic acid, selected from the group consisting of geraniol, ethyl methyl phenyl glycidate, vanillin, ethyl pelargonate, isoamyl acetate, ethyl butyrate, naphthyl ethyl ether, ethyl acetate, isoamyl butyrate, diacetyl, cinnamic acid, oil of cinnamon and decalactone.

(The claim numbers assigned correspond to the numbers in the reissue patent.)

The examiner rejected the claims as obvious based primarily on two references that confirmed that 2-methyl-2-pentenoic acid was a known component of strawberries. One of the references simply identified the material via an analytical procedure without physically isolating the material, and the other reference was the applicant's own technical report. The cited references were not listed as prior art per se, but rather as evidence of scientific fact that the strawberry component 2-methyl-2-pentenoic acid was found naturally in strawberries. The USPTO examiner rejected the claims, stating:

The claimed compound is a natural constituent of strawberries. The analysis of natural constituents of foods is now conventional as it was at the time of the filing of this application. To analyze strawberries, identify the claimed constituent and use it in its obvious environment, e.g. food or flavoring, would be obvious to the ordinary worker in the art. Attention is drawn to the Kratz paper, cited above, which sets forth that inventor Kratz did not do anything more than analyze strawberries so that information derived from the analysis could be used to improve the imitation strawberry flavor. There is nothing in the record to indicate unobvious subject matter. On the contrary, the record establishes that applicants have done no more than do the expected analysis for the

---

[131]*In re Kratz*, 592 F.2d 1169 (CCPA 1979).

expected objective. The recitation in the claims of the use of the "substantially pure" or "synthetically produced substantially pure" are not seen to influence the conclusions reached. The synthetic or substantially pure compound would be obvious over the natural constituent.

After failing to overcome the examiner's rejection, the applicants appealed to the USPTO Board of Appeals. The applicants argued that the required element of their claims, 2-methyl-2-pentenoic acid, was isolated only after a very lengthy and arduous analysis of the strawberries, which required the assistance of "advanced" analytical techniques including nuclear magnetic resonance, gas chromatography, and infrared spectroscopy.[132] The applicants further argued that strawberries contain a wide variety of components, and it was unexpected to find that one compound could lend the fragrance and taste of the strawberry. A supporting declaration was also submitted by the applicants that purported to demonstrate unexpected superiority of flavorings when treated with just a small amount of the isolated, claimed material.[133] The declaration also was relied on to show the low levels of 2-methyl-2-pentenoic acid in natural strawberries versus the levels required in the claims. However, the Board agreed with the examiner's position and discounted the difficulty of the separation as well as the sophistication of the instrumentation used to assist in that endeavor, arguing that these conditions were not contributive to patentability. The Board also pointed out that the compound itself was not novel to this application but had been previously prepared.[134]

The applicants then appealed to the CCPA where they had better luck. The CCPA began its written opinion by reminding the USPTO Appeal Board that it was impermissible to negate inventiveness based on the method used by the applicants for discovering their invention.[135] Second, they noted that the obviousness determination must be based on a comparison of the claimed matter and the prior art. As such, the prior art at the time of the invention was the "natural strawberry and its attendant taste." From this, there was no basis for selecting the claimed compound and using it

---

[132]Not so advanced these days.

[133]In this case, it was ultimately determined that no prima facie case of obviousness existed, so there was no need to decide the question based on the strength of unexpected results. Nevertheless, it is typical in a legal venue to argue in the alternative. Thus, for example, one is permitted to argue that there is no prima facie case of obviousness. But this does not prevent one from also submitting secondary evidence of nonobviousness, such as unexpected results. The examiner or court is obligated to consider all of the evidence together when arriving at its ultimate conclusion.

[134]It's helpful to understand that because the applicants are claiming the compound *with* further limitations, the absolute novelty of the compound (by itself) is not a prerequisite to the absolute novelty of the claim. Claim 13 is directed to a process that involves using the compound as a component to effect a particular end, and claim 14 describes a composition including a certain ratio of the claimed substance together with another additive selected from a list of possible components.

[135]Recall from the examiner's rejection the following: "Attention is drawn to the Kratz paper, cited above, which sets forth that inventor Kratz did not do anything more than analyze strawberries so that information derived from the analysis could be used to improve the imitation strawberry flavor." In contrast, 35 U.S.C. §103 (obviousness statute) states that "[p]atentability shall not be negatived by the manner in which the invention was made."

in compositions as claimed by the applicants. But the court did not stop there. Even if the components of the strawberry were identified in the prior art as components of strawberries, such a list would not direct a person to any particular compound for use in any particular purpose. Pointing to the requirement that obviousness requires at least at least some predictability, the court considered that even if 2-methyl-2-pentenoic acid was known in the art as a strawberry component, the prior art must provide some forseeability that the component would be "a significant strawberry flavor ingredient." The CCPA thus concluded the claims were patentable.

In the instant case, the USPTO appeared to assert an obviousness rejection based on how the discovery itself was made, rather than giving weight to the failure of the prior art to motivate one to specifically arrive at the claimed invention. In effect, the USPTO took the position that the isolation, extraction, identification and use for compositions or methods was essentially a technical endeavor lacking in inventive insight. This can be an easy mistake to make, as we all carry ideas that confuse inventive insight with the requirement of invention vis-à-vis obviousness. This distinction is of such importance that the 35 U.S.C. §103 statute itself reminds us that "patentability shall not be negatived by the manner in which the invention was made." Focusing on the claimed substance itself, the Court could have asked: If it was not known that 2-methyl-2-pentenoic acid was in strawberries (or any other flavorful substance), what would have motivated one of ordinary skill to claim that very same compound in a flavor-enhancing composition? Alternatively, even if it were known that 2-methyl-2-pentenoic acid is a component of strawberries, why would one of ordinary skill in the art have thought that it would lend a strawberry flavor and, furthermore, what sort of composition would such a flavor have been useful in? Finally, the fact that the compound itself was known in the art (but not as a strawberry component) would not motivate one to provide for a flavor-enhancing composition containing the same compound. There are literally many millions of organic compounds that are known in the art. What would motivate one to test this particular compound, out of the many millions, for this particular purpose? The disclosure in the art, absent an accompanying utility is not considered sufficient motivation to include the compound in a composition with a demonstrated utility. In the instant case, not only was the compound first isolated by the inventors from strawberries, they also found utility for that compound. Furthermore, they provided for compositions useful for specific purposes where those compositions included concentrations of the material that were not the same as they existed in the natural state, composition ratios and amounts that would be capable of imparting the strawberry flavor to foods. The CCPA was able to divorce the routine nature of the method used to find 2-methyl-2-pentenoic acid from the *nonobviousness* of the 2-methyl-2-pentenoic acid as claimed.[136]

Although the case just presented focused on the isolation of molecules and proteins from complex mixtures found or derived from nature, the same principles can also

---

[136]Notice that the applicants were not attempting to claim the process they used to isolate the 2-methyl-2-pentenoic acid. If they had been trying to claim a process for isolating flavorants from strawberries, for example, it might have been more appropriate to focus on the obviousness/nonobviousness of the techniques used since those might lend patentable distinction to such a process.

apply even where the prior art is not what is found in nature but rather compounds previously prepared by man, but only in a different, less pure form. Such was the situation in the next case, *In re Cofer*, arising from an appeal to the CCPA of a USPTO rejection of the two claims shown here:[137]

1. As a manufacture, free-flowing crystals of 2,2-bis(2,3-epoxypropoxyphenyl) propane.

8. As a manufacture, free-flowing crystals of substantially pure 2,2-bis(2,3-epoxypropoxyphenyl)propane characterized by a sharp melting point of about 43.5 degrees C, a weight to epoxide ratio of about 170 grams per equivalent epoxide, total chlorine content of less than 0.1 percent by weight, saponifiable chlorine content of less than 0.01 percent by weight, total hydroxyl content of less than 0.01 gram equivalents per 100 rams, each, and a viscosity, when a supercooled liquid, of less than about 40 poises at 25 degrees C.

The claimed compound of the invention is produced by the reaction of epichloro-hydrin and bisphenol A and was well known in the art and was also known to be useful in the preparation of thermosetting epoxy resins. The claim preamble "As a manufacture ...," is equivalent to saying "As a product of manufacture ...," which in turn means that these two claims are compound claims. Although the compound was well known in the literature, it was reported in the context where it made up to 70% to 90% of a complex liquid mixture. According to the applicant's patent application specification, no method had been described that would allow the production of pure 2,2-bis-[(4)-2,3-epoxypropoxyphenyl]propane directly from a reaction mixture. Prior attempts to recover the product of the reaction resulted in, at best, a viscous liquid that adversely affected the usefulness of the product. In contrast to the prior art, applicant's methods resulted in material of sufficient purity that the compound was isolated as a free-flowing, crystalline powder. This material could be easily handled and allegedly could be used to prepare epoxy resins equal or superior to the liquid material prepared previously.

In the course of rejecting the claimed crystalline material of applicant's invention, the examiner concluded that simply claiming this compound in a more pure and different physical state was obvious. The USPTO Board of Patent Appeals upheld the examiner's rejection and *attempted* to lay down what looked like a broad, per se rule on the issue, saying, "merely changing the form, purity or another characteristic of an old product, the utility remaining the same as that for an old product, does not render the claimed product patentable." However, the CCPA soundly rejected the Board's reasoning, noting that: "Merely stating a compound or composition is obvious, without factual support, is not sufficient." In reversing the Board's decision, the court reminded the Board that the obviousness analysis must consider the subject matter *as a whole*. The invention as a whole includes the fact that the claimed compound was a solid, which the court found was not suggested in the art and furthermore, the court held that those of ordinary skill in the art *would not have*

---

[137] *In re Cofer*, 354 F.2d 664 (CCPA 1966).

known how to obtain such crystals. As we have discussed previously, the prior art must be enabled. In this case, the Court found that the prior art did not describe the claimed compound as a pure crystalline solid nor would it have enabled its production. Accordingly, they reversed the USPTO obviousness rejections to the two claims at issue.[138]

In summary, at its core the obviousness determination is subjective in nature. What would be obvious to one person might not be to another. To make the inquiry more objective, techniques are used such as asking the fact-finder to attempt to stand in the shoes of one of ordinary skill in the art at the time of the invention and judge the inventiveness based on that criterion. But in reality, this is difficult to do because, ultimately, if the invention is not explicitly disclosed in the art, then one can only at best surmise what one of ordinary skill would or wouldn't do in any given situation. Putting oneself in another's shoes is difficult enough under any circumstance but probably even more so when one is asking the fact-finder to put himself in the position of a person having ordinary skill in an art of which the fact-finder may in fact know little or nothing about. Further complicating matters is the tendency for humans to engage in hindsight reconstruction, unable to fully appreciate the many choices an inventor had when he decided to go the particular way that he ultimately did choose. In hindsight, good choices often seem obvious but when confronted with the thicket of choices available at the time of the invention, the inventor does not have the benefit of his own success to guide him. Perhaps in answer to these myriad difficulties, the court in *Graham* set out to create objective criteria for defining obviousness but left available numerous ways that nonobviousness might be demonstrated through secondary considerations. These considerations provide proxies that can allow one to draw inferences regarding the nonobviousness of the invention itself. The definition of obviousness remains far from set in stone, as exemplified by the recent Supreme Court opinion in *KSR v. Teleflex*, and we should content ourselves to accept a certain amount of uncertainty no matter how much we might wish otherwise.

---

[138]The two claims would issue as claims 1 and 2 in U.S. 3,413,320.

CHAPTER 9

Basic Requirements of Patentability: Written Description, Enablement, and Best Mode

By this point in this text, we have learned a great deal about the requirements of patentability in view of what the prior art has to offer. We have focused especially on the novelty and nonobviousness requirements and the respective place for each in the patentability analysis. Furthermore, we have also examined the utility requirement—the invention must be new and nonobvious and must be useful as well. Although these three requirements together might already seem like an awful lot for inventors to deal with, we are not quite done yet.

The requirements of 35 U.S.C. §112 are the last of the requirements for patentability that will be discussed in this text. As part of the patentee's grand bargain with the public, they are required to provide exchange of a full disclosure of their invention in return for a federally enforceable right of exclusion. Since the main thrust of patent law is the advancement of the useful arts, it is not unreasonable to require the inventor to make their invention fully available to the public. Remember that the public is giving something of potentially very high value to the patentee, the right to exclude the public from using the claimed invention for a period of time approaching 20 years. After this lease period is over, the claimed invention becomes part of the public domain. The requirements of 35 U.S.C. §112 are in place so that the public does not give up too much in exchange for too little.

Viewed from the other side, one can consider the alternative. If there were no possibility of exclusivity, what inducement would exist for the full disclosure of one's invention? If the invention had any value at all, the full disclosure in the absence of exclusivity would prompt a great deal of copying by others not saddled with the cost of research and development. In such a scenario, there would be little inducement for research and even less inducement for disclosure of the research; the things most likely to be disclosed would be those with the least value. As a result of these considerations, the system works best when the equation is balanced.[1]

---

[1] The patent bargain is not forced—if the inventors do not wish to disclose the invention, they might, in some circumstances, attempt to keep their invention secret. However, many inventions for which the ultimate

The Chemist's Companion Guide to Patent Law, by Chris P. Miller and Mark J. Evans
Copyright © 2010 John Wiley & Sons, Inc.

Collectively, the requirements of written description, enablement and best mode are captured by 35 U.S.C. §112 first paragraph (§112 ¶1) which provides:

> The specification shall contain a *written description* of the invention, and of the manner and process of making and using it, in such full, clear, concise, and exact terms as to *enable* any person skilled in the art to which it pertains, or with which it is most clearly connected, to make and use the same, and shall set forth the *best mode* contemplated by the inventor of carrying out his invention (emphasis added).

In the section emphasis has been placed on the three key components of written description, enablement, and best mode. Each of these three requirements is separate and distinct, so that is how they will be treated in this chapter. Though it might seem obvious, each of the three sections of the written description requirement applies to the *claimed invention only*. It is surprisingly easy to get caught up with peripheral aspects of the invention that do not directly involve claimed elements and, as a result, overinterpret the ambit of §112 ¶1. Likewise, it is easy to forget that written description is required for all of the elements in a claim and as a result, attempt to make claim amendments during patent prosecution that do not find adequate written description support in the patent specification. Making the claim elements the focus will help to keep one on track with respect to §112 ¶1 issues.

In Chapter 2 we learned that the filing of a U.S. patent application serves as a constructive reduction to practice as of the date that the application is filed. However, we also learned that the patent application as filed needed to satisfy the requirements of §112 ¶1. Similarly we learned that an applicant is awarded a priority date as of their date of filing the patent application, but this assumes that the filed patent application satisfies the requirements of §112 ¶1 as well. This means that the value of the patent application to the patentee is largely premised on the ability of the patentee to fulfill the requirements of §112 ¶1.

9.1 THE WRITTEN DESCRIPTION REQUIREMENT

The first part of §112 ¶1 provides that the specification "shall contain a written description of the invention." According to current federal law jurisprudence, this requirement is separate from the enablement requirement, meaning that an analysis aimed at determining whether the requirements of §112 ¶1 are met requires a separate inquiry for both written description and enablement. In turn, there are two aspects

---

goal is commercial in nature cannot be kept secret. In the chemical arts, analytical tools for reverse engineering a chemical composition are very powerful. In contrast, a process for making a chemical product could more readily be kept a trade secret; reverse engineering a molecule or composition might not tell the process by which that material was prepared. In the pharmaceutical and consumer health arena, myriad regulations would prevent one from marketing a product without providing significant information regarding the contents. For this reason, intellectual property beyond patents can be of critical importance for protecting a product; trademarks are a critical component of intellectual property for many branded products.

to the written description requirement. The first aspect of the written description requirement is of significant practical importance, especially for the patent practitioner, while the second aspect relates to a more controversial interpretation of the written description requirement arising in a couple of important and recent CAFC decisions.

The first aspect of the written description requirement is often (but not always) encountered during the patent examination process when claims are amended or added during patent prosecution such that they are different from what was in the original patent application. In order for these amended or newly added claims to meet the written description requirement, the patent application *as filed* must provide written description support for any of the newly added subject matter. This means that the patent specification as originally filed must be able to provide a level of support sufficient to demonstrate the limitations and elements of each and every added or amended claim. An attempt to add new subject matter to a patent specification after it has been filed, whether such new subject matter is added into the claims during prosecution or into the specification by way of attempting to correct the original specification, is not allowed if that subject matter is not fully supported in the original patent specification. Any attempt to make such an addition during the patent prosecution process should draw a "new matter" rejection, and a request will be made that the newly added subject matter be removed.

Before proceeding further, it might be helpful to briefly review the patent prosecution process to better envision how new matter rejections typically occur. When an applicant files a patent application, that patent application is good for all the subject material that it contains—whether the subject matter is in the patent disclosure or in the original claims as filed. During the course of patent prosecution, it is very common for the patent practitioner to amend claims, remove claims, and add new claims to the patent application. The ability to make claim amendments (or even substitute entire claims) is important because such changes may be necessary to respond to objections or rejections that a patent examiner raises during the examination process. The rejection of a claim can be made for many different reasons, such as over questions of novelty, obviousness, or even the written description requirement itself, but whatever the reason, the appropriate amendment of a claim, or substitution of the claim with a new claim may circumvent the patent examiner's objection or rejection of that claim. There are additional reasons an applicant may wish to amend or add claims during the patent prosecution process. Since the patent prosecution process usually unfolds over a period of several years, the applicants for a patent often learn more about their invention after the application is filed, meaning they may have learned more about the critical features of their invention, what they are and how to exploit them, and wish to make sure that those critical features become claimed features of the invention. The patent applicant is allowed to do this provided the claims, as amended or added, are supported by the patent application as originally filed. For this reason (among others), patent specifications often contain very detailed listings of reagents, excipients, ranges, structural embodiments, and the like. The required written support can also come from the description of the invention and all relevant embodiments, or be taught from the examples, drawings, formulae, etc. Such detailed disclosures can help supply the support necessary to amend and add claims at a later date.

In a straightforward example of a written description requirement violation involving new matter, let's consider the example where a patent specification is filed that discloses specifically only the molecules A and B. During the course of the patent prosecution, the inventor contacts the agent prosecuting the application and informs her that he has prepared an additional example, compound C, and wishes to also claim it in the application that was already filed. He asks if it is too late to add the additional compound to the patent claims. What do you think? If you said, "No way!" You're right! This amendment will not be allowed. The application as filed must provide complete support for the invention claimed. In this example, if the inventor wished to get literal patent coverage for the new compound then a new patent application containing a written and enabled description of compound C would need to be filed.[2] If a new patent application or a continuation-in-part application is filed containing compound C which the earlier-filed specification did not include, then compound C will be considered to have been filed, for purposes of establishing a filing date of priority for that compound, as of the date the new application or continuation is filed. This of course can affect the patentability of compound C since it is possible that relevant prior art might have appeared between the filing of the two patent applications (including possibly the publication of the applicant's own first application). In the situation just discussed, the inventor did not include compound C anywhere in the first patent application, and thus an attempt to claim compound C in that patent application will most likely fail for lack of written description support.

There would have been a different outcome if the inventor had included compound C in the original patent application, even if the compound were not part of the original claim set that was presented when the patent application was filed. Where compound C is included in the patent disclosure (but not in the claims), the compound can be claimed later during the patent prosecution process because the later-claimed subject matter (compound C) has the necessary written description support in the specification. As a result, compound C would be afforded a priority date from the date the application was filed since the compound was part of the patent application as of that date, even though it was not specifically claimed until later. For this reason, the initial claim set in a patent application or even a subsequent claim set introduced during patent prosecution is not necessarily a complete picture of the potential of that patent application to issue additional claims. Yes, it is the issued claims that are enforceable against potential infringers, but it is the entire specification that serves as a wellspring from which those claims arise.[3] If we recall the discussion back in

---

[2]Later in this section, we will learn more about what constitutes new matter, but for now it is sufficient to appreciate that the addition of a compound species to a claim where the specification did not disclose that species is almost always going to be treated as new matter.

[3]This concept is always important to keep in mind when analyzing freedom to operate issues because it is not only the issued claims we need to consider when looking at potentially dominating patents. In circumstances in which a patent application is still active through continuations or the possibility of a broadening reissue, there still exists the very real potential that the applicants will add additional claims to subject matter for which the written description requirement is met in that patent application. In other words, if the compound, composition, process, etc. that you are worried about is disclosed but not claimed in an issued patent, you are not out of the woods yet if there are still pending patent applications in the family (e.g., continuations) that disclose the embodiment you are worried about.

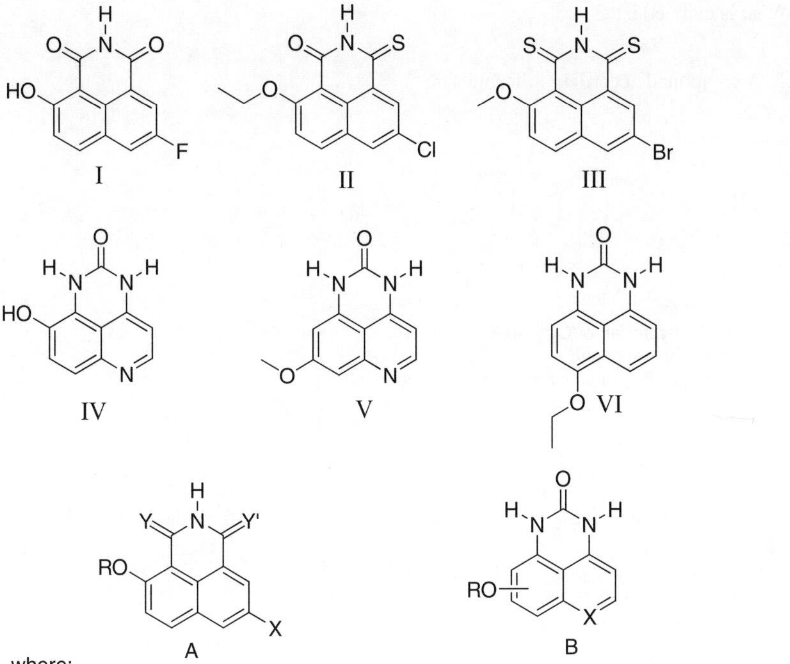

FIGURE 9.1 Markush structures and compound examples disclosed in the 10/690,732 application.

Chapter 2, we know that patent applications can be considered like links in a chain. In the patent chain, filing date priority can be followed back through the chain until the earliest date that the particular subject matter was included in the patent application that provides the necessary written description support for the later set of claims, no matter how much later those claims occurred, provided that the support of the claimed subject matter was present in each one of the linking applications and the later-filed patent applications properly claimed priority to the earlier-filed cases.[4]

Let's consider the following example. An applicant files a regular U.S. patent application (application number 10/690,732) on June 1, 2001, disclosing the six compounds I–VI and the 2 Markush structures A and B shown in Figure 9.1.

---

[4]As a drafting matter, each application in the chain must properly claim priority to each of the patent applications in the chain. You have probably seen claims to priority in patent applications and/or on the cover sheet to the patent application or patent. The claim to priority lists the applications in the chain and their relationship one to the other. The process can also be thought of as something akin to a relay race in which the baton must be properly handed from one runner to the next runner. To properly claim priority to one or more earlier filed patent applications, the later filed patent application must cite the patent application or patent applications that are being claimed for priority date purposes. Moreover, each

What is claimed is:

1. A compound according to formula A:

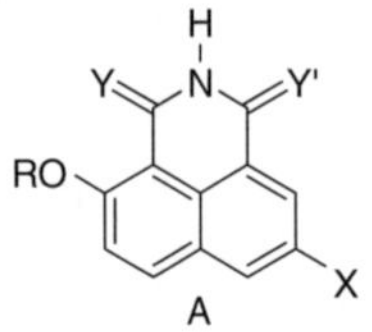

where:
R = hydrogen or C_{1-2} alkyl;
X = halogen; and
Y and Y' are each independently oxygen or sulfur.

2. A compound according to claim 1, said compound chosen from the group consisting of:

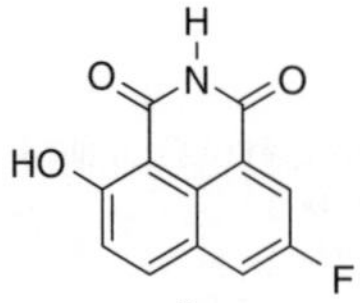

FIGURE 9.2 Claims issuing from first application.

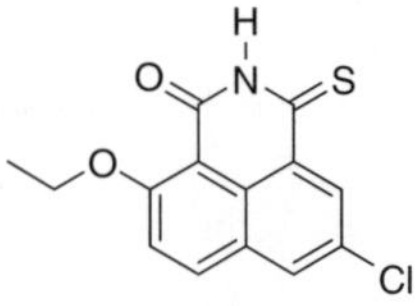

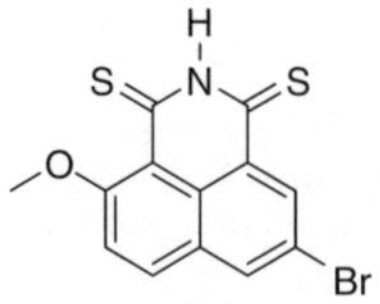

During the course of the patent prosecution, the applicants were successful in getting the two claims shown in Figure 9.2 issued in U.S. 6,992,424 on July 15, 2004. Before to the issuance of the U.S. 6,992,424 patent, the applicants file a continuation application (application number 10/999,621). In the continuation application, two more claims to the remaining subject matter (Figure 9.3) were issued in U.S. 7,532,923 on September 29, 2008.

For the examples illustrated in Figures 9.2 and 9.3, both sets of claims were supported in the original application as filed, but the claims that issued in each patent described a different genus of compounds, and moreover, they issued in different patents on dates separated by more than 4 years. Despite this fact, both sets of claims were supported by the original disclosure, and consequently, the priority date for

patent application that is filed in the chain that claims priority to an earlier patent application or patent applications, must be co-pending with at least one of the patent applications at the time the later application is filed. If the earlier patent application goes abandoned or issues fully into a patent before the later patent application is filed, then the priority chain can be broken and the ability to claim priority to the earlier filed application(s) lost.

1. A compound of formula B:

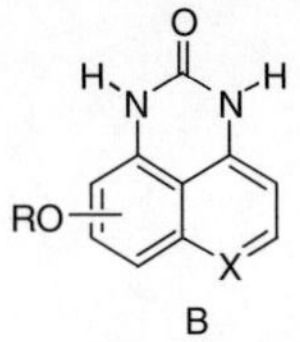

B

where:
R = hydrogen or C_{1-2} alkyl; and
X = C or N.

2. A compound according to claim 1, said compound selected from the group consisting of:

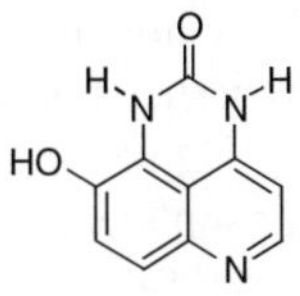

IV 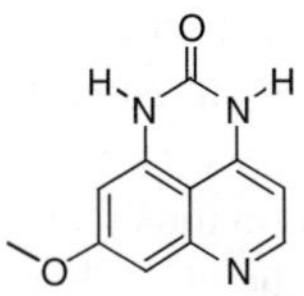   V   o VI

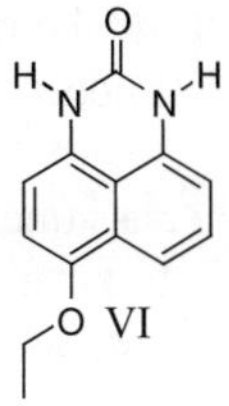

FIGURE 9.3   Claims issuing from the continuation application.

both sets of claims traces back to the original patent application: June 1, 2001.[5] This example is meant to demonstrate the importance of the disclosure as well as the claims for determining what can be properly patented. As long as the disclosure meets the patentability requirements (including the written description requirement) as of the date that the disclosure is filed, claims to that subject matter can be made in a patent application that properly claims priority from the one or more provisional patent applications, regular patent application(s), continuation application(s), or divisional application(s) containing the requisite disclosure.

One aspect in all this is that the claims that are originally filed with the specification are part of the specification, and thus should not qualify for new matter rejections. This means that an applicant may submit a list of claims with the original specification and during the prosecution may withdraw one or more of the claims, only to resubmit them later in the prosecution of that same patent application (or one

---

[5] Another possible outcome from the scenario just described is where a restriction requirement would be issued by the patent examiner at the USPTO. Compounds I–III together with genus A would make up one group, and compounds IV–VI together with genus B would make up the other group. The applicant would be requested to select one of the groups for prosecution and cancel the claims to the other. The claims to the other could then be taken up in a divisional patent application. In any event, the result would be the same with regard to priority date and patent life (both the same and dependent on the underlying patent application).

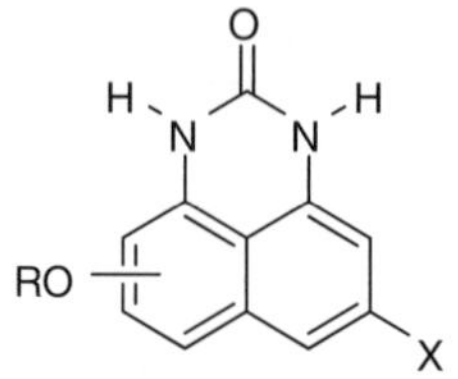

where:
R = hydrogen or C_{1-2} alkyl; and
X = halogen.

FIGURE 9.4   Claim 1 from third patent application (second continuation).

of its progeny). The newly introduced claim should not draw a new matter rejection because it was in the original patent application; the patentee had literal possession of those claims at the time the patent application was filed. Those claims must still be enabled and the best mode taught, but there should not be any issues related to new matter.

However, let's further assume from the previous scenario that our applicant wishes to add a claim to the Markush structure shown in Figure 9.4 in yet a third patent application (second continuation) that contains the identical disclosure as the previous applications.

Referring back to the patent disclosure in Figure 9.1, we see that there was no written description support for the Markush of formula C; it was not disclosed in the patent application as filed. Attempting to add the claim to the compound C in a continuation patent application and claiming priority to the original patent application filing date will fail. If the applicant wished to add the claimed subject matter of the Markush structure shown in Figure 9.4, she would have to file a new patent application (or a continuation-in-part) for which her priority date would be the new filing date, and any analysis of patentability in view of the prior art would have to take the new filing date into account.

The example just shown for genus C was a fairly straightforward matter. However, the question of whether claimed subject matter complies with the written description requirement is often clouded by the fact that the claim language does not have to be literally mirrored in the patent specification. For example, some elements may be inherent and although they were not literally present in the specification, they can still be explicitly claimed, provided such an element or elements was necessarily present in the thing being claimed, even though its presence was not overtly recognized.[6] A patent specification might be filed describing a step in which a material is combusted but the requirement that oxygen be required for the combustion is not stated explicitly in the original patent application. Nevertheless, it should be allowable to amend a

---

[6]This should sound familiar as inherent elements also may be found in the context of prior art where those inherent elements can be used to support novelty rejections even though the elements lack specific disclosure provided they meet the same requirement that they are *necessarily* present (and not just a matter of *possibly* being present).

claim to the process so that oxygen is included explicitly as a required element in the claimed process since oxygen is an inherently present reactant (by definition) in the combustion process.

Corrections to the specification of obvious mistakes can be made when one of ordinary skill in the art would be able to recognize the mistake as well as the correction that is to be made.[7] What does this have to do with new matter or a failure to support the change? If an applicant corrects an application that has already been filed, the result is not unlike amending a claim during the patent prosecution process. The difference in the language in the claim from what is disclosed in the original specification must be fully supported by the specification as filed, or the application is new subject matter and needs to be part of a new patent application. A change to the specification is no different. However, where the mistake and correction are both obvious, then no substantive change has been attempted; typographical errors easily fall within this category. Similarly, where an U.S. application is originally filed in a language other than English and a subsequent English translation is filed that contains an error, that error can be corrected using the originally filed foreign language as support for the correction.[8]

In some instances, a patent applicant may wish to include large amounts of information in the patent specification but to do so would make the specification impractically long. For example, it might be known from one or more prior art references that an extensive group of compounds can act as polymer formation initiators. Instead of explicitly listing each and every reagent in the patent specification, the applicant may choose to incorporate those reagents by reference to one or more prior art document(s) that contain the information. The act of incorporating another reference or portion of another reference, if done properly, will have the same effect as if the reference or portion of the reference were explicitly present in the patent specification.[9]

---

[7]MPEP 2163.07 citing *In re Oda*, 443 F.2d 1200 (CCPA 1971).

[8]MPEP 2163.07.

[9]Material incorporated into a patent application can be classified as *essential* or *nonessential*. Essential material is necessary to describe the claimed invention, provide an enabling disclosure of the claimed invention, or describe the best mode of the invention. If the incorporated material is essential, then it can incorporate only material from a U.S. patent, a U.S. patent application, or a pending U.S. patent application. The U.S. patent, patent application, or pending application cannot, in turn, incorporate essential matter itself. Nonessential subject matter is subject matter that is included for background purposes and can be incorporated by reference even when the original document is a non-U.S. patent document or a nonpatent document, provided it is not a reference to browser-executed code. A mere reference to another document is not sufficient for an incorporation by reference but rather the patent application must specifically identify the document that is being incorporated including the specific portions of the document that are being incorporated. A sample incorporation by reference might read something to the effect, "this application herein incorporates U.S. 1,234,567, columns 11 through 26." Very often one typically finds broad attempts at incorporation by reference such as "this document further includes U.S. published application 2007/0012345, which is herein incorporated in its entirety." Given the MPEP's instruction for direction to include the specific portions being referenced, it is not entirely clear how effective such broad incorporations as demonstrated by the latter example are likely to be (MPEP 608.01(p)). It is interesting that the advice for specific incorporation of the relevant portions appears to be more honored by patent drafters in its breach than in the observance.

These are the relatively easy examples that lie on either extreme: Either the added material was clearly new matter or not. As we saw previously, an attempt to add a new compound or Markush structure in a claim that was not disclosed in the patent application disclosure as filed will likely draw a new matter rejection. In contrast, obvious mistakes with obvious corrections can be corrected without introducing new matter into the claims or specification. In reality, chemical and pharmaceutical cases are seldom as simple as these two extremes. In chemical patent cases, the core substance is often a Markush generic structure that the applicant wishes to claim some subcombination or a chemical composition or process where the reagents or reaction variables (temperature, time, concentrations) are described in a patent specification that teaches acceptable ranges but where the applicant wishes to claim one or more subranges or overlapping range. As a first approximation, one should expect that any patent claims that lack literal support in the specification are likely to face an uphill struggle should the applicant wish to claim such material. Likewise, when one argues that literal support in a patent application can be found but that support requires cobbling together different pieces of the patent, there will also be major challenges.

These issues will be demonstrated by our next case, *Purdue Pharma v. Faulding Services*.[10] The Purdue Pharma Company (Purdue) owns U.S. 5,672,360 (the '360 patent) which is drawn to methods of treating pain by the administration of opiates (e.g., morphine) on a once-a-day schedule. A competitor company, Faulding began marketing a long-acting morphine formulation that included instructions that it could be administered once or twice a day. Purdue brought suit against Faulding alleging that Faulding's marketing of their product infringed the '360 patent.[11] Claim 1 (which was not present in the original patent application as filed but rather was included during patent prosecution) of the '360 patent is presented here:

We claim:

1. A method of effectively treating pain in humans, comprising orally administering to a human patient on a once-a-day basis an oral sustained release dosage form containing an opioid analgesic or salt thereof which upon administration provides a time to maximum plasma concentration (T_{max}) of said opioid in about 2 to about 10 hours and *a maximum plasma concentration (C_{max}) which is more than twice the plasma level of said opioid at about 24 hours after administration of the dosage form*, and which dosage form provides effective treatment of pain for about 24 hours or more after administration to the patient (emphasis added).

---

[10] *Purdue Pharma. v. Faulding Inc.,* 230 F.3d 1320 (CAFC 2000).

[11] When Faulding began selling its sustained-release formulation, the '360 patent had not yet issued. Purdue first brought suit alleging infringement of an issued patent U.S. 5,478,577 (the '577 patent). When this suit was filed, Purdue had pending before the USPTO an application that claimed priority to the '577 patent application. During prosecution of this second application, Purdue canceled all the original claims and added all new claims. The USPTO examiner stated "the new claims are supported by the specs" and the application issued as the '360 patent. Purdue then amended the infringement suit against Faulding by asserting infringement of the '360 patent.

In the District Court where the infringement action was heard, it was held that Faulding's production and sale of their drug did infringe the asserted claims of the '360 patent but that those claims were invalid because they lacked the written description requirement required by §112 ¶1. On appeal, the CAFC panel hearing the case clarified the issue as whether the claim limitation "a maximum plasma concentration (C_{max}) which is more than twice the plasma level of said opioid at about 24 hours after administration of the dosage form" ($C_{max}/C_{24} > 2$) was adequately described in the patent specification as filed. Noting that the District Court did not find support for that limitation in the patent specification as filed, the CAFC cited the standard level of support required as being that one of skill in the art, upon reading the specification, "must immediately discern the [claim] limitation at issue."[12]

On its behalf, Purdue argued that support for the critical limitation $C_{max}/C_{24} > 2$ was found in the specification. In particular, Purdue argued that the portions of the specification reproduced below, describing their discovery as providing "24 hour oral opioid formulations which do not exhibit a substantially flat serum concentration curve" supported their specific claim language requiring a $C_{max}/C_{24} > 2$. Excerpt from specification:

> The state-of-the-art approach to controlled release opioid therapy is to provide formulations which exhibit zero order pharmacokinetics and have minimal peak to trough fluctuation in opioid levels with repeated dosing. This zero order release provides very slow opioid absorption, and a generally flat serum concentration curve over time. A flat serum concentration is generally considered to be advantageous because it would in effect mimic a steady-state level where efficacy is provided but side effects common to opioid analgesics are minimized. . . .

> It has now been surprisingly discovered that quicker and greater analgesic efficacy is achieved by *24 hour oral opioid formulations which do not exhibit a substantially flat serum concentration curve*, but which instead provide a more rapid initial opioid release so that minimum effective analgesic concentration can be more quickly approached in many patients who have measurable if not significant pain at the time of dosing. . . . Also surprising and unexpected is the fact that while the methods of the present invention achieve quicker and greater analgesic efficacy, there is not a significantly greater incidence in side effects which would normally be expected as higher peak plasma concentrations occur (emphasis added).

They buttressed their contention that one of skill in the art would understand that "do(es) not exhibit a substantially flat concentration curve" means a C_{max}/C_{24} >2 by providing evidence in the form of examples from the scientific and/or patent literature to make their case. The essential thrust of their arguments was that one of ordinary skill in the art would find their specification language equivalent to what they were trying to claim. Unfortunately for Purdue, none of the examples provided were persuasive to the CAFC because none of those cited references had the type of specificity needed to prove the correlation between the specification language and the claim language. For example, one of the cited references taught that a morphine

---

[12]Quoting from *Waldemar Link Gmbh & Co. v. Osteonics Corp.*, 32 F.3d 556, 558 (CAFC 1994).

formulation with a peak-to-trough morphine ratio of 60% to 100% was "substantially flat." However, the CAFC pointed out that the reference was silent on the question of whether a peak-to-trough ratio of more than 100% would be not substantially flat; the CAFC was unwilling to accept Purdue's implication from that statement.[13] Purdue further argued that the specific examples contained in the specification supported the more generic relationship as claimed. In this regard, the CAFC noted that while two examples out of the seven presented in the specification possessed the claimed trait, other examples in the specification did not. More important, there was nothing in the specification that would indicate that this relationship was a feature of their invention; there simply was no highlighting of the relationship at all, and nothing to indicate to one of ordinary skill that this specific relationship was their invention.

Citing a much earlier federal court case, the CAFC made the analogy that finding support for claim language in the specification was something akin to finding one's way through a forest. Clear blaze marks needed to be left on the trees to point the way.[14] The CAFC explained that it does not suffice to provide a forest of trees in the specification and then later pick out a single tree and say "here is my invention." The direction for picking out those single trees must be present in the original specification. Any attempt to claim particular trees or groups of trees retrospectively that are not marked prospectively is overreaching by the patent applicant. Even where specific examples are present in the invention that do possess certain traits or numeric relationships, this does not, by itself, allow one to later claim any of those ratios, unless the specification clearly identifies the later claimed ratio as a clear feature of the invention. Regarding Purdue's attempts to claim the key relationship from the mere presence of two examples that inherently contain the relationship, the CAFC summarized Purdue's attempt as picking "a characteristic possessed by two of their formulations, a characteristic that is not discussed even in passing in the disclosure, and then making it the basis of claims that cover not just those two formulations, but any formulation that has that characteristic. This is exactly the type of overreaching the written description requirement was designed to guard against."

The *Purdue* case can be described as a failed attempt to claim a broader generic relationship from specific examples described in the specification. The desire to claim subject matter broader then what is supported in the specification can also occur in cases that are more chemical in substance. In particular, one can envision different scenarios where an applicant describes specific examples of a process, a molecule, or a composition. In the case of a process where an actual example is provided, several variables (reaction time, temperature, solvent volume, etc.) will be set to specific values, the values that the process was actually conducted at. However, absent additional language in the original specification (including the originally filed

[13]Even if they were willing to accept the implication that the reference in evidence did support the equivalence of the claim language, this would not necessarily be dispositive of the final question. Remember that Purdue is trying to prove that one of ordinary skill in the art would understand the specification language to be equivalent to the claim language. However, the understanding of one of ordinary skill in the art may or may not be the same as what the single cited reference supports. There may have been many other references and expert witnesses that supported the opposite conclusion.

[14]Citing *In re Ruschig*, 379 F.2d 990 (CCPA 1967).

claims) explaining how such variables might be modified, a later attempt to broaden the examples by including ranges for any of the variables will likely fail for lack of proper support. For example, consider a patent application as filed that describes a specific example of a chemical process in which a reaction is heated at 30°C. Absent the description of a temperature range in the original specification, the applicants would not be allowed to add claims to the process covering a range, for example, of 20° to 60°C. An attempt to add such a claimed range during the prosecution of the patent should be rejected as improperly claiming new matter. However, if the original specification disclosed the process examples properly spanning a range of 20° to 60°C, either in the disclosure or in the originally filed claims, then a new matter rejection is not appropriate.

A similar situation occurs when an original specification contains single molecule (species) examples only, and the applicant nevertheless attempts to add claim(s) to a broader genus that would contain one or more of the presented examples. For example, the disclosure of the compounds in the original specification as filed shown in Figure 9.5 *would not* provide sufficient support for adding a claim during the patent prosecution to the new Markush genus (also shown in the figure).

Not only might applicants want to broaden their original filing by adding claims that exceed the scope of the originally filed patent application, the reverse situation can also occur; an applicant might attempt to add or amend claims to subject matter in a way that results in a narrower claim than what is supported by the original specification. Perhaps somewhat counterintuitively, this scenario will also fail for inadequate written description support. We previously discussed an applicant's attempt at inappropriately broadening a single disclosed temperature for a process step of 30°C into a previously *undisclosed* range of 20° to 60°C. In the reverse scenario, the specification would describe a range of 20° to 60°C, and the applicant would attempt

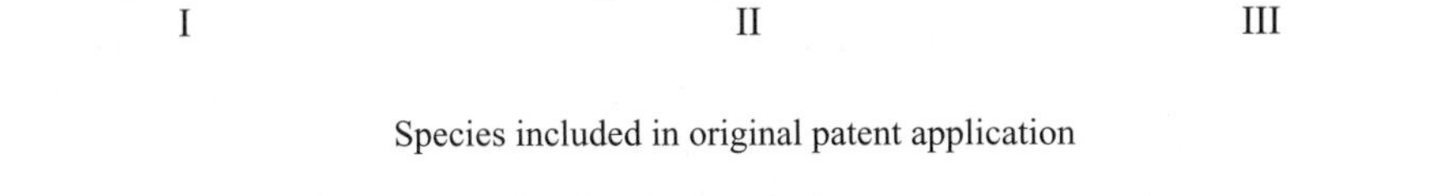

Species included in original patent application

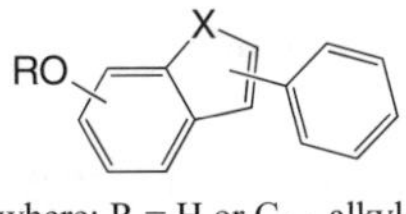

where: R = H or C$_{1-3}$ alkyl; and
X = O, NH os S

Markush structure not supported by originally filed species (shown above)

FIGURE 9.5   Species disclosed in application and later attempted Markush claim.

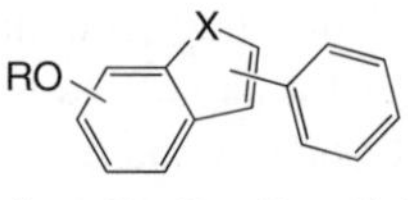

where: R = H or C$_{1-3}$ alkyl; and
X = O, NH os S

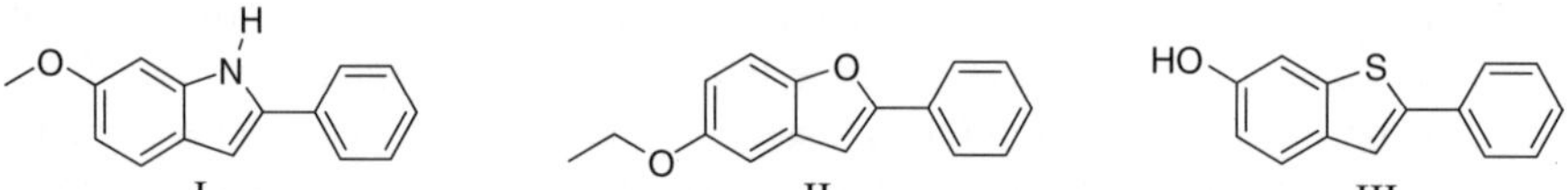

I   II   III

FIGURE 9.6 Disclosure of genus does not provide written description support to species.

to include a claim limitation containing a single temperature point (30°C) that falls within the range. The amendment would fail because the applicant is trying to add new matter to the patent application, even though the claim is narrower than the range disclosed in the specification.[15]

Likewise, the disclosure of *only* the genus shown in Figure 9.6 in the original patent specification would not support the later attempt to separately claim any of the species that were not originally disclosed (the reverse of Figure 9.5). The broader disclosure of either a range or a Markush does not, by itself, allow one to separately claim the subcombinations or species components that make up the range or Markush structure. The forest and trees analogy from the *Purdue* case is most appropriate in those circumstances since the original specification can be thought of as disclosing a generic forest and the later added claim as attempting to pick out a single tree.[16]

This discussion begs the question of why an applicant might wish to narrow his description during patent prosecution to beyond what he originally filed. Whereas the advantages of a broad claim are more intuitively obvious in terms of a broadened claim's greater area of intellectual property real estate, the ability to narrow a claim can serve several critical purposes as well. First, a claim as filed in the initial specification might not be valid in view of certain prior art, whereas a narrower claim might avoid the effect of that prior art. One scenario where this can occur is during the course of patent prosecution when the applicant becomes aware of prior art that renders his earlier filed claims not novel, whereas a narrower claim might avoid that problem. Another area where support for narrowing claim amendments is put into issue is in the arena of patent interferences. Sometimes when attempting to provoke an interference, one or more of the prospective parties to the interference will need to change his patent

---

[15]See MPEP 2163.05 for discussion of changing claim scope during patent prosecution.
[16]Broadening of a composition range beyond what is supported in the specification is also impermissible. The court in the case of *In re Wertheim* 541 F.2d 257 (CCPA 1976) found that a specification describing a range of 25% to 60% and specific examples containing 36% and 60% did not support a claim to "at least 35%" because such a claim did not provide an upper limit although a claim to "between 35% and 60%" did (MPEP 2163.05).

We claim:

1. A compound according to formula I:

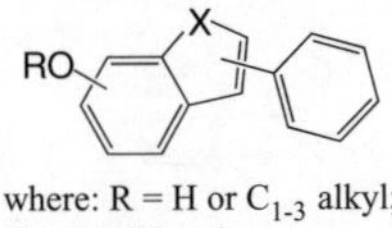

where: R = H or C_{1-3} alkyl; and
X = O, NH os S

I

2. A compound according to claim 1, wherein said compound is selected from the group
consisting of:

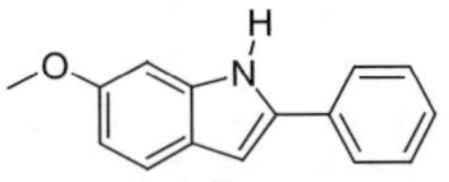

Ia 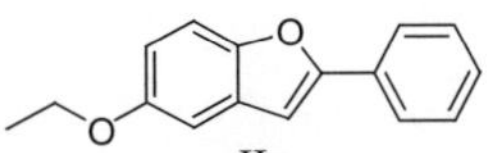   IIa 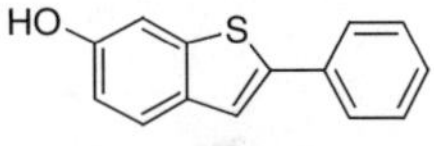   IIIa

FIGURE 9.7   Claims to Markush and species.

claim language to match the language of the other patent application or patent so that
an interference can be properly declared.

To tie these new matter issues together and place them into a patent prosecution
context, let's return to our previous Markush example from this section but augment
the facts slightly. Envisage the scenario where the applicant has filed, in his original
patent application, claims to the Markush and species shown in Figure 9.7. No
other Markush structure was included in the originally filed patent application. Very
soon after the applicant *files* his original patent application, a competitor's patent
application *publishes* with the compounds shown in Figure 9.8 disclosed.

A quick look at the competitor patent application's cover sheet indicates that the
original application, from which the published patent application properly claims
priority to and which contains support for the claimed compounds, was filed about

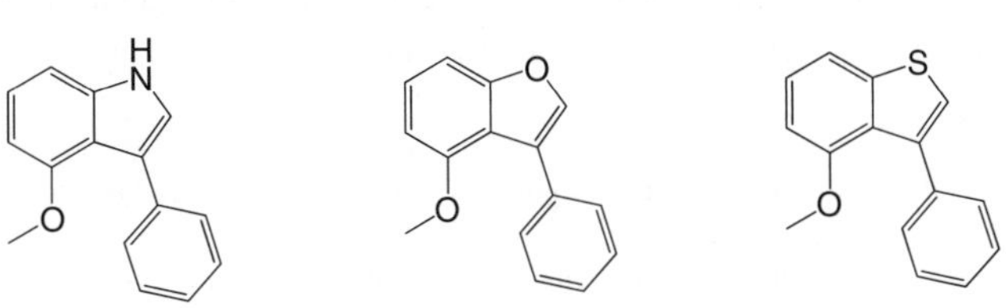

FIGURE 9.8   Intervening prior art species.

1. (Canceled.)

2. A compound selected from the group consisting of:

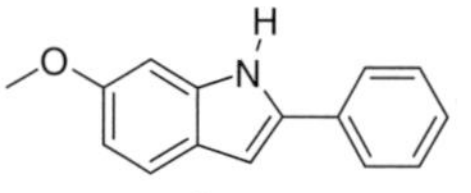

Ia 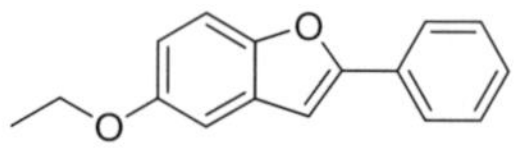   IIa 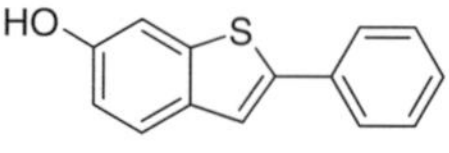   IIIa

3. (New) A compound according to structure II

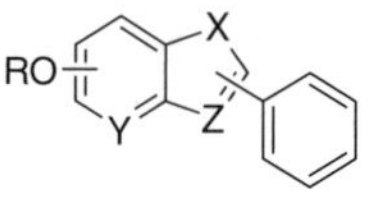

II

where: R = H or C_{1-3} alkyl;
X = O, NH or S;
Y = C;
Z = C; and
when OR is at position Y, and the phenyl group is at position Z; then
OR is not = OCH_3.

FIGURE 9.9   Amended Markush structure.

18 months before its publication. The applicant realizes that he *probably* can no longer claim the Markush genus because each one of the species from the competitor's patent application falls within the scope of the Markush genus that our applicant is claiming.[17]

When the applicant begins prosecution of the patent application, the first thing he does is submit an amended claim set where the claim to the Markush of formula I is canceled and the new Markush, shown in Figure 9.9, is introduced as claim 3.

Why would the applicant wish to make the amendment? First, if he doesn't, then the originally claimed Markush of claim 1 would be anticipated (rendered not novel) as it is currently written; recall that a single earlier disclosed species falling within a later claimed Markush will anticipate the later claimed Markush structure. By redrawing the Markush definition carefully, it is possible to exclude the anticipating species while minimizing the amount of claim scope lost. Unfortunately the applicant in this case is almost certain to get a new matter rejection because at the time of filing,

---

[17]The word *probably* is used because §102(e) prior art can be antedated by showing prior invention. On the surface, the applicant's chances don't look especially promising because he just filed the patent application, whereas the competitor's patent application just published. Since the normal time from filing a provisional patent application to publication of an application properly claiming priority to that application is approximately 18 months in the United States, there is at least a good chance the competitor actually invented the subject matter before the applicant as well.

the applicant did not have possession of the later added Markush of claim 3.[18] Absent the necessary direction in the specification as filed, one of ordinary skill in the art would not have envisioned the Markush genus as added in claim 3. Moreover, it did not matter that the added claim was to a narrower genus of compounds; the written description requirement applies the same regardless of whether the subject matter is narrowed or expanded.

It is reasonable then to ask how the applicant might have avoided the problem of not being able to narrow his genus scope. While no plan is failsafe (especially in patent law), an applicant can increase his ability to claim varying, narrower slices of his Markush structure by providing support in his original specification as filed for those narrower embodiments. In the example we just analyzed, the applicant might have provided a number of alternate Markush structures of varying scope in the original patent application. Even without a crystal ball, the provision of narrowing Markush embodiments will increase the probability of finding fallback positions that are broader than the species embodiments but possibly still narrow enough to avoid intervening or nondiscovered prior art.

For purposes of quick demonstration, let's briefly return to the problem we just discussed in order demonstrate how the inclusion of various intermediate Markush structures might have helped the applicant. A reasonable way to narrow the Markush structure would be to provide a couple of embodiments wherein the species examples are still covered by each one of the included genera but where additional Markush scope is progressively whittled down. For example, the set of Markush embodiments might have been provided in the initial patent application as filed (either in the claims, the disclosure or both) as shown in Figure 9.10. If the applicant in the previous example had provided the second and third embodiments shown in the figure, he would have been able to present alternate Markush structures that would not read on the prior art species disclosure while still providing broader coverage than the individual species embodiments.[19] Note that these genera can be constructed rather systematically without having to see into the future but rather based on narrowing the broadest Markush so that alternate views of the invention will be provided in the specification as filed so that they can be claimed later if necessary.

Beyond the prior art issues like those just discussed, the presentation of the more narrowly tailored genera can also *help* avoid potential enablement issues. Although we will be discussing enablement in depth in the next section, suffice for now to appreciate that what is claimed must be enabled throughout the scope of the claim. It is not surprising that it is often more difficult to demonstrate sufficient enablement for a broader claim than for a narrower claim because there is more subject matter to enable. By providing additional claims of varying scope, a failure to enable the

---

[18]"Any claim containing a negative limitation which does not have basis in the original disclosure should be rejected under 35 U.S.C. 112, first paragraph, as failing to comply with the written description requirement." MPEP 2173.05(i).

[19]This does not mean he is out of the woods yet with respect to the prior art. Potential issues of obviousness *could* still remain though, as we learned earlier, prima facie obviousness can be rebutted through secondary evidence, whereas a lack of novelty cannot.

First embodiment and broadest (same as previous):

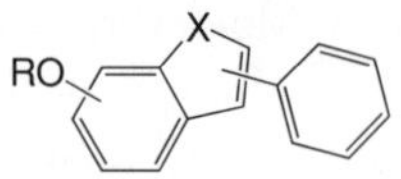

where: R = H or C$_{1-3}$ alkyl; and
X = O, NH os S

Second embodiment:

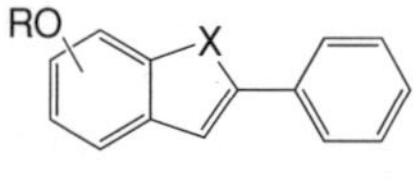

where: R = H or C$_{1-3}$ alkyl; and
X = O, NH os S

Third embodiment:

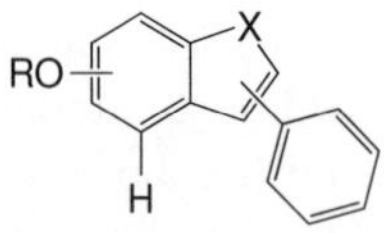

where: R = H or C$_{1-3}$ alkyl; and
X = O, NH os S

Fourth embodiment:

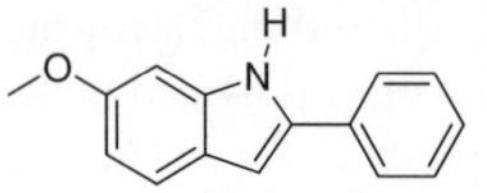

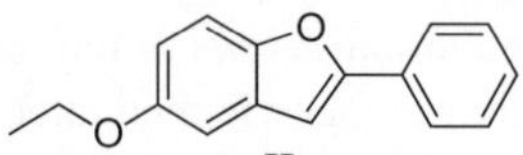

FIGURE 9.10   Progressively narrowed embodiments.

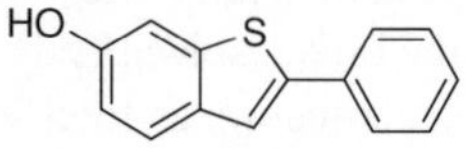

broadest genus does not necessarily lead to a failure to enable alternate descriptions of varying scope.

Although the written description requirement is most typically discussed in the context of when an applicant attempts to amend claims or the patent specification after the patent application has already been filed, there is a second, broader aspect to consider. This second aspect of the written description requirement relates not to the amendment process per se but rather to whether the claims, even where present in the original specification (such that there is no new matter issue), have written description support in the patent application. This written description requirement is separate from the new matter context discussed earlier, and is often discussed in the context of a separate requirement that the inventors demonstrate that they have "possession of the invention" at the time the patent application is filed by the disclosure in the specification itself. Asking an applicant to demonstrate possession of the invention is to require that she actually invented what she is claiming. An applicant could claim a broad genus of compounds or compositions that are well within the ability of one of ordinary skill in the art to make and use, without the inventor having made any actual examples of the compounds that are being claimed. The synthetic techniques available to one of ordinary skill in the art are quite extensive, and viable schemes to many previously unprepared compounds can often be rapidly generated without resorting to extraordinary creativity or skill. If an applicant were to claim a broad genus of compounds or compositions as well as a method for identifying the utility of those compounds or compositions, the enablement requirement could arguably be met if a credible assertion of utility has been made. You might have an almost visceral reaction against such broad claims, even though they might arguably be enabled. One's instincts should allow one to sense that broad claims need broad support (and vice versa) even if they are enabled. One shouldn't be able to reap what she has not sowed.

A recent case heard on appeal by a panel of the CAFC, *Univ. of Rochester v. G.D. Searle & Co.* provides an example of a written description, possession of the invention controversy.[20] Although this is a pharmaceutical science rather than a pure chemistry case, the general discussion in the opinion is instructive for its reflection of the current judicial policy values in regard to whether a broad claim finds adequate written description support in the specification (outside of the new matter context already discussed).

Traditional nonsteroidal anti-inflammatory drugs (NSAIDs) such as ibuprofen, naproxen, ketoprofen, and aspirin exert their effects by inhibiting cyclooxygenase enzymes (cyclooxygenases) responsible for the production of prostaglandins. In the early 1990s scientists discovered that there were two distinct cyclooxygenases, which were subsequently labeled COX-1 and COX-2. It was determined that COX-1 was responsible for the production of prostaglandins in the gastrointestinal tract. The prostaglandins produced in the gastrointestinal tract by the COX-1 enzyme serve primarily a protective role, and the inhibition of COX-1 is believed to result in stomach irritation and ulcers. In contrast, COX-2 has been shown to be responsible for the formation of prostaglandins involved in pain and inflammation. Selective inhibition of the COX-2 enzyme would thus reduce pain and inflammation but produce less gastrointestinal irritation than nonselective inhibition of COX-1 and COX-2.[21]

Scientists at the University of Rochester filed patent applications describing and claiming numerous aspects related to the COX-1/COX-2 pathway. One of the patent

---

[20] *Univ. of Rochester v. G.D. Searle & Co.*, 358 F.3d 916 (CAFC, 2004).

[21] Drugs specifically designed for COX-2 selectivity have been marketed quite successfully, especially in the United States. The drugs that were part of this litigation were celecoxib (Celebrex®) and valdecoxib (Bextra®). COX-2 selective inhibitors have been shrouded in controversy since their use has been in some cases linked with increased risk of heart attacks and stroke. Celebrex and Bextra were both marketed by Pfizer, but Bextra was pulled from the United States market in 2005 out of safety concerns. The COX-2 selective inhibitor rofecoxib (Vioxx®) was marketed by Merck and was voluntarily pulled from the market by Merck in 2004. Celebrex remains on the market, and worldwide sales in 2007 were over $2 billion.

applications issued as U.S. 6,048,850 (the '850 patent) and contained several broad claims to methods of administering to a person a nonsteroidal COX-2 selective compound. For the purpose of the present discussion, claim 1 is representative of the asserted claims from the '850 patent and is shown here:

> 1. A method for selectively inhibiting PGHS-2 activity in a human host, comprising administering a non-steroidal compound that selectively inhibits activity of the PGHS-2 gene product to a human host in need of such treatment.

> (PGHS-2 is an acronym for prostaglandin H synthase, which is another name for the COX-2 enzyme.)

The '850 patent described the COX-1 and COX-2 enzymes as well as assays useful for distinguishing whether a compound was a selective inhibitor for the COX-2 enzyme. The '850 patent also described methods of using COX-2 selective compounds as well as their formulation, routes of administration, possible effective doses, and suitable dosage forms. On the day that the '850 patent issued, the University of Rochester sued Pfizer, G. D. Searle, and related parties in a federal district court alleging that Pfizer's marketing of Celebrex and Bextra infringed several claims of the patent (including claim 1).[22]

Pfizer alleged that the claimed invention was invalid for, among other things, failure to comply with the written description requirement. The District Court found for Pfizer, holding that Rochester's claimed invention was invalid for failure to comply with the written description because the Rochester patent neither disclosed a nonsteroidal compound that selectively inhibits the COX-2 enzyme nor provided any idea how such a compound could be made other than by "trial and error" research.[23] Instead of a specific description of the invention all the public received was a method of identifying and using compounds without any description of the compounds themselves. As the District Court explained in its decision,

> Tellingly, ... what plaintiff's experts' [sic] do not say is that one of skill in the art would, from reading the patent, understand what compound or compounds—which, as the patent makes clear, are necessary to practice the claimed method—would be suitable, nor would one know how to find such a compound except through trial and error.... Plaintiff's experts opine that a person of ordinary skill in the art would understand from reading the '850 patent what method is claimed, but it is clear from reading the patent that one critical aspect of the method—a compound that selectively inhibits PGHS-2 activity—was hypothetical, for it is clear that the inventors had neither possession nor knowledge of such a compound.

The claims were therefore held to be invalid. The University of Rochester appealed the decision to the CAFC.

---

[22] Both Celebrex and Bextra were developed by Searle, which was subsequently purchased by Monsanto. Monsanto later merged with Pharmacia and Upjohn, and the resultant company later merged with Pfizer.
[23] *University of Rochester v. G.D. Searle & Co.*, 249 F.2d 216 (W.D. N.Y. 2003). The District Court also held the claims to be invalid due to lack of enablement, an issue we will discuss further later in this chapter.

The CAFC upheld the District Court opinion, citing from an earlier decision: "A description of what a material does, rather than what it is, usually does not suffice" to satisfy the written description requirement.[24] The failure to disclose any actual compounds that would work for the claimed methods meant that the inventors failed to adequately describe their invention; they did not have possession of the subject matter that they were claiming. In effect, if the asserted claims had been held enforceable, then the patentees would have effectively captured any COX-2 selective compounds, including compounds not yet made or even imagined at the time the application was filed. Claim 1 of the patent effectively covers the entire mechanism of action of what could theoretically amount to a very large and potentially incredibly diverse set of molecular structures, none of which was described in the patent itself. Instead of structural description of any compounds or substances that would accomplish the claimed methods, the public was offered the instructions as to how one could recognize such a substance if such a substance in fact existed. The essential test characteristic of the claimed method relied on a functional definition of any compounds to be used to accomplish the claimed method but did not provide the bridge between the required function and a structural corollary that any chemist could possibly understand. In the District Court decision, the federal judge ruling on the case considered the claimed method in this case as being similar to an attempt to claim a method of turning lead into gold:

> In effect, the '850 patent claims a method that cannot be practiced until one discovers a compound that was not in the possession of, or known to, the inventors themselves. Putting the claimed method into practice awaited someone actually discovering a necessary component of the invention. In some ways, this is reminiscent of the search for the so-called "philosopher's stone," eagerly sought after by medieval alchemists, which supposedly transmutes lead into gold. While the Court does not mean to suggest that the inventors' significant work in this field is on a par with alchemy, the fact remains that without the compound called for in the patent, the inventors could no more be said to have possessed the complete invention claimed by the '850 patent than the alchemists possessed a method of turning base metals into gold.[25]

None of this is to say that a structural component must be explicitly laid out where one of ordinary skill in the art could readily equate the given function with a structure (which was not the case here). For example, if one listed a DNA sequence in a patent, it could be reasonable for one to also claim sequences that hybridize with the listed sequence since the base-pair complementarily is readily appreciated by one of skill in the art.

---

[24] *Univ. of Rochester v. G.D. Searle & Co.*, *id.*, referring to *Regents of the Univ. of Cal. v. Lilly*, 119 F.3d 1559, 1568 (CAFC 1997).

[25] The District Court judge went on to explain that a patent for the philosopher's stone was actually issued during the reign of Edward III. Apparently, they were not as keen on the written description requirement back in those days. Neither were they very keen on the utility requirement—a *credible* utility must be alleged (though perhaps it was a credible utility at that time). However, there probably would be no rejections based on novelty for such a claim.

The broader issue in this case is not whether the patentees had made a discovery worth patenting but how far the federal courts were willing to allow the patentees go in exploiting that discovery. As a policy matter, we wish to reward inventors for their efforts and encourage innovation at the same time. If the patent-in-suit had been upheld, then its holders would have the right to tap into the royalty stream for any COX-2 selective drugs that might someday be discovered. Would the requirement of paying downstream royalties to the holders of a mechanism of action patent increase the pursuit of drugs acting on that mechanism? The answer is probably not. In the best case, companies would have the option of licensing the technology and would find it still financially viable to go after the targeted mechanism with novel therapeutic compounds. In the worst case, the owner of the mechanism of action patent would not allow others to work in the area at all, or alternatively, they might price its use so high that it was not financially viable for others to license. In these latter scenarios, the enforcement of the broad patent rights would decrease the actual amount of useful pharmaceutical research that could be done in the area.

Surely you are thinking that since the University of Rochester researchers had made such an important discovery that it merited some reward. As a matter of public policy, we undoubtedly wish to encourage these types of breakthroughs in basic research. As we have already come to appreciate, patent law is a delicate balancing act between encouraging inventors through the awarding of the exclusive right to practice the invention while at the same time ensuring that the net result is not harmful to the overall level of research and discovery. One could argue that "the patent law is the patent law" and that it should not matter what the policy might be for any given circumstance, but the reality is that patent law is almost always subject to some degree of interpretation, meaning that there is often no strictly black or white answer to any particular issue before the courts. Where matters of interpretation must be made, the public policy behind the various interpretations cannot be ignored. In the instant case, we have a policy that says we should reward inventors, and we also have a policy that says we need to encourage innovation. Sometimes these go hand in hand, and at other times they are at odds. For example, when an inventor is rewarded with a patent that essentially can block or otherwise hinder all activity in a given area, the public policy might direct us to consider whether rewarding such inventors, as an inducement to future inventors, is worth the cost of blocking the application of that research in the present. Sometimes the answer is no; in such instances, the policies behind not allowing one patent to block the efforts of many might be honored by a court finding that the invention that was actually made is much narrower than the invention that was claimed. Accordingly, the interpretation of the written description requirement is sometimes the strategic ground where this policy meets patent law in the federal courts. This does not mean that the inventors do not deserve a patent but only that their patent should not overreach the scope of their discovery. Perhaps, for example, if claims directed to specific assays useful for determining whether a compound is selective for one of the cyclooxygenase enzymes had been asserted, their validity in the federal courts might have been upheld. In effect, it appears that is what their patent provided written description support for. In such a circumstance, one could plausibly argue that the courts would allow the University of Rochester to reap what it sowed without significantly hampering effort in the area of pharmaceutical research.[26]

9.2 ENABLEMENT

The second portion of the §112 ¶1 is referred to as the *enablement requirement* and is typically considered to be made up of two separate requirements. The first part is the "make and use" part and requires that one provide in the specification enough instruction to teach one how to make and use the *claimed* invention without having to resort to undue experimentation. The second part of the enablement requirement requires that the use taught be credible; this requirement is co-extensive with the utility requirement.[27]

The make and use portion of the enablement requirement means that one of ordinary skill in the art should be able to prepare and practice the claimed invention without having to engage in "undue experimentation" to do so. A patent application is no place to play hide and seek but rather should provide the level of instruction to prepare and practice the invention described in the claims. This does not mean that the applicant needs to understand or explain the theory behind the invention or precisely how the invention works, unless such explanation is required to make the utility of the invention credible (second part of the enablement requirement). Furthermore, an applicant does not need to include every detail already known in the prior art and in fact it is preferred that he does not.[28] As was the case for determining whether the applicant demonstrates possession of the invention, the issue of whether an invention meets the enablement requirement is a mixed question of fact and law. The determination of whether an invention is enabled depends on a number of factors related to the invention and knowledge of the art. A number of these factors were delineated and explained in a case opinion by the CAFC.[29] These are the so-called *Wands* factors and serve as the general template the USPTO examiners use when determining whether a patent application undergoing examination satisfies the enablement requirement. The eight *Wands* factors are listed and explained briefly in the following paragraphs.

First, one needs to consider *the breadth of a claim*. Generally, the broader the claim scope, the greater the amount of disclosure required. This is logical from a technical perspective as well as a policy perspective. From the technical perspective, a more broadly claimed invention will capture more embodiments or a wider range of a single embodiment, or a combination of both. A larger number of claimed embodiments will often require more exemplification or a greater level of description because *the claim must be enabled throughout its entire scope*, meaning each of the

---

[26]For another recent case with a similar subject and holding, see *Ariad Pharmaceuticals v. Eli Lilly* 2008–1248 (CAFC 2009).

[27]We discussed in Chapter 6 that a patent that lacks utility will receive rejections under both §101, the utility requirement, and §112 ¶1, the enablement requirement.

[28]Though one person's tedious detail might be another person's essential fact.

[29]*In re Wands*, 858 F.2d 731 (CAFC 1988).

claimed embodiments must be enabled. For example, a granted claim to a chemical Markush structure grants its holder the right to exclude others from making, using, or selling compounds falling within its scope. If the patent specification does not teach one of ordinary skill in the art how to make and use the compounds falling within the Markush, then the patent holder has been awarded a broader right to exclude then what he has actually enabled. It is very easy to draw chemical Markush structures on a piece of paper—with the description of just a few broad variables, one can in a few minutes of drafting time define and claim a huge amount of chemical space, thus potentially blocking others from working in that space without providing sufficient working instruction in how to make or use the invention. From a policy standpoint, such a dichotomy between disclosure and enablement upsets the balance of interests between the inventor and public. The right to exclude is a valuable right that needs to be proportionate to the information given to the public in return. Finally, if broad intellectual turf were surrendered so easily, there would be no inducement for inventor's to actually "do the work." Allowing structures on paper to substitute for the real thing would simply encourage paper land rushes more than actual research and testing.[30] Clearly the latter is of much more value. Therefore, broader claim scope will generally (and should) require more instruction on how to make and use the claimed invention.

The second factor considers *the nature of the invention* and the type of disclosure contained in the application. Some inventions by their very nature may be more conceptual in nature, and others require more instruction on how to actually make and use the claimed invention. For example, some mechanical inventions might need nothing more than a blueprint to allow one to make and use the invention.[31] In other words, sometimes little more than a look or inspection of a drawing can allow one to comprehend not only the invention and that it would work but also that it can be made and used without additional comment or instruction from the inventors. The sum total of the disclosure plus the knowledge already available in the art enable the invention. In contrast, many areas of chemistry are more empirical in nature. Absent specific instruction regarding exactly what reagents are used, reaction times required, temperatures employed, etc., a chemical process may fail and/or a composition may not be produced at all or, even if successful, the final result may require a large amount of experimentation to get it to work.

The third *Wands* factor, *the state of the prior art* refers essentially to the teachings in the prior art at the time of the invention. Prior art, since it is available as teachings to those of skill in the art, can help fill certain gaps in the disclosure that were left out. By way of hypothetical example, consider a patent application that leaves out an extraction procedure for removing an organic acid impurity from an organic solvent. This might not be a problem, as there exists considerable prior art regarding the extraction of acids from organic solvents using aqueous bases, etc. Such an omission is unlikely to cause one to have to resort to undue experimentation to solve the problem, since such problems are solved as a matter of routine in the prior art.

---

[30]Not that this doesn't sometimes occur anyway.
[31]*In re Gay*, 309 F.2d 768 (CCPA 1962).

The fourth factor requires an ascertainment of the *level of ordinary skill in the art*. Where the level of skill in the art is high, the ordinary practitioner is assumed to be able to practice a complex invention with less likelihood of requiring undue experimentation, as she is presumably more skilled in getting experiments to work in her particular field of endeavor. In contrast, some inventions might be aimed at a layperson audience, (i.e., the general public). In such an example, one would assume less competence for any technical subject matter required by the invention; therefore, more detailed instruction on how to make and use the invention would be required. For example, the patent drafter could reasonably assume the general public would know how to screw off the top of a screw-top bottle but probably would need instruction on how to put the threads into one. In typical patent prosecution or chemical and pharmaceutical patent litigation where a finding on this topic must be made, the level of ordinary skill in the art is often determined to be very high. The generic skill level for a given art is considered to be a combination of education and working experience. It is not uncommon for a federal court to find the level of ordinary skill in the chemical or pharmaceutical arts to be at least a bachelor's degree in the discipline together with several years of working experience. With a high level of ordinary skill in the art, less basic instruction needs to be provided. As a practical matter, this might mean that a chemical worker with ordinary problem-solving skills probably would not need specific instruction on how to make a solution of standard concentration, pH, etc.

The fifth Wands factor is the *level of predictability in the art*. If an art is generally unpredictable, then the amount of disclosure will need to be higher than that required for a more predictable art. This is true because one of ordinary skill in the art would not be able to extrapolate easily from a single example or limited disclosure and thus more instruction might be required. As was already mentioned, chemistry is often very unpredictable and highly empirical; small changes in chemical structure or composition can often lead to large changes in chemical or biological properties. Accordingly, this consideration generally supports the requirement that chemical or pharmaceutical patents provide a greater level of detail. Remember that enablement requires that we teach how to make and use the invention throughout the claimed scope of the invention. If one thinks of the specification like a map, then the map will need more detail when the terrain is rougher and more unpredictable.[32]

The sixth *Wands* factor relates specifically to *the direction actually provided by the inventor* in the patent application. Where the inventor has provided a lot of direction on how to make and use the invention, it is less likely that undue experimentation will be required to practice the invention.

The seventh *Wands* factor relates to whether there are *working examples of the invention*—these are usually helpful to one who wishes to make and use the invention. This should not be a surprise, especially when the working examples fall within the

---

[32]The general unpredictability of the chemical arts was specifically commented on in the opinion of *In re Marzocchi*, 439 F.2d 220, 223-224 (CCPA 1971) where it was stated: "[In] the field of chemistry generally, there may be times when the well-known unpredictability of chemical reactions will alone be enough to create a reasonable doubt as to the accuracy of a particular broad statement put forward as enabling support for a claim."

claim scope (and they actually work!) because one essentially should have to simply repeat the examples to produce the invention, or at least the part of it the example represents. The presence of working examples is typical in chemical composition cases and thus, for example, a patent claiming a genus of compounds that provides several examples of the preparation of compounds having utility within the genus will be considered to have working examples of the invention. If an invention claimed a method of using the compound to treat a disease in a mammal, then a working example of that method might be demonstration of the compound being tested in a relevant mammalian model or a prophetic example explaining how such a treatment could be carried out. As a general matter, applications directed toward compounds having utility as drugs typically contain a section describing how the compounds may be formulated, dosed, and delivered to humans. Although these experiments have often not been done as of the date of the filing of the application, they can be written prophetically to help provide the necessary guidance.[33]

This last consideration from the *Wands* factors directly assesses the *amount of experimentation required to make or use the invention in view of the direction provided in the disclosure*. At first blush, this might appear to be the final determination masquerading as the eighth factor, but it's not. The ultimate determination of whether the invention is enabled or not is not whether significant experimentation is required but whether the experimentation required to make and use the invention is an *unreasonable or undue* amount. Some experimentation may still be required after reading the disclosure but it is not per se unreasonable if the experimentation is routine or if the disclosure provides considerable guidance as to the direction the experimentation should proceed.

The question of whether a disclosure is enabling is a broad inquiry that is fact intensive. While certain factors are outside of the applicant's control when the application is drafted (such as *Wands* factors 2–5), for those within the applicant's (and/or her representative's) control, one should keep these factors in mind. For example,

---

[33] Prophetic or constructive examples refer to examples that are hypothetical in nature but provide the applicants a mechanism to explain how they think the example could be made or performed. It is sometimes true that an applicant will have sufficient knowledge of the invention and technical area to provide explicit instruction without having carried out the representative example. Often times, such examples are used to augment the presence of other, actual examples to demonstrate how the invention can be made or used beyond the scope of the work that has actually been done. The use of prophetic examples must be approached cautiously. In particular, it is very important that an applicant clearly identify examples or procedures that have not actually been performed. Failure to clearly identify work that has actually been done from work that has not actually been done has been used to support findings of inequitable conduct in more than one case. In *Hofmann-La Roche v. Promega Corp,*. 323 F.3d 1354 (CAFC 2003), a patent was invalidated because the use of the past tense instead of the present tense implied that the experiments described had actually been performed as they were described; in patent parlance, examples written in the present tense indicate a prophetic example, whereas those written in the past tense indicate that the experiments were actually carried out. It's a thin but deep line. To stay on the right side, one should clearly delineate what actually has been done from what could be done or what has sort of been done. The biggest challenge is in describing experimental procedures where most of it has occurred as written but perhaps different experiments have been combined, since multistep procedures are often not done in a linear fashion. The impact of omissions or inaccuracies will depend, in part, on the criticality/materiality of those procedure to the claimed invention.

if one wishes to claim broadly, then one might be advised to provide representative description commensurate with the claim scope. Likewise, if one is making a broad claim, the provision of several working examples within and representative of the scope of the claim is helpful. In contrast, if one were claiming a single species or specific composition per se, then possibly just a single working example demonstrating how the compound is prepared might be sufficient (together with a credible assertion of utility).

In some instances, the issue of inoperable embodiments within a claim arises. Strictly speaking, one could plausibly argue that a claim containing one or more inoperable embodiments was not enabled throughout the claim scope since a person could not be expected to be able to make and use an inoperable embodiment of the claim. However, if we examine the inoperable embodiment issue in view of the overarching requirement that one can make and use the claimed invention without *undue experimentation*, then it can be appreciated that the issue is not whether there is an inoperable embodiment in the claim but whether the operable embodiments of the claim can be discerned without undue experimentation. This does not mean to imply that attempting to knowingly claim a large amount of inoperable subject matter is recommended, it's not. It means simply that the presence of one or more inoperable embodiments of a claimed invention is not always a harbinger of nonenablement.

The enablement requirement is difficult to understand and apply because it requires that we not only understand what the patent application actually teaches, but we also need to ascertain how that knowledge in combination with what is already known in the art would be applied by one of ordinary skill in the art to determine if the practice of the claimed invention requires undue experimentation. To further compound the challenge, there is very little enablement federal case law in the chemical and pharmaceutical arts to light our way.

Fortunately for us, however, the chemists at Miracles in a Bottle have been extra busy of late since their management has imposed a new time deadline for their stated goal of curing the world's Alzheimer disease epidemic. Unfortunately for the employees, it seems that "Discovering yesterday's drugs, tomorrow ... maybe" is no longer soon enough. To properly motivate the employees, management has determined that patent applications must be filed immediately to cover all recently completed as well as ongoing work. The chemists in the Alzheimer disease research group have been working on small peptide-aldehyde inhibitors of the β-amyloid secretase enzyme (BACE), which they hope can function to prevent the accumulation of the neurotoxic peptide fragment, β-amyloid fragment 1-42 ($A\beta_{1-42}$). Upon hearing of their new directive, the chemists promptly stop all ongoing research activities in order to write up their records of invention, including the experimental description of how they made the peptides of interest and tested them for inhibition of BACE activity. In particular, they have prepared the six modified peptides shown in Figure 9.11.

In their record of invention, the chemists explain that they have discovered a novel series of hexapeptides bearing an aldehyde at the C-terminal end. While they have not had the opportunity to fully characterize the activity through extensive structure–activity investigations as is their usual preference, they have done some

H₂N−Val−Ala−Leu−Asp−Glu−Gln−CHO
 1   2   3   4   5   6

H₂N−Ile−Ala−Leu−Asp−Glu−Glu−CHO

H₂N−Val−Leu−Leu−Asp−Glu−Gln−CHO

H₂N−Val−Ile−Leu−Asp−Glu−Ser−CHO

H₂N−Leu−Ala−Leu−Asp−Glu−Gln−CHO

H₂N−Val−Val−Ile−Asp−Glu−Thr−CHO

FIGURE 9.11   Amino acid sequence of modified peptides prepared by chemists at Miracles in a Bottle pharmaceutical company.

molecular modeling and preliminary co-crystallization studies with the BACE enzyme and the six peptide aldehydes. The chemists observed a general unpredictability in regard to which peptides function as inhibitors and which are inactive, although there was a trend where hydrophobic amino acids at the one to three-positions and a polar amino acid at the six-position were preferred. In addition, an aspartic acid at the four-position and a glutamic acid at the five-position were preferred for optimal activity. The chemists believed the necessary pharmacophore for inhibition at the BACE active site is likely be a pretty narrow set because, as they say, "that BACE enzyme is one picky eater." Unfortunately, they were not able to test any of the compounds in vivo using their mouse β-amyloid model because they did not have the required time with management on their backs. Just 2 weeks after the chemists presented their record of invention and discussed their results with the suddenly very busy patent attorney, a patent application was drafted and submitted to the USPTO.

The patent application as drafted contains a description of how to make the specifically disclosed peptides as well as additional description explaining that it was "well within the purview of one of ordinary skill in the art to make hexapeptide amino acid sequences by standard synthesis, solid phase peptide synthesis or synthesis by recombinant techniques." The patent application also described the in vitro BACE inhibition assay and the data generated from it. The application further explains:

> Despite the challenges associated with discovering potent BACE inhibitors due to the very particular active site of BACE and the general unpredictability of designing specific inhibitors for this enzyme, the examples contained herein all demonstrate an IC$_{50}$ of inhibition for BACE in the low micromolar range. Accordingly, the peptides of this invention are useful for the treatment of Alzheimer's disease. Furthermore, one of ordinary skill in the art appreciates that one could take the peptides of this invention, optionally combine them with excipients and compress them into tablets suitable for oral delivery to patients in need thereof. The dosages may vary from 1 mg to 500 mg, depending upon the patient's weight and need. In addition to these uses, the peptides of this invention may be used to ascertain the activity of BACE enzyme preparations. Incubation of an active enzyme preparation with a compound of this invention and measurement of a changed substrate turnover rate indicates that the enzyme preparation has specific BACE activity. The peptides of this invention may be prepared according to the methods described herein but may also be prepared by other standard peptide synthesis methods well-known to those of ordinary skill in the art.

What we claim:

1. A hexapeptide according to the following formula:

NH_2-AA-AA$_2$-AA$_3$-Asp-Glu-AA$_6$-CHO

where: AA, AA$_2$, AA$_3$ and AA$_6$ are each and independently any natural L-amino acid;

Asp is L-aspartic acid; and

Glu is L-glutamic acid.

2. The hexapeptide according to claim 1 where: AA, AA$_2$ and AA$_3$ are each independently selected from the group consisting of Val, Ala, Leu and Ile, and AA$_6$ is selected from the group consisting of Gln, Glu, Ser and Thr.

3. A hexapeptide selected from the following:

H_2N−Val—Ala—Leu−Asp−Glu−Gln−CHO
　　　　1　 2　 3　 4　 5　 6

H_2N−Val—Leu-Leu−Asp−Glu−Gln−CHO

H_2N−Leu−Ala−Leu−Asp−Glu−Gln−CHO

H_2N−Ile—Ala—Leu−Asp−Glu−Glu−CHO

H_2N−Val—Ile—Leu−Asp−Glu−Ser−CHO

H_2N−Val−Val—Ile—Asp−Glu−Thr−CHO

4. A method of treating Alzheimer's disease, comprising the daily oral administration of a tablet containing from between 1 to 500 mg of a peptide according to claim 4 to a patient in need thereof.

FIGURE 9.12　Claims to anti-amyloidogenic peptides and their uses.

The claims to the patent application are shown in Figure 9.12. After some time, the first substantive response to the patent application is received by the Miracles in a Bottle patent attorney. Before opening the USPTO response, the patent attorney decides to break for lunch thinking, "I'd better eat lunch now before I open this thing, otherwise I might lose my appetite."

Back from lunch and after a very brief time for quiet reflection (i.e., nap), the attorney screws up his courage and takes the envelope from his assistant's desk when his assistant is not looking. He then quickly ducks back into his office, pulling the door shut behind him. "After that last fiasco, I think I better keep these USPTO letters to myself from now on," he thinks. Inhaling deeply, he opens the correspondence from the USPTO and reads the relevant portions:

Claims 1–4 are pending in the instant application. Claims 1 and 4 are rejected for failure to satisfy 35 U.S.C. §112 ¶1. Claims 2 and 3 are objected to as being improperly dependent from a rejected claim. They will be allowed if rewritten in proper form.

The USPTO examiner further explained his reasoning:

> At the outset, we note that claim 1 is drawn to an extremely broad set of hexapeptide C-terminal aldehyde sequences. Claim 1 requires only that amino acids 1-3 and 6 be a naturally occurring L-amino acid. A claim must be enabled such that one of ordinary skill in the art can make and use the claimed invention throughout its entire scope without resorting to undue experimentation. Even if an invention can be made, a credible utility must still be taught and moreover, that use must be credible throughout the scope of the claim as well. The primary inquiry the USPTO undertakes when examining claims for compliance with 35 U.S.C. §112 ¶1 is not whether one of ordinary skill in the art would have to engage in any experimentation to make and use the claimed invention but whether that experimentation is *undue* (*In re Wands*, 858 F.2d 731, 737 (CAFC 1988). Whether experimentation is "undue" can be ascertained by considering the eight Wands factors:
>
> 1. Breadth of the claims;
> 2. Nature of the invention;
> 3. State of the prior art;
> 4. Level of ordinary skill in the art;
> 5. Predictability of the art;
> 6. Amount of direction provided in the specification;
> 7. Any working examples; and
> 8. Quantity of experimentation needed relative to the disclosure.
>
> Before applying the Wand's factors to claim 1, we stipulate at the outset that the peptides of claim 1 can be *made* without undue experimentation. It is accepted that the state of the art with regard to peptide synthesis coupled with the applicants' teaching are sufficient to satisfy this aspect of 35 U.S.C. §112 ¶1. In contrast, however, a credible utility is not established throughout the claimed scope. The Wand's factors have been applied as follows:
>
> 1. The breadth of the invention. The applicants have claimed an extraordinarily broad array of peptide sequences in claim 1. Using just the standard 20 natural amino acids in claim 1 covers 160,000 peptide sequences. It is axiomatic that a broader invention requires broader enablement.
> 2. The nature of the invention. The present invention is directed towards peptide inhibitors of the BACE enzyme. The binding and inhibition prerequisites for active peptide inhibitors of this enzyme are not adequately understood nor explained in such a way that one can extrapolate from a very limited set of examples to a very broad set of claimed compounds.
> 3. The state of the prior art. Numerous studies with small molecule inhibitors of BACE have been published and some studies with peptide inhibitors have also been disclosed. However, with regard to peptide aldehyde inhibitors of BACE, very little is known. Accordingly, the state of the prior art with regard to claims 1 is relatively undeveloped.
> 4. The level of ordinary skill in the art. The level of ordinary skill in the chemical and pharmaceutical arts is high.

5. The level of predictability in the art. Medicinal chemistry is generally an unpredictable and largely empirical science. Small changes in a molecule's structure often lead to large and unpredictable changes in its activity. Despite significant advances in structure-based design, the make-and-try approach is often still the rule of the day. But these are generalities and a lower or higher level of predictability against certain targets with certain scaffolds is sometimes known or can be demonstrated. In such instances, the more specific information can be used to make a finding in place of the more general assumption. In the instant case, the applicants themselves have described both the difficulty of finding active site inhibitors of the BACE enzyme as well as a lack of predictability for making BACE inhibitory molecules. The general lack of predictability in the medicinal chemistry arena coupled with the applicants' own statements support a finding of a high level of unpredictability for the identified utility (i.e. inhibition of the BACE enzyme).

6. The level of direction provided. The applicants have provided sufficient direction for making the explicit examples of the invention and it is stipulated that one of ordinary skill in the art could make the claimed peptides. However, the applicants have not provided a clear path or direction towards understanding which of the more than one hundred thousand peptides covered by claim will actually have the desired activity.

7. The presence of working examples. The applicants have provided six working examples of the invention. The examples have been demonstrated to have inhibitory activity against *in vitro* BACE preparations. Determining whether this utility for these compounds is sufficient to demonstrate *in vivo* activity against Alzheimer's disease is not necessary to the present inquiry. The applicants have posited that the compounds of this invention are also useful for measuring the specific BACE enzyme activity of a preparation and this is sufficient to establish that working examples are present in the application.

8. The amount of experimentation required to make or use the invention in view of the direction provided in the disclosure. The direction provided in the disclosure is sufficient for the production of the compounds of the invention.

No single Wand's factor should be considered by itself but rather the collective effect of the total should be balanced carefully in the final determination. Claim 1 fails to satisfy the enablement requirement primarily because the breathtaking scope of the claim coupled with the highly unpredictable nature of the underlying science is clearly insufficient. The disclosure of little more than a handful of active compounds pales next to the unclimbed mountain of peptides remaining to be made but yet still claimed. While the level of skill in the art is high, it is not high enough to traverse this towering obstacle without undue experimentation and accordingly, claim 1 is rejected. It should be noted that claim 2 does not suffer the same fatal defect because the claim scope is considerably attenuated and moreover, a reasonable correlation between the disclosed active compounds and the claimed compounds has been established—the claim scope is commensurate to what is taught in the application.

Claim 4 is drawn to a method of treating Alzheimer's disease in a patient in need thereof. The treatment of Alzheimer's disease is highly speculative even under the best of circumstances, but here the applicants are claiming a method of treating this CNS disease by the oral administration of a peptide. The difficulty in achieving systemic delivery of peptides after oral administration is well known The rapid gut metabolism of peptides coupled with generally poor intestinal absorption pose very significant hurdles. (see, e.g. *Foye's Principles of Medicinal Chemistry*, Lippincott Williams & Wilkins; Sixth Edition edition (September 1, 2007)). Further compounding this challenge is the presumed requirement that these synthetic peptides effectively cross the blood brain barrier where the BACE inhibition needs to take place. Despite these significant challenges, the applicants have provided limited *in vitro* data only. The in vitro data indicate that micromolar concentrations of the drug are required for inhibition of the enzyme. While these concentrations might be achievable in a test tube and therefore suitable for laboratory diagnostic work, it is simply not credible to think that such concentrations could be reached after oral administration of the peptides of the invention In order to achieve the method of claim 4, one of ordinary skill in the art would have to engage in undue experimentation and even then any chance of success would be remote. Therefore, claim 4 is also rejected for lack of enablement.

After reading the entire office action, the patent attorney leans back in his chair and breathes a heavy sigh of relief, "Two claims out of four, that's 50%. That was a passing grade for me in law school and so I'll take it as a passing grade now. This calls for a celebration!"

9.3 BEST MODE

The last required gear in the written description machinery is the best mode. The best mode requirement exists to ensure that the applicants do not reap the benefits of a patent while hiding from the public the best way to practice the invention. How could they do that? Wouldn't the patentee want to make sure not only to disclose the best method of practicing the invention but also specifically to claim the best mode to the extent possible? The answer is often but not always, and here is why. Applicants are primarily interested in obtaining a right to exclude others from practicing the invention they wish to claim. At the same time, the applicants may be reluctant to give away all of their secrets. Absent the best mode requirement, applicants might try to have the best of both worlds by providing enough written description (enablement, written description) to get the claim issued but withhold from disclosing their best way of practicing the invention that they have claimed. In this way, they can retain at least some of their trade secrets and at the same time have a valid enforceable right of exclusion against the public for the entire claim scope, including their best mode of practicing the claimed invention. Does this seem fair? I don't think so either. The best mode requirement serves to prevent the just-described scenario from happening.[34]

[34]The best mode requirement is not mirrored outside the United States and has been the subject of a fair amount of criticism. The primary complaints appear to be centered on the inordinate amount of litigation energy and resources directed toward finding best mode violations. Congressional acts targeted toward reforming U.S. patent law have specifically provided for the elimination of the best mode requirement. For

The best mode requirement is a two-prong inquiry that first asks if the *inventor(s)* contemplated a best mode of practicing the *claimed* invention and second, whether the written description in the patent application disclosed the best mode such that one reasonably skilled in the art could practice it.[35] The first prong of the inquiry addresses *whose* best mode we are talking about. The use of the word *best* implies subjectivity—ask any group of people what the best of anything is, and you are likely to get more than just one answer. Since the *inventors* of the claimed invention are the most knowledgeable about the invention, it is reasonable to require the inventors to provide what they think is the best mode of practicing the claimed invention. It is possible that another way of practicing the invention is arguably a "better best mode," but if the inventors did not reasonably contemplate it as such, they have not violated the requirement.[36] The first prong of the inquiry is a subjective standard, meaning the judicial body evaluating the evidence must make the determination whether the inventor(s) did not provide what he believed to be the best mode of practicing the claimed invention. Figuring out somebody's actual belief can be very difficult because the ultimate keeper of that information may not be the most reliable of tour guides. Who is going to admit hiding the best mode of practicing their claimed invention when they know that admission might invalidate the claim or even the entire patent?[37] This is not meant to imply that inventors are fundamentally dishonest, but there can be many versions of the same story and time and personal interests have a way of recoloring details long since faded. However, proving somebody's state of mind or

example, HR 2795 (the Patent Reform Act of 2005) would have eliminated the best mode requirement, but this bill did not pass. The Patent Act of 2009 is currently pending and would retain the best mode requirement but would remove it as a basis of patent invalidity. This of course begs the question of what sort of compulsion would accompany the requirement if the act did pass.

[35] See *Chemcast Corp. v. Arco Indus. Corp.*, 913 F.2d 923, 927-28 (CAFC 1990); *Fonar Corp. v. General Elec. Co.*, 107 F.3d 1543, 1548 (CAFC 1997); *United States Gypsum Co. v. National Gypsum Co.*, 74 F.3d 1209, 1212 (CAFC 1996).

[36] In *Glaxo v. Novopharm*, 52 F.3d 1043 (CAFC 1995), the CAFC upheld the District Court's ruling that the failure to include a preferred process for making the claimed form of the drug was not a violation of the best mode because that mode was not known to the inventor at the time of the filing of the patent application, despite the fact that others at Glaxo were aware of the preferred process and had even brought it to the attention of the agent filing the case. In other words, the assignee knew of a different best mode than did the inventor. In upholding the District Court's decision, the CAFC stated:

> Glaxo knew of the azeotroping process and knew that this process would be used commercially to produce pharmaceutical forms of the claimed product. The record also indicates that these individuals as well as their English patent agent were concerned that failure to disclose the azeotroping process may present a best mode problem. However, in neither instance did Glaxo nor its patent agent appropriately consider that inventor Crookes knew nothing of the azeotroping process. That Glaxo thought it may have a best mode problem either because of its incorrect or incomplete consideration of U.S. patent law does not make it so.

[37] Failure to meet the best mode requirement results in the invalidation for the one or more claims that the inventor's best mode was not adequately provided. The failure to provide the best mode can also result in the invalidation of the entire patent if the best mode was withheld with an intention to deceive (i.e. a finding of inequitable conduct renders the entire patent nonforceable).

knowledge does not require a signed confession or direct admission but rather can be gleaned from all of the circumstances including the witness's personal connection with the case. The inventor has the advantage of being the person familiar with the relevant facts but also *may* have a strong personal interest in seeing the patent validity upheld.[38] The fact-finder will take into account not only what a witness says but also that witness's credibility; statements are measured in their overall context. This overall context can include an inventor's publications, notebook entries, and grant applications. Where joint inventors are listed on a patent, each inventor's perceived best mode must be included in the patent. In such a circumstance, there may be more than one best mode, and accordingly, each of the best modes will need to be included in the patent application.[39]

While we have properly focused on the inventor's perception of what the best mode is, the best mode does not have to be part of the applicant's invention per se but can be part of what is known in the art or even provided to the inventor by, for example, a particular commercial supplier. For example, if a claimed process requires a catalyst and the inventors have found that a particular commercial supplier's catalyst is best for practicing the claimed process, then that detail will need to be disclosed in the patent application. However, the potential inconvenience of this latter requirement is significantly tempered by the more fundamental requirement that the best mode applies to the claimed invention only.

The focus on the claimed invention is sensible from both policy and practical enforcement standpoints. It is not fair to grant a right to exclude that is greater than the scope of disclosure, but it's also not fair to require disclosure beyond the granted right of exclusion. If the patent claims only a composition then it is usually not necessary to describe the best way to make the compound; the process to make the compound might even be the subject of a separate patent application or kept as the applicant's trade secret. In such a circumstance, the patent would need only to disclose those details that materially affected the invention as claimed. This issue will be specifically covered in our next case.

The case of *Bayer v. Schein Pharm.* arose when Bayer sued Schein for patent infringement.[40] Bayer Corporation is the assignee of the U.S. 4,670,444 patent (the '444 patent) that includes the potent, broad-spectrum antibiotic ciprofloxacin (Cipro®).

---

[38]Where inventors have assigned their rights in the invention to another party (e.g., employer), their personal financial interest in the case is not as strong as the assignee though they still may have an interest in seeing the patent upheld even without a direct financial stake. One's personal integrity and professional reputation must also be considered when evaluating the inventor's interest.

[39]During the prosecution of patent applications before the USPTO, the patent examiners assume the best mode has been satisfied unless there is clear evidence to the contrary. The USPTO will not subject the inventors to questioning regarding their beliefs or intent regarding best mode compliance during the prosecution of the patent application. The examiners do not have the means or the mandate to make such investigations. However, in a patent interference or patent litigation context, questions of best mode can be raised since questions to the inventors may be asked through written interrogatories and depositions and evidence may be subpoenaed.

[40]*Bayer Corp. v. Schein Pharm. Co.* 301 F.3d 1306 (CAFC 2002).

The '444 patent claimed priority to U.S. application number 292,560 (the '560 application), an earlier patent application. The defendants (Schein) argued that earlier, foreign-issued Bayer patents to Cipro were prior art against the '444 patent under §102(d). If the '444 patent could properly claim priority to the '560 patent application, then the earlier, foreign-issued Bayer patents would no longer be prior art under §102(d) and no longer the basis for an anticipation rejection. Schein further asserted that Bayer had failed to provide the best mode of practicing the claimed invention (Cipro) in the '560 application because the Bayer inventors had failed to disclose the best mode of preparing the claimed compounds (including Cipro). If the '560 patent application did not meet the requirements of the written description requirement, including the best mode, then Bayer could not rely on the earlier filed parent patent application for a priority date. The foreign patents would provide a §102(d) basis for invalidation of the claim and Bayer would lose.[41]

The synthesis of the target compounds could be accomplished by more than one method, though the inventor involved in the claimed subject matter had a preference to make the target compounds by a procedure using a particular intermediate that was not disclosed in the '560 patent application. The particular class of intermediate preferred by the inventor *was not claimed* in the '560 application, although that particular class of intermediates was claimed separately in a different patent application by Bayer. It was not contested by Schein that the procedure supplied in the '560 patent application enabled one of ordinary skill in the art to make the claimed compounds, only that the '560 patent did not indicate the preferred intermediate. The majority panel of the CAFC held that the best mode applies only to preferred embodiments of the *claimed* invention itself or to practices of making or using the claimed invention that *materially affected* the properties of the claimed invention. In the instant litigation, the claims of the invention were directed to compounds and not to methods of making those compounds. The preparation of the claimed compounds could be accomplished by different routes, some preferred over others. However, since the inventors were not claiming the synthetic route or the intermediate used in the preferred route, it follows that they need not disclose that route or intermediate. In fact, the preferred route became the subject matter of a separate patent application.

According to the Bayer court, however, we still need to determine whether the failure to provide the inventor's best mode for making the compound "materially affected" the properties of the claimed compound. As an abstract legal matter, one can at least appreciate the CAFC panel's position because it is conceivable that inventors might not disclose the best mode for making a composition where that best mode positively affected the properties of the claimed composition itself. For example, one might envision a best mode crystallization process as a final synthetic step where that crystallization process yields a particularly preferred crystalline form

---

[41] In Chapter 2 we briefly discussed the possibility of claiming a priority date from an earlier filed patent application provided, among other things, that the earlier filed patent application met the requirements of §112, including providing the best mode of practicing the claimed invention.

that cannot be achieved by any of the other methods disclosed in the patent application. Nevertheless, if the claim is drawn broadly to the composition in any form, then the inventors will be afforded the right to exclude others for making, using or selling all crystalline (and noncrystalline) forms of the composition, without actually providing direction as to the best form. However, phrases such as *affect the properties* of the claimed subject matter can be tortuous in trying to pin down. To what extent must it *affect the properties*? Must the properties be affected in a predictable way? Does the inventor need to investigate how the properties are affected once he realizes they are affected? What if the properties are affected in both positive and negative ways from the inventor's perspective? While such a rule sounds good in theory, it actually has the potential to open a legal Pandora's box, and partly for this reason was heavily criticized in a concurring opinion in the decision.[42]

A violation of the best mode requirement has severe consequences. Any claim that fails to meet the best mode requirement is invalid, and the entire patent can possibly be invalidated as well (if the best mode was intentionally withheld). Despite the severity of consequences for not meeting the best mode requirement, the overall impact on patent litigation outcomes is minimal. In fact, the CAFC in *Bayer v. Schein* noted that in the history of the CAFC and its predecessor court (the CCPA), only seven patents had had one or more claims invalidated as a consequence of a best mode violation. Despite the fact that best mode violations are found rarely, a discussion of the requirement is still warranted. First, the legal landscape can change over time and patents can be litigated long after the application is filed. The closer one tries to walk next to the edge of the best mode cliff, the better the chance of falling over should the legal landscape eventually shift. Second, you as an inventor control the best mode because it is your belief that determines what it is though it is sometimes very difficult to determine what the claimed invention will be while you are still drafting the technical description of the invention. The best course is one of full disclosure to your agent or attorney. Drafting and filing decisions should be then made in view of both the ultimate invention likely to be claimed as well as the strategy for dealing with any related material that will not be part of the claimed invention.

---

[42] Judge Rader who concurred on the opinion to the extent they got the overall decision right ("no best mode violation") and to the extent that the majority focused on the claimed invention only he was also satisfied. He did object, however, to the idea that "unclaimed" properties of the invention could apparently be subjected to the analysis as well. The desire to contain the best mode requirement is consistent with the legislative and judicial trends towards limiting (or even eliminating) the requirement. Judge Rader opined:

> When extended beyond the scope of the claimed invention, the best mode requirement becomes as insidious and destructive as a hidden landmine. One of the cases emphasized by this opinion, *Dana*, 860 F.2d 415, illustrates those disturbing implications. In *Dana*, the inventor claimed a seal apparatus, not any method at all, let alone a method of treating elastomeric material to ensure its longevity. Having invented a unique seal apparatus, the inventor could not have guessed that the best mode would reach out to encompass a process to increase the useful life of one component of the invention—a process that was already well known in the prior art to boot. *Dana*, 860 F.2d at 419. Nonetheless, this court invalidated the patent because the undisclosed method affected the life of the elastomeric material and thus the satisfactory performance of the seal.

This related but potentially not claimed material may be dealt with by keeping it as a trade secret, filing a separate patent application if feasible, or still including it in the patent specification as a defense to others filing on the subject matter, and thus potentially preventing you from practicing your own best mode.[43]

[43]This is sometimes referred to as a *defensive filing* or a *defensive disclosure* because the publication of the best mode in the patent application can serve as prior art under §102(a), §102(b), and §102(e). Defensive publications need not be in filed patent applications but also may be accomplished in other ways. One possibility is publishing the unclaimed material in a scientific journal or alternatively, in the *United States Patent Gazette*. Some additional aspects of the best mode that are worth keeping in mind are (1) although the best mode of practicing the claimed invention must be included in the patent application, it does not need to be designated as such (*Ernsthausen v. Nakayama*, 1 USPQ2d 1539 (Bd. Pat. App. & Inter. 1985); (2) updating the best mode is not required after the first application in the chain is filed unless the new best mode relates to a continuation-in-part application; and (3) a best mode defect cannot be cured by adding new matter to an already pending application. See MPEP 2165.01.

■■■■■ AFTERWORD AND SOURCES

> Lately it occurs to me,
> what a long strange trip it's been.
>
> —The Grateful Dead

For those of you who managed the entire journey through to this point, congratulations—you are finished! Maybe you've enjoyed yourself (not too much though, I suspect) and perhaps—better still—you've learned a few new things. Not to take the shine off, but one issue with learning patent law is that unlike scientific laws, patent law changes course endlessly. The changes may be less frequent but more sweeping where Congress enacts direct changes to the patent laws, or they may be subtler, where the federal court's doctrines on issues evolve over time. Either way, one should be careful in interpreting any issue through a lens cast many years before. All interpretations of law need to be checked against any recent legislation or federal court decisions. While this text should provide you with a good working foundation for future use, it is not a substitute for consulting more current sources and of course, consulting with a qualified patent agent or attorney for consequential matters. To assist in future endeavors, a short list of related resources that you may find amusing and perhaps even helpful is provided here. Good luck!

1. *The Manual of Patent Examination Procedures* (the MPEP). This is available for free at www.uspto.gov/web/offices/pac/mpep/mpep.html. Fortunately, the entire manual is searchable as well as printable so your ability to use it should not be limited in any way. The MPEP contains a wealth of details concerning almost any patenting topic you might be interested in. Incidentally, if you are interested in taking the patent agent's certification exam (yes, you too can become a patent agent!) then you and this manual need to become the very best of friends. Every answer to the exam can be found within its four square walls and perhaps, more important, you are allowed access to the MPEP while taking the test. If you are seriously interested in taking the U.S. patent agent exam then you are also encouraged to take a patent bar study course.

2. The Court of Appeals for the Federal Circuit. As you may have noticed, much of the quoted case law in this text reference opinions from the Court of Appeals of

The Chemist's Companion Guide to Patent Law, by Chris P. Miller and Mark J. Evans
Copyright © 2010 John Wiley & Sons, Inc.

the Federal Circuit. If you go to the CAFC's website (www.cafc.uscourts.gov) you will have access to the latest court opinions in downloadable and printable form (free, too). The opinions can be found by clicking on the "Opinions and Orders" link near the top of the page. Unfortunately, the decisions presently only go back to 2004, so if you need to go back further then you can check with Findlaw (www.findlaw.com/casecode/courts/fed.html), where free CAFC case law is also available to 1997.

3. The USPTO website (www.uspto.gov). The USPTO website is particularly valuable for finding up-to-date fee information on most aspects related to the USPTO's dealings, including up-to-date fee information, published patents and patent applications and online file histories (through the "Public PAIR" option).

4. Patent blogs. There are numerous patent blogs that are typically frequented by patent practitioners and where myriad topics are presented and discussed. In many blogs you will find a large number of links to other blogs as well as additional resource pages. The blogs make for interesting reading and can make for a good jumping-off point for any topic you wish to get more current on. Some of the more popular blogs include (but are no means limited to): Patently-O Blog, Patent Docs, Patent Prospector, Patent Barista, I/P Updates, IPWatchdog, and Patently Silly (for the lighter side of things).

5. The Internet. Probably the single most useful research tool for starting a patent law topic search is the Internet. There are a seemingly infinite source of information available and most of it is just a Google search or two away (including Google patents!). From the general to the specific, many (if not most) of your questions can be answered if you are persistent enough. General sources that are very helpful on the Internet include Wikipedia as well as Google patents. For a reminder, many of the court opinions referred to in the case citations in this book can be retrieved simply by pasting the citation into the Google search tool and then analyzing the hits, just make sure the case title is the same as in the citation, often you will be led to other cases that are citing the case you are attempting to find. Harder to find and very often older case law may have to be obtained through a pay service such as WestLaw or Lexis-Nexus. Most likely you should not need to reach this level of detail in your own work but if for some reason you have the need, you should at least know where to look.

■ ACKNOWLEDGMENTS

Chris Miller: I would like to thank the many patent attorneys who helped me learn patent law early in my career. I'd like to specially thank Dr. Michael Straher Esq. who was my first patent boss and was incredibly patient and kind in dealing with my many questions. Despite what you may have heard about law firm partners, not all of them are difficult to work for. In a similar vein, I'd like to acknowledge the entire Marc DeLuca Esq. group for all that they did to assist me in my early development as a chemical patent attorney.

Since my chemistry career is a bit longer in tooth, there simply is no way to acknowledge all of the persons who mentored me but that won't stop me from naming at least a few. First and foremost is Professor Peter Wipf (University of Pittsburgh). Professor Wipf served as my graduate research adviser and taught me a tremendous amount of chemistry and science in general. I will forever be grateful to him for his incredible patience and generosity with his time. I'd also like to thank Dr. Ivo Jirkovsky, Dr. John Yardley, and Dr. Jay Wrobel for mentoring my industrial chemistry career. I'd like to further acknowledge my undergraduate chemistry research adviser, Dr. John Peterson, who introduced me to organic chemistry research and set me on a path I have enjoyed ever since.

Finally, I'd like to express my sincerest gratitude to my family for putting up with having their husband and father distracted with this project. Without their support (or at least benign neglect) I would have been hard-pressed to finish this book. Skye, I wonder what your first book will be about?

Mark Evans: I would like to thank Brother Robert Conley, Ph.D, F.M.S, for introducing me to the rigors of science in general and the wonders of chemistry in particular. I would also like to thank Dr. Paul Hartig, Dr. David Olton, and Dr. Richard Scarpulla for nurturing that interest throughout my early career. Finally, I would like to thank Mr. Joseph Sauer, J.D., for sparking my interest in patent law and providing the necessary resources and guidance to dive in deeply.

Ex parte A, 17 USPQ2d 1716 (Bd. Pat. App. & Inter. 1990), 185

Abbott Laboratories v. Geneva Pharmaceuticals, Inc. 182 F.3d 1315 (CAFC 1999), 98

In re Adamson, 275 F.2d 952 (CCPA 1960), 242

A.F. Stoddard & Co. v. Dann, 564 F. 2d 556 (D.C. Cir. 1977), 121

AK Steel Corp. v. Sollac, 344 F.3d 1234 (CAFC 2003), 146

In re Albrecht, 514 F.2d 1389 (CCPA 1975), 210

Allen Eng'g Corp. v. Bartell Industries, Inc., 299 F.3d 1336 (CAFC 2002), 97

In re Aller, 220 F.2d 454 (CCPA 1955), 266

Amgen Inc. v. Hoechst Marion Roussel, 314 F.3d 1313 (CAFC 2003), 170

In re Antonie, 559 F.2d 618 (CCPA 1977), 267

Ariad Pharmaceuticals v. Eli Lilly (CAFC 2009), 304

Atofina v. Great Lakes Chem. Corp., 441 F.3d 991 (CAFC 2006), 185

In re Baird, 16 F.3d 380 (CAFC 1994), 252

Bayer v. Housey, 340 F.3d 1367 (CAFC 2003), 12

Bayer Corp. v. Schein Pharm. Co. 301 F.3d 1306 (CAFC 2002), xiii, 315

In re Bergy, 596 F.2d 952 (CCPA 1979), 158

In re Blaisdell, 242 F.2d 779 (CCPA 1957), 97

Ex parte Bland, 3 USPQ2d 1103 (Bd. Pat. App. & Inter. 1986), 202

Ex parte Blattner, 2 USPQ2d 2047 (Bd. Pat. App. & Inter. 1987), 222

In re Brana, 51 F.3d (Fed. Cir. 1995), 107, 165

Brenner v. Manson, 383 US 519 (1966), 161

Bristol-Myers Squibb Company v. Rhone Poulenc Rorer, Inc. et al., 326 F.3d 1226 (CAFC 2003), 66

In re Burckel 592 F.2d 1175 (CCPA 1979), 233

In re Burhans, 154 F.2d 690 (CCPA 1946), 269

Burlington Indus. v. Dayco Corp., 849 F.2d 1418, 1422 (CAFC 1988), 78

Burroughs Wellcome Co. v. Barr Labs, Inc., 40 F.3d 1223 (CAFC 1994), 127

The Chemist's Companion Guide to Patent Law, by Chris P. Miller and Mark J. Evans
Copyright © 2010 John Wiley & Sons, Inc.

The Chemist's Companion Guide to Patent Law, by Chris P. Miller and Mark J. Evans
Copyright © 2010 John Wiley & Sons, Inc.